AF581854

Land Use and Water Quality

Land Use and Water Quality

Editors

Brian Kronvang
Dico Fraters
Frank Wendland

MDPI • Basel • Beijing • Wuhan • Barcelona • Belgrade • Manchester • Tokyo • Cluj • Tianjin

Editors
Brian Kronvang
Aarhus University
Denmark

Dico Fraters
National Institute for Public Healt and the Environment
Centre for Environmental Quality
The Netherlands

Frank Wendland
Forschungszentrum Juelich, Institute of Bio- and Geosciences (IBG)
Germany

Editorial Office
MDPI
St. Alban-Anlage 66
4052 Basel, Switzerland

This is a reprint of articles from the Special Issue published online in the open access journal *Water* (ISSN 2073-4441) (available at: https://www.mdpi.com/journal/water/special_issues/LUWQ2019).

For citation purposes, cite each article independently as indicated on the article page online and as indicated below:

LastName, A.A.; LastName, B.B.; LastName, C.C. Article Title. *Journal Name* **Year**, *Article Number*, Page Range.

ISBN 978-3-03943-503-6 (Hbk)
ISBN 978-3-03943-504-3 (PDF)

Cover image courtesy of Danish Centre for Environment (DCE), Aarhus University.

Contents

About the Editors

Brian Kronvang, Professor. He graduated with an MSc (1985) in Geoscience, Aarhus University, Denmark, where he also completed his Ph.D. in 1996. He worked as scientist at the Freshwater Laboratory, Environment Protection Agency, Ministry of Environment, Denmark, during 1986–1988 and at the National Environmental Research Institute (NERI), Denmark, during 1989–1996. He was appointed Senior Scientist at the National Environmental Research Institute (NERI), a position he held from 1996 to 2008. He served as Research Professor from 2008 to 2014 at the Department of Bioscience, Aarhus University, Denmark. From 2014 onward, he has served as Professor of Catchment Science and Environmental Management at the Department of Bioscience, Aarhus University, Denmark. He has been Guest Professor at Nanjing University, Inst. Geogr. and Limnol., China Academy of Sciences (CAS) from 2009 to 2013.

Dico Fraters, Senior Researcher. After his graduation with a BSc in Agronomy from HAS University of Applied Science (Den Bosch) and Soil Science from Wageningen University & Research, he worked at the International Institute for the Semi-Arid Tropics in Niger from 1983 to 1986. He graduated with an MSc in Soil Science (Wageningen) in 1988, and since 1989, has been employed as scientist at the National Institute for Public Health and the Environment in Bilthoven, the Netherlands. From 1993 to 2013, he was responsible for the national program for monitoring the effectiveness of the minerals policies, and since 2000, for the Netherlands Member State reports, obligatory under the EU Nitrates Directive. He has been involved in several international projects and organized international workshops and conferences.

Frank Wendland, Professor. After a master's degree in Geology from the University of Cologne in 1988, he completed his doctorate degree in Geosciences at this same university in 1992. Since 1991, he has worked as a scientist at Research Centre Juelich, Germany, where he became head of the research group "Modelling and Management of Catchments" in 1995 at the Program Group Systems Analysis and Technology Evaluation, and in 2006, at the Institute of Bio- and Geosciences. From 2014 onward, he has been Honorary Professor of Hydrogeological Systems Analyses at the Brandenburg University of Technology Cottbus-Senftenberg, Germany.

Preface to "Land Use and Water Quality"

This reprint of the special issue on Land Use and Water Quality in WATER is from the 4th International Conference 'Land Use and Water Quality—LUWQ' held at Aarhus University from 3–6th June 2019. The Danish Centre for Environment (DCE) and Energy and Department of Bioscience, Aarhus University was the Danish organizers together with RIVM National Institute for Public Health and the Environment, The Netherlands, Geological Survey of Denmark and Greenland (GEUS), Copenhagen, Denmark and Umweltbundesamt (UBA), German Environment Agency, Dessau, Germany.

Brian Kronvang, Dico Fraters, Frank Wendland
Editors

Editorial

Land Use and Water Quality

Brian Kronvang [1,*], Frank Wendland [2], Karel Kovar [3] and Dico Fraters [3]

[1] Department of Bioscience, Aarhus University, Vejlsoevej 25, DK-8600 Silkeborg, Denmark
[2] Forschungszentrum Jülich, Institute of Bio- and Geosciences (IBG), Institute 3: Agrosphere, 52425 Jülich, Germany; f.wendland@fz-juelich.de
[3] National Institute for Public Health and the Environment, Centre for Environmental Quality, P.O. Box 1, 3720 BA Bilthoven, The Netherlands; karel.kovar@rivm.nl (K.K.); dico.fraters@rivm.nl (D.F.)
* Correspondence: bkr@bios.au.dk

Received: 24 June 2020; Accepted: 26 August 2020; Published: 28 August 2020

Abstract: The interaction between land use and water quality is of great importance worldwide as agriculture has been proven to exert a huge pressure on the quality of groundwater and surface waters due to excess losses of nutrients (nitrogen and phosphorous) through leaching and erosion processes. These losses result in, inter alia, high nitrate concentrations in groundwater and eutrophication of rivers, lakes and coastal waters. Combatting especially non-point losses of nutrients has been a hot topic for river basin managers worldwide, and new important mitigation measures to reduce the input of nutrients into groundwater and surface waters at the pollution source have been developed and implemented in many countries. This Special Issue of the Land use and Water Quality conference series (LuWQ) includes a total of 11 papers covering topics such as: (i) nitrogen surplus; (ii) protection of groundwater from pollution; (iii) nutrient sources of pollution and dynamics in catchments and (iv) new technologies for monitoring, mapping and analysing water quality.

Keywords: land use; water quality; nutrients; surplus; management; mitigation measures; monitoring and mapping technologies

1. Introduction

Agriculture provides food, fibre, energy and, last but not least, a living for many people around the world [1–3]. A drawback of agricultural production is pollution of the terrestrial environment with nitrogen via atmospheric deposition [4] and the aquatic environment by nutrients, pesticides and trace elements [5–9]. Thus, growth of agricultural production, as has occurred in Europe and North America since the 1950s and more recently in many other parts of the world, threatens the quality of groundwater and surface waters or has already led to deterioration of the quality of these waters [10–12]. Problematic hotspot areas are found in all countries around the world with typical examples in Denmark, the Netherlands, Germany, Belgium, United Kingdom, northern Italy, France, China, the United States and New Zealand.

Policies to abate the deterioration of water quality have been developed and programmes to improve water quality implemented around the world. For example, the European Union has adopted directives (the Nitrates Directive in 1991 [13]; the Water Framework Directive in 2000 [14]; the Groundwater Directive in 2006 [15]) with the objective of obtaining good quality in all waters by 2027.

Experience from the last 15 to 25 years shows that it will be a great challenge to realise the objectives of these policies in the remaining years of this decade, not only because the easy, low-cost measures have already been implemented, but also because the need for food for a growing world population creates a pressure to increase agricultural production. Since the easy, low cost measures generally have already been introduced, there is a need for answers to questions like 'Which measures are most effective and at the same time the most cost-effective for water quality improvement'? As most of the additional measures will have a larger effect on farm management, their implementation may be less straightforward. This raises the question 'Should measures be enforced by law or implemented on a voluntary basis'?

Schoumans et al. [16] developed a new system for implementing mitigation measures in agricultural systems to reduce phosphorus losses to water. Moreover, within the EU Cost-Action 869 programme, a wide range of mitigation measures for reducing nutrient emissions from agriculture was collected [17]. Mitigation of nonpoint source nutrient pollution can be grouped into eight categories: (1) nutrient management; (2) crop management; (3) livestock management; (4) soil management; (5) water management within agricultural land; (6) land use change; (7) landscape management and (8) surface water management. These mitigation measures can be further divided into source and transport measures (Figure 1) [18,19].

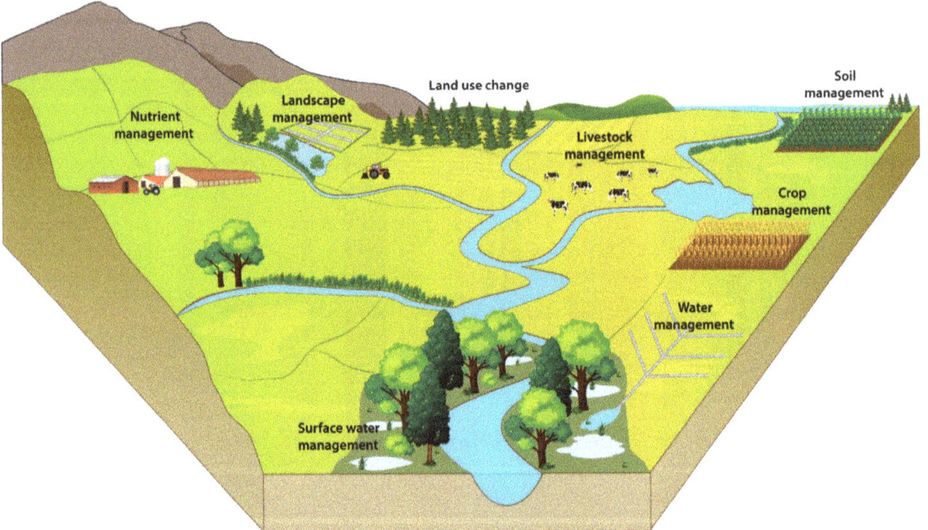

Figure 1. Categories of mitigation options for nonpoint source nutrients in a landscape and catchment context.

Source measures typically belong to groups 1–4 and 6, whereas transport measures belong to groups 5, 7 and 8 (Figure 1). Mitigation measures can be adapted as mandatory requirements for farmers as was the case during the period 1987–2015 in Denmark. This strategy was successful as demonstrated by a reduction of nitrogen and phosphorus concentrations and loadings in streams and rivers of nearly 50% and 70%, respectively, since 1990 [20]. A new era of management has, however, evolved in Denmark involving the application of targeted mitigation measures in agriculture, including local N-retention in groundwater and surface waters [21], as well as identification of phosphorus loss hotspots in agricultural catchments (P-risk mapping) [22]. Currently, development, scientific testing and application of new mitigation measures are ongoing worldwide. Research into new and more engineered types of mitigation measures assisting in removing and capturing nutrients during the transport from source to recipient waters is in progress [18,19].

Some of the above issues are touched upon in this Special Issue from the Land use and Water Quality (LUWQ) conference series, of which the most recent, LuWQ2019, was held at Aarhus University from 3 to 6 June 2019.

2. History and Themes of the LuWQ Conference Series 2013–2021

The Land use and Water Quality conferences have their origin in the series of so-called international MonNO3 workshops focusing on monitoring the effectiveness of the European Nitrates Directive action programmes. The first MoNO3 workshop was held in The Hague in 2003 and the second in Amsterdam in 2009. The workshops were limited to invited researchers and policy advisors from north-west and central European countries. At the latter workshop, the importance of the European Water Framework Directive for nutrients and other agriculture-related pollutants became clear. Furthermore, in countries outside the European Union, interesting developments in agri-environmental research, water management and policies have taken place. Therefore, we decided to broaden the scope of the meetings and to continue them in the form of conferences.

Thematically, LuWQ currently covers a wide range of topics covered by the nine themes. Three themes relate to more fundamental research:

(a) to increase our knowledge about 'system functions', i.e., basic hydrogeological and biogeochemical processes and related tools and methodologies,
(b) water quality monitoring which is about improving the effectiveness and increasing the added value of monitoring,
(c) impact of weather variability and climate change on water quality.

Three themes deal with the impact of policy and measures on water quality on plot, field, catchment and national scales:

(d) assessing national or regional policy, e.g., with regard to the effectiveness of programmes of measures on water quality,
(e) improving water quality by farm management practices (monitoring and modelling) and changes in land use,
(f) improving water quality by establishing eco-technological mitigation measures and discussing development, testing, implementation and operation to quantify the effects of such measures.

The last three themes cover management and social–economic aspects:

(g) managing protected areas for water supply and nature conservation including risk assessment techniques, monitoring and modelling,
(h) decision-making on Programmes of Measures with topics which look into the role of stakeholder input and science in policy decision-making, and
(i) implementation of Programmes of Measures that focusses on social and economic incentives and regulatory mandates that drive implementation (carrots and sticks).

At the first LuWQ conference in The Hague in 2013, we had a total of 170 participants from 30 different countries from all continents. Since then, the number of participants rose to 175 at the conference in Vienna in 2015, to 195 at the conference in The Hague in 2017 and to 240 participants in Aarhus in 2019. Feedback of participants resulted in a separate theme about monitoring in 2017 and distinguishing explicitly separate themes on farm management and technical measure for improving water quality in 2019. Furthermore the setup of the conference developed, for example, by introducing special sessions and workshops giving more room for in-depth discussion. However, the key strength of the LuWQ conferences remained which is twofold: (1) LuWQ has a well-defined narrow focus on 'agriculture and water quality' and (2) the conference is broadly oriented with regard to the various professional disciplines related to the conference topics.

3. Contributions

This Special Issue from the LuWQ2019 Conference hosted by Aarhus University, Danish Centre for Environment and Energy and Department of Bioscience, includes a total of 11 papers related to land use and water quality. The papers are grouped into four sections according to their main aims. The first section includes three papers discussing the use of the nitrogen surplus as an indicator to assess the impact of agri-environmental policies and climate change on water quality. Protection of groundwater from pollution is the guiding theme of the second section. The common theme of the three papers in the third section is sources and dynamics of nutrients in catchments, and the last section contains four papers that discuss new technologies for monitoring, mapping and analysing water quality.

3.1. Nitrogen Surplus

The nitrogen (N) surplus is an appealing indicator for assessing agri-environmental policies. Klages et al. [23] used this indicator to demonstrate the effect of the sunny and dry summer of 2018 on the environment. The 2018 drought resulted in yield reductions between 10 and 35% of the main crops grown in Germany, which led to an additional average N surplus of 30 kg/ha. Their calculations showed that this surplus would have been higher (43 kg/ha) if farmers had not changed their regular fertilisation practices. Klages et al. [23] referred to data showing an increase in mineral N in soil (upper 0.9 m) in southern Germany of 36 kg/ha between spring 2018 and spring 2019. They therefore concluded that water shortages in a changing climate not only impacted agricultural production and yields but also created further challenges and threats to nutrient management and the environment. To provide farmers with tools to reduce these risks, they discussed measures for bringing down N losses in autumn and winter, counteracting the risk of crop failure and to increase N-efficiency during the cropping season.

However, in a second paper, Klages et al. [24] questioned the use of the N surplus as a unified indicator for water pollution at farm level in Europe. They analysed the methods of fertilisation planning and nitrogen budgeting at farm level in fourteen European countries and came to the conclusion that there is no unified indicator to the assist management of nutrients and water quality at farm level. Although fertilisation planning is defined as "good agricultural practice" in the Nitrates Directive, the directive does not accept the N surplus as an indicator for the success of its implementation into national legislation. The authors argued that this is an obstacle to harmonising the methodology of using nitrogen budgets as agri-environmental indicators at farm level. Additionally, the lack of farm specific data for calculating budgets makes the N surplus at farm level less reliable. Nevertheless, Klages et al. [24] state that nitrogen budgets at farm level can serve as a simple benchmark instrument, next to autumn mineral nitrogen values of the soil, to identify the best management practices under defined environmental and farming conditions.

A model-based assessment of the nitrogen surplus can also be used to analyse the fate of agricultural nitrogen inputs into groundwater and nitrate concentrations in leachate as described in the paper by Wendland et al. [25]. They coupled a water balance model, an agro-economic model and a reactive transport model to calculate nitrate concentrations in leachate from N emissions from agriculture, urban systems and small sewage plants in North Rhine-Westphalia with a spatial resolution of 100 by 100 m. Wendland et al. [25] compared the modelled nitrate concentrations in leachate of different land uses with measured nitrate concentrations in groundwater from 1500 monitoring wells. The model results corresponded well with the measured results, with the exception of wells located in aquifers with high denitrification capacities due to the fact that denitrification in groundwater was not accounted for in the modelled nitrate concentrations in the leachate. Wendland et al. [25] highlight the importance of considering systematically not only agricultural N-emissions when modelling nitrate concentrations in densely populated regions, but argued that N-inputs from all sources (including small sewage plants, urban systems and nitrogen oxides deposition) should be included, and such a model seems promising for evaluating the relative contribution of each pollutant to the overall N-pollution. The latter has supported the selection and adequate dimensioning of regionally adapted agricultural N reduction measures.

3.2. Protection of Groundwater from Pollution

The paper by Nicholson and colleagues [26] is a comprehensive review carried out within the framework of the FAIRWAY project where approaches to protect drinking water from nitrate and pesticide pollution have been assessed. The authors reviewed more than 150 decision support tools (DSTs) applied by farmers, advisors, water managers and policy makers across the European Union with the aim to identify tools applicable at farm scale. Out of these, 12 DSTs were selected for practical testing at nine case study sites across Europe, based on their pertinence to the local challenges at the nine locations. It turned out that the exchange of DSTs across Europe is still challenging because of the different legislative environments, advisory frameworks, country-specific data/calibration requirements, regional climate/soil differences, and language issues. Specifically, the authors found that although many countries have developed comparable DSTs for similar problems, these—albeit being based on the same input data—may deliver very different results. The authors conclude that the potential for exchange of DSTs between countries is limited at present but necessary in order to promote a more effective drinking water protection within EU, for instance by collecting the required data in such a way that any transnational DST can be used consistently and thereby increase the comparability between regions. Although the paper focuses on Europe, the concept may also be applicable in other regions experiencing problems with high nitrate and pesticide concentrations in groundwater sources.

3.3. Nutrient Sources and Dynamics in Catchments

The paper "Influence of Farming Intensity and Climate on Lowland Stream Nitrogen" by Goyenola et al. [27] elucidates the patterns and driving factors behind N fluxes in lowland stream ecosystems differing in land use, soil characteristics and climatic-hydrological conditions. For the purpose of the paper, three complementary monitoring sampling characteristics were applied to 43 selected streams showing contrasting farming intensities in a humid subtropical climate with flashy conditions in Uruguay and a humid cold temperate with stable discharge conditions in Denmark. The results proved that, in both study regions, farming intensity determined the concentrations in and loss to streams of total dissolved N, nitrate and total N in spite of the differences in soil and climatic–hydrological conditions between the case study areas. Ammonium and dissolved organic N concentrations did not show significant responses to the farming intensity or climatic/hydrological conditions, whereas a high dissolved inorganic N to total dissolved N ratio was associated with the temperate climate and high base flow conditions but not with farming intensity. As for nitrate pollution of surface waters, the authors touched upon an issue of general importance in many lowland areas. In the Danish catchments, artificial drainage through tile drains is widespread, these acting as a shortcut pathway for the transport of nitrate from the soils to surface waters. As part of the nitrate leached from soils is transported directly in tile drains to streams and thus escapes from the denitrification processes in the groundwater of lowland areas, connected streams draining intensively farmed catchments are characterised by higher nitrate concentrations than streams in intensively farmed catchments without drainage systems such as the subtropical streams in Uruguay.

Chen et al. [28] draw attention to the source apportionment and spatiotemporal dynamics of nitrogen distribution and transport in hilly areas with a monsoon climate. Specifically, the paper presents the results of applying a SWAT model to the Qiandao Lake Basin (China) to identify drivers and transport processes in order to facilitate effective nutrient management. The simulation showed that the basin's annual average total nitrogen load between 2007 and 2016 of ca. 11,500 tons was represented realistically by the model. The authors conclude that most of all available optimised localised SWAT model parameters suitable for hilly areas with a monsoon climate, such as shallow soil layers, were needed to provide reliable estimates, e.g., with regard to realistic representation of the N loss peak from fertilised tea plantations in the subsurface flow in the period before the rainy season. For such areas with intensive agriculture and high total nitrogen (TN) yields, the authors suggest that water conservation practices should be implemented. Another main finding was that the largest contributors to the N loads were domestic sewage (ca. 22%) and livestock production

(21%), which highlights the need for not only constructing more wastewater treatment plants in areas identified as sites of high domestic and industrial pollution but also for increasing their operational efficiency. Although the paper focuses on a Chinese catchment, the concept of optimised localised SWAT model parameters may be relevant for application in similar monsoon basins elsewhere that experience problems with nitrate pollution of surface waters.

Ta et al. [29] performed a distributed model analysis on the gap between the current water status and the good ecological status stipulated by the European Water Framework Directive (WFD) for an entire region of Germany (state of Schleswig-Holstein) by applying the model system GROWA-MEPhos with high spatial resolution. The model results on water balances and phosphorus sources and emissions showed that phosphorus target concentrations were exceeded in 60% of the 6407 analysed sub-catchments. The authors also found that a 269 tons P year-one reduction of the phosphorus input to rivers and lakes in Schleswig-Holsten is required, corresponding to approximately 31% of the total emissions to surface waters. As most of the phosphorus emissions into surface waters are derived from agricultural diffuse sources, Ta et al. [29] concluded that the implementation of measures directed against fertilisation and erosion control is needed. Moreover, improved treatment of waste in sewage treatment plants may be an option, especially when they play a major emission role in urban areas. Finally, the authors found that technically sound and plausible methods are required to better describe the phosphorus retention processes in rivers and lakes in order to improve the model results.

3.4. New Technologies for Monitoring, Mapping and Analysing Water Quality

Song et al. [30] investigated the effects of land use on water quality across multiple spatial scales based on both circular and riparian buffers in a rapidly urbanised region in Hangzhou City, China. Their results showed that total nitrogen (TN) and total phosphorus (TP) concentrations were closely related to land use at circular buffer scale, while relatively strong correlations were found between land use and algae biomass at riparian buffer scale. At circular buffer scale, significant positive correlations appeared between land use and TN concentrations when areas of industrial land, urban greenspace and roads increased, whereas negative correlations were found with forests, lakes, ponds and rivers. In the case of TP, significant correlations emerged between land use and industrial areas at circular buffer scale, the correlation improving with increasing buffer size (100 m to 1000 m). In addition, significant correlations existed between land use and algae biomass (total chlorophyll a), with the clearest correlation occurring at the riparian buffer scale, and the authors therefore concluded that riparian buffer zones might play a key role in the conservation of aquatic ecosystems. Finally, the variation in water quality explained by landscape metrics increased with increasing buffer size, implying that land use patterns may have a closer correlation with water quality at larger spatial scales than those of this investigation (maxima of 1000 m for circular buffers and 300 m for riparian buffers).

Zhang et al. [31] investigated a water quality predictive model based on a combination of Kernal Principal Component Analysis (kPCA) and a Recurrent Neural Network (RNN) to forecast the trend of dissolved oxygen based on sensor data on electrical conductivity, pH, turbidity and chlorophyll-a in the Burnett River, Australia. Enormous amounts of water quality data are collected by advanced sensors, and the authors therefore foresee that use of data-driven models for predicting trends in water quality will increase in the future. The water quality variables used in their investigation were reconstructed based on the kPCA method, which aims to reduce the noise from raw sensory data and preserve usable data. Due to the RNN's recurrent connections, their model was able to include previously gathered information to predict a trend for the future of water quality data. The kPCA-RNN model achieved R2 scores up to 0.908, 0.823 and 0.671 in predicting the concentration of dissolved oxygen in the upcoming 1, 2 and 3 h, respectively, using data from the previous 24 h time intervals as the model input. The predictive accuracy of the kPCA–RNN model was at least 8-17% better than that of other comparative models. Zhang et al. [28] concluded that the model inputs can be improved to include extra information such as rainfall and to cover longer periods of time and, also, that the water quality predictive model can be extended to support simultaneous prediction of multiple water variables.

Hashemi et al. [32] investigated nutrient concentration–discharge (C–Q) relationships for hysteresis behaviours and export regimes in 87 Nordic small streams draining mini-catchments from three Nordic countries (Finland, Sweden and Denmark). The classification applied is based on a combination of stream export behaviour (dilution, constant, enrichment) and hysteresis rotational pattern (clockwise, no rotation, anticlockwise). The classification is intended to assist in the development of pollution monitoring and management in catchments as it requires an in-depth understanding of spatial and temporal factors controlling nutrient dynamics in streams. Hashemi et al. [32] documented that the dominant nutrient export regimes were enrichment for NO3− and constant for total organic N (TON), dissolved reactive P (DRP), and particulate P (PP). Nutrient hysteresis patterns were primarily clockwise or no hysteresis. A Principal Component Analysis (PCA) considering the effects of catchment size, land use, climate, and dominant soil type showed that land use and air temperature were the dominant factors controlling nutrient C–Q types. Therefore, the nutrient export behaviour in streams draining Nordic mini-catchments seems to be dominantly controlled by their land use characteristics and, to a lesser extent, their climate.

Kim et al. [33] investigated lag time as an indicator of the link between nitrogen (N) surplus in agriculture and groundwater nitrate concentrations utilizing a cross correlation analysis method. The authors used groundwater monitoring data from three case study sites with groundwater-based drinking water abstraction: Tunø and Aalborg-Drastrup in Denmark and La Voulzie in France. In these case study sites, soil surface N surplus and long-term groundwater monitoring data were available for the analysis. Kim et al. [33] found that for Tunø and La Voulzie, where matrix flow is the dominant groundwater pathway, the N lag times continuously increased with an increasing distance from the agricultural N source (in Tunø: from 0 to 20 years between 1.2 to 24 m below the land surface (mbls); at La Voulzie: from 8 to 24 years along the groundwater pathway). However, in Aalborg-Drastrup where both matrix and fracture flows are important groundwater pathways, the N lag times showed a greater variability with depth: for instance, a 23-year lag time at 9–17 mbls compared to a 4-year lag time at 21–23 mbls. The N lag times estimated in the study were generally found to be comparable to groundwater ages measured by chlorofluorocarbons (CFCs). Kim et al. [33] concludes that the lag time may be a useful indicator to reveal the hydrogeological links between the agricultural pressure and water quality state, which is fundamental for a successful implementation of drinking water protection plans.

To sum up, this Special Issue clearly fills in a gap in the knowledge about the many linkages between land use and water quality. Special thanks are due to all the authors for accepting the call and to the publishers for their professional assistance at all stages.

Author Contributions: All authors have read and agreed to the published version of the manuscript.

Funding: This study was part of the Nordic Centre of Excellence BIOWATER project and funded by Nordforsk (project number 82263).

Conflicts of Interest: The authors declare no conflict of interest.

References

1. Tilman, D.; Cassman, K.G.; Matson, P.A.; Naylor, R.; Polasky, S. Agricultural sustainability and intensive production practices. *Nature* **2002**, *418*, 671–677. [CrossRef] [PubMed]
2. Erisman, J.W.; Sutton, M.A.; Galloway, J.; Klimont, Z.; Winiwarter, W. How a century of ammonia synthesis changed the world. *Nat. Geosci.* **2008**, *1*, 636–639. [CrossRef]
3. Jensen, L.S.; Schjoerring, J.K.; van der Hoek, K.W.; Damgaard-Poulsen, H.; Zevenbergen, J.F.; Pallière, C.; Lammel, J.; Brentrup, F.; Jongbloed, A.W.; Willems, J.; et al. Benefits of nitrogen for food fibre and industrial production. In *The European Nitrogen Assessment*; Sutton, M.A., Howard, C.M., Erisman, J.W., Billen, G., Bleeker, A., Grennfelt, P., van Grinsven, H., Grizzetti, B., Eds.; Cambridge University Press: Cambridge, UK, 2011.

4. Bobbink, R.; Hicks, K.; Galloway, J.N.; Spranger, T.; Alkemade, R.; Ashmore, M.; Bustamante, M.; Cinderby, S.; Davidson, E.A.; Dentener, F.; et al. Global assessment of nitrogen deposition effects on terrestrial plant diversity: A synthesis. *Ecol. Appl.* **2010**, *20*, 30–59. [CrossRef] [PubMed]
5. Pimentel, D.; Houser, J.; Preiss, E.; White, O.; Fang, H.; Mesnick, L.; Barsky, T.; Tariche, S.; Schreck, J.; Alpert, S. Water resources: Agriculture, the environment, and society source. *BioScience* **1997**, *47*, 97–106. [CrossRef]
6. Peierls, B.L.; Caraco, N.F.; Pace, M.L.; Cole, J.J. Human influence on river nitrogen. *Nature* **1991**, *350*, 386–387. [CrossRef]
7. Townsend, A.R.; Howarth, R.W.; Bazzaz, F.A.; Booth, M.S.; Cleveland, C.C.; Collinge, S.K.; Dobson, A.P.; Epstein, P.R.; Holland, E.A.; Keeney, D.R.; et al. Human health effects of a changing global nitrogen cycle. *Front. Ecol. Environ.* **2003**, *1*, 240–246. [CrossRef]
8. Reid, W.V.; Mooney, H.A.; Cropper, A.; Capistrano, D.; Carpenter, S.R.; Chopra, K.; Dasgupta, P.; Dietz, T.; Duraiappah, A.K.; Hassan, R.; et al. *Millennium Ecosystem Assessment: Ecosystems and Human WellBeing: Synthesis*; Island Press: Washington, DC, USA, 2005.
9. Billen, G.; Silvestre, M.; Grizzetti, B.; Leip, A.; Garnier, J.; Voss, M.; Howarth, R.W.; Bouraoui, F.; Lepistö, L.; Kortelainen, P.; et al. Nitrogen flows from European watersheds to coastal marine waters. In *The European Nitrogen Assessment*; Sutton, M.A., Howard, C.M., Erisman, J.W., Billen, G., Bleeker, A., Grennfelt, P., van Grinsven, H., Grizzetti, B., Eds.; Cambridge University Press: Cambridge, UK, 2011.
10. Sims, J.T.; Sharpley, A.N. *Phosphorus: Agriculture and the Environment*; Book Series: Agronomy Monographs; American Society of Agronomy, Inc.; Crop Science Society of America, Inc.; Soil Science Society of America, Inc.: Madison, WI, USA, 2005; Volume 46, p. 1121, ISBN 9780891181576, Online ISBN 9780891182696. [CrossRef]
11. Schlesinger, W.H. On the fate of anthropogenic nitrogen. *Proc. Natl. Acad. Sci. USA* **2009**, *104*, 203–208. [CrossRef]
12. Voß, M.; Baker, A.; Bange, H.W.; Conley, D.; Cornell, S.; Deutsch, B.; Engel, A.; Ganeshram, R.; Garnier, J.; Heiskanen, A.-S.; et al. Nitrogen processes in coastal and marine systems. In *The European Nitrogen Assessment*; Sutton, M.A., Howard, C.M., Erisman, J.W., Billen, G., Bleeker, A., Grennfelt, P., van Grinsven, H., Grizzetti, B., Eds.; Cambridge University Press: Cambridge, UK, 2011.
13. European Commission. *Directive of the Council of December 12, 1991 Concerning the Protection of Waters against Pollution Caused by Nitrates from Agricultural Sources (91/676/EEC)*; European Commission: Brussels, Belgium, 1991; pp. 1–8.
14. European Commission. *Directive 2000/60/EC of the European Parliament and of the Council of 23 October 2000 Establishing a Framework for the Community Action in the Field of Water Policy*; European Commission: Brussels, Belgium, 2000; pp. 1–72.
15. European Union. Directive 2006/118/EC on the protection of groundwater against pollution and deterioration Directive *2006/118/EC* of the European Parliament and of the Council. *Off. J. Eur. Union* **2006**, *372*, 19–31.
16. Schoumans, O.F.; Chardon, W.J.; Bechmann, M.E.; Gascuel-Odoux, C.; Hofman, G.; Kronvang, B.; Rubæk, G.H.; Ulén, B.; Dorioz, J.-M. Mitigation options to reduce phosphorus losses from the agricultural sector and improve surface water quality: A review. *Sci. Total Environ.* **2014**, *15*, 468–469. [CrossRef]
17. Bechmann, M.; Gascuel-Odoux, C.; Hofman, G.; Kronvang, B.; Litaor, M.I.; Lo Porto, P.A.; Newell-Price, G.R. *Mitigation Options for Reducing Nutrient Emissions from Agriculture. A Study Amongst European Member States of Cost Action 869*; Schoumans, O.F., Chardon, W.J., Eds.; Wageningen, Alterra, Alterra-Report 2141; 2011; p. 144. Available online: https://www.researchgate.net/publication/254832937_Mitigation_options_for_reducing_nutrient_emissions_from_agriculture_a_study_amongst_European_member_states_of_Cost_action_869 (accessed on 3 June 2020).
18. Hoffmann, C.C.; Zak, D.; Kronvang, B.; Kjaergaard, C.; Carstensen, M.V.; Audet, J. An overview of nutrient transport mitigation measures for improvement of water quality in Denmark. *Ecol. Eng.* **2020**. [CrossRef]
19. Carstensen, M.V.; Hashemi, F.; Hoffmann, C.C.; Zak, D.; Audet, J.; Kronvang, B. Efficiency of mitigation measures targeting nutrient losses from agricultural drainage systems: A review. *Ambio* **2020**. [CrossRef] [PubMed]
20. Dalgaard, T.; Hansen, B.; Hasler, B.; Hertel, O.; Hutchings, N.J.; Jacobsen, B.H.; Jensen, L.S.; Kronvang, B.; Olesen, J.E.; Schjoerring, J.K.; et al. Policies for agricultural nitrogen management—Trends, challenges and prospects for improved efficiency in Denmark. *Environ. Res. Lett.* **2014**, *9*, 115002. [CrossRef]

21. Højberg, A.L.; Windolf, J.; Børgesen, C.D.; Troldborg, L.; Tornbjerg, H.; Blicher-Mathiesen, G.; Kronvang, B.; Thodsen, H.; Ernstsen, V. National nitrogen model, catchment model for loading and mitigation. In *National Geological Survey for Denmark and Greenland Report*; 2015; p. 111. ISBN 978-87-7871-417-6. (In Danish)
22. Andersen, H.E.; Kronvang, B. Modifying and evaluating a P index for Denmark. *Water Air Soil Pollut.* **2006**, *174*, 341–353. [CrossRef]
23. Klages, S.; Heidecke, C.; Osterburg, B. The Impact of Agricultural Production and Policy on Water Quality during the Dry Year 2018, a Case Study from Germany. *Water* **2020**, *12*, 1519. [CrossRef]
24. Klages, S.; Heidecke, C.; Osterburg, B.; Bailey, J.; Calciu, I.; Casey, C.; Dalgaard, T.; Verloop, K.; Velthof, G. Nitrogen Surplus—A Unified Indicator for Water Pollution in Europe? *Water* **2020**, *12*, 1197. [CrossRef]
25. Wendland, F.; Bergmann, S.; Eisele, M.; Gömann, H.; Herrmann, F.; Kreins, P.; Kunkel, R. Model-Based Analysis of Nitrate Concentration in the Leachate—The North Rhine-Westfalia Case Study, Germany. *Water* **2020**, *12*, 550. [CrossRef]
26. Nicholson, F.; Laursen, R.K.; Cassidy, R.; Farrow, L.; Tendler, L.; Williams, J.; Surdyk, N.; Velthof, G. How Can Decision Support Tools Help Reduce Nitrate and Pesticide Pollution from Agriculture? A Literature Review and Practical Insights from the EU FAIRWAY Project. *Water* **2020**, *12*, 768. [CrossRef]
27. Goyenola, G.; Graeber, D.; Meerhoff, M.; Jeppesen, E.; Teixeira-de Mello, F.; Vidal, N.; Fosalba, C.; Ovesen, N.B.; Gelbrecht, J.; Mazzeo, N.; et al. Influence of Farming Intensity and Climate on Lowland Stream Nitrogen. *Water* **2020**, *12*, 1021. [CrossRef]
28. Chen, D.; Li, H.; Zhang, W.; Pueppke, S.G.; Pang, J.; Diao, Y. Spatiotemporal Dynamics of Nitrogen Transport in the Qiandao Lake Basin, a Large Hilly Monsoon Basin of Southeastern China. *Water* **2020**, *12*, 1075. [CrossRef]
29. Ta, P.; Tetzlaff, B.; Trepel, M.; Wendland, F. Implementing a Statewide Deficit Analysis for Inland Surface Waters According to the Water Framework Directive—An Exemplary Application on Phosphorus Pollution in Schleswig-Holstein (Northern Germany). *Water* **2020**, *12*, 1365. [CrossRef]
30. Song, Y.; Song, X.; Shao, G.; Hu, T. Effects of Land Use on Stream Water Quality in the Rapidly Urbanized Areas: A Multiscale Analysis. *Water* **2020**, *12*, 1123. [CrossRef]
31. Zhang, Y.; Fitch, P.; Thorburn, P.J. Predicting the Trend of Dissolved Oxygen Based on the kPCA-RNN Model. *Water* **2020**, *12*, 585. [CrossRef]
32. Hashemi, F.; Pohle, I.; Pullens, J.W.M.; Tornbjerg, H.; Kyllmar, K.; Marttila, H.; Lepistö, A.; Kløve, B.; Futter, M.; Kronvang, B. Conceptual Mini-catchment Typologies for Testing Dominant Controls of Nutrient Dynamics in Three Nordic Countries. *Water* **2020**, *12*, 1776. [CrossRef]
33. Kim, H.; Surdyk, N.; Møller, I.; Graversgaard, M.; Blicher-Mathiesen, G.; Henriot, A.; Dalgaard, T.; Hansen, B. Lag Time as an Indicator of the Link between Agricultural Pressure and Drinking Water Quality State. *Water* **2020**, *12*, 2385. [CrossRef]

Article

The Impact of Agricultural Production and Policy on Water Quality during the Dry Year 2018, a Case Study from Germany

Susanne Klages *, Claudia Heidecke and Bernhard Osterburg

Coordination Unit Climate, Johann Heinrich von Thünen-Institute, Bundesallee 49, 38116 Braunschweig, Germany; claudia.heidecke@thuenen.de (C.H.); bernhard.osterburg@thuenen.de (B.O.)
* Correspondence: susanne.klages@thuenen.de; Tel.: +49-531596-1111

Received: 14 March 2020; Accepted: 18 May 2020; Published: 26 May 2020

Abstract: The hot summer of 2018 posed many challenges with regard to water shortages and yield losses, especially for agricultural production. These agricultural impacts might further pose consequent threats for the environment. In this paper, we deduce the impact of droughts on agricultural land management and on water quality owing to nitrate pollution. Using national statistics, we calculate a Germany-wide soil surface nitrogen budget for 2018 and deduce the additional N surplus owing to the dry weather conditions. Using a model farm approach, we compare fertilization practices and legal restrictions for arable and pig breeding farms. The results show that, nationwide, at least 464 kt of nitrogen were not transferred to plant biomass in 2018, which equals an additional average nitrogen surplus of 30 kg/ha. The surplus would even have amounted to 43 kg/ha, if farmers had continued their fertilization practice from preceding years, but German farmers applied 161 kt less nitrogen in 2018 than in the year before, presumably as a result of the new implications of the Nitrates Directive, and, especially on grassland, owing to the drought. As nitrogen surplus is regarded as an "agri-drinking water indicator" (ADWI), an increase of the surplus entails water pollution with nitrates. The examples of the model farms show that fertilization regimes with high shares of organic fertilizers produce higher nitrogen surpluses. Owing to the elevated concentrations on residual nitrogen in soils, the fertilization needs of crops in spring 2019 were less pronounced than in preceding years. Thus, the quantity of the continuously produced manure in livestock farms puts additional pressure on existing storage capacities. This may particularly be the case in the hot-spot regions of animal breeding in the north-west of Germany, where manure production, biogas plants, and manure imports are accumulating. The paper concludes that water shortages under climate change not only impact agricultural production and yields, but also place further challenges and threats to nutrient management and the environment. The paper discusses preventive and emergency management options for agriculture to support farmers in extremely dry and hot conditions.

Keywords: climate change; nitrogen budget; nitrogen balance; water quality; land management; extreme weather events

---

1. Introduction

The summer of 2018 was one of the hottest and driest in Europe [1]. Like in other European countries, precipitation in 2018 fell far below the long-term mean (Figure 1a). This refers to all months between February and November 2018 [2].

In comparison with the long-term mean, extended sunny periods between April and November 2018 dominated [1] (Figure 1b). These dry and sunny weather conditions led to a drastic reduction in soil moisture in autumn 2018 [3]. In September and October 2018, the usable field capacity dropped to around 30%, with the exception of south-east Bavaria in southern Germany, where values above

100% were attained. In November 2018, usable field capacity still remained at levels between 35% and 55% [3]. In some areas in the eastern German federal states, the values were remarkably low, at 0–25% [3].

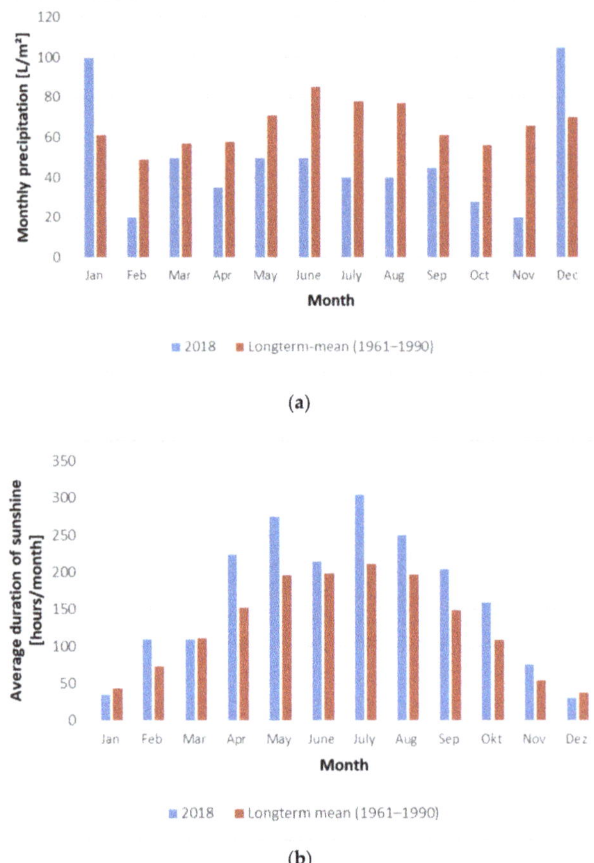

Figure 1. Conditions in Germany in 2018: (**a**) average monthly precipitation compared with long-term means (in L/m^2) [2] (source: [2], translated); (**b**) average monthly sunshine in 2018 in Germany compared with long-term means (in hours/month) [4] (source: [4], translated).

This had an enormous impact on the agriculture of European Union (EU) Member States in northern Europe, like Germany, especially owing to decreasing crop yields. In comparison with the long-term level, yield reductions in 2018 nationwide were significant for all arable crops and for grassland (Figure 2).

The year 2018 can thus be regarded as an example of the impact of extreme weather on agriculture, and serves as a basis for discussion on how extreme events also influence land use and water quality issues. Globally, anthropogenic climate forcing has doubled the probability of years that are both warm and dry in the same location relative to their baseline data [6]. Further, they find an increased joint probability that key crop and pasture regions simultaneously experience unusually warm conditions in conjunction with dry years. That extreme events severely impact agricultural production has been analyzed, for example, by Rosenzweig et al. [7]. Research has also been conducted to analyze the impact on extreme events on nutrient concentrations in groundwater using modelling approaches or statistical regression analysis [6,7].

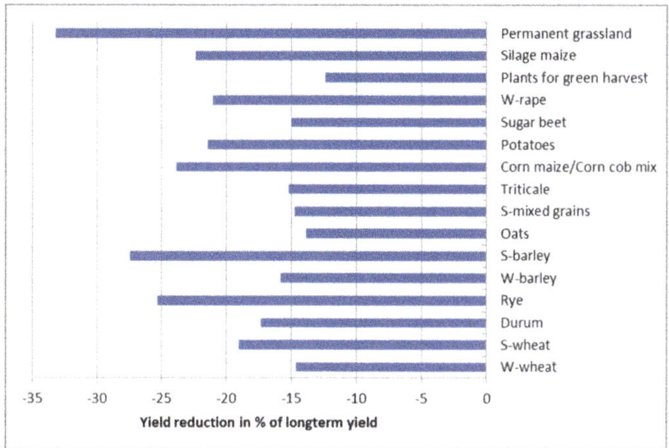

Figure 2. Yield reduction in Germany in 2018 in % of long-term yield (2012–2017) [5] (source: [5], translated).

According to the Organisation for Economic Co-Operation and Development (OECD) [8,9], the gross nitrogen budget (GNB) is an appropriate method to calculate comparable indicators on the regional and national scale. The GNB, as national farm gate budget, is the only European agri-environmental indicator (AEI) for nitrogen efficiency, whereby low efficiency goes along with high nitrogen surplus [10]. Budgets are used in numerous studies as a tool to monitor changes in nitrogen and phosphorus leaching to groundwater and surface water [11–23]. As investigated in the Horizon 2020 European Union-funded project FAIRWAY (farm systems management and governance for producing good water quality for drinking water supplies), running from 2017 to 2021, nitrogen surplus can be regarded as an "agri-drinking water indicator" (ADWI) [24], which is applied on different scales, from national to farm scale, within the European Union; an increase in N surplus entails an increase in water pollution with nitrates [25]. Different budget types were categorized and visualized by means of diagrams, and their implementation at different scales in Europe has been investigated [25].

Gu et al. [26] analyzed the impact of climate change on nutrient surpluses in China on an average perspective until 2050. Bouwman et al. [27] calculated global nitrogen budgets and included global change analysis for the future. The direct relationship between an extreme weather event like a drought and its impact on the nitrogen surplus as indicator for water pollution has so far not been investigated.

This article further discusses the impact on agriculture regarding weather and growing conditions in 2018, the impact on yield and nitrogen uptake, as well as economic aspects. Furthermore, we analyze the potential water quality impact by agriculture according to farm type and discuss possible measures for reducing losses in autumn/winter, for reducing risk of crop failure, and for increasing the overall N-efficiency. We thus discuss nitrogen flows and surpluses that could potentially influence the water quality. The specific impact of agricultural production on nitrogen concentrations in ground water and surface water is not directly analyzed in this paper.

The effects of extreme weather conditions on nitrogen supply due to climate change can be summarized as follows [28]:

- Relative/absolute lack of water and of the resulting lack of nutrients owing to missing precipitation and increased evaporation, especially in spring and early summer. This can result in reduced nitrogen efficiency.
- Depending on the individual level of the infiltration capacity of soils, excess water due to high precipitation may lead to a reduction of air-filled pores, which will affect root growth. Moreover, nitrogen losses are likely to occur as a result of leaching and/or aerial emission. This would also lead to low nitrogen efficiency.

- Owing to an increase in temperature, the vegetation period is prolonged. This has an effect on the nitrogen dynamics in soil, stimulating microbial mineralization of soil biomass. This affects soil nitrogen dynamics.

Climate change, therefore, indicates a need for the implementation of adapted nitrogen fertilization strategies and optimized land use systems.

2. Materials and Methods

We calculated a simplified nationwide nitrogen soil surface budget in order to estimate changes in the average nitrogen surplus and the total value of nitrogen surplus as a result of the dry weather in 2018. In addition, we calculated nitrogen budgets for two model farms in order to highlight the effects of fertilization practices in certain farm types.

2.1. Simplified Nationwide Nitrogen Soil Surface Budget

A simplified nationwide nitrogen soil surface budget for Germany for the average of the preceding years (2012–2017), as well as with the data specifically available for the dry year 2018, was established. As our focus is on the impact of the drought on agricultural plant production, we used a simplified soil surface budget as the budgeting method. We estimated the nitrogen inflow by the quantities of all relevant types of fertilizers used and deduced the nitrogen removal by crop according to average yields in the preceding years (2012–2017) and the dry year 2018. For this nationwide estimation, we did not consider N-deposition, as this was assumed not to change much between succeeding years. As there is no precise method available for the quantification of legume N-fixation in order to estimate differences between years [18,29], we did not include nitrogen inflows from this source. The N-uptake by grazing animals was assumed to be negligible; as most farm animals are now kept indoors, methods for quantification are not precise and statistics are not available.

2.1.1. Estimation of Nitrogen Fertilization

Nitrogen fertilization was calculated separately for the different fertilizer types:

- Manure: The number of farm animals, grouped in animal categories, was taken from the last national Farm Structural Survey in 2016 [30]. Animal counts were multiplied with nitrogen excretion data for each animal category as listed in the annex of the Fertilization Ordinance [31], the national legal basis for implementing Nitrates Directive [32]. For each animal category, the percentage of nitrogen incorporated in liquid and solid manure was estimated. For each manure type, gaseous losses of nitrogen in the stable and in storage, respectively, during application of the animal manure on the field were calculated according to standard values of the Fertilization Ordinance [31], and not in accordance with the German emission inventory [33]. The plant availability of nitrogen in organic fertilizers like manure applied to soil was calculated in the year of application and the following year as percentage of the applied nitrogen, in accordance with the standards of the German Fertilization Ordinance [31].
- Imports of manure: Total amounts of imported manure from the Netherlands were taken from an agricultural news platform [34]. The corresponding nitrogen amount was calculated by multiplication with standard data of the Fertilization Ordinance [31].
- Compost: The total amount of nitrogen in compost used in agriculture was estimated by the national Scientific Advisory Board on Fertilizer Issues (Wissenschaftlicher Beirat für Düngungsfragen—WBD) [35].
- Sewage sludge: The total amount of nitrogen in agriculturally used sewage sludge was estimated by WBD [35].
- Digestate of plant origin: Total nitrogen from digestates of plant origin from biogas plants was estimated using information from the website of the German Biogas Association [36]. The overall installed electrical output is currently 4,200 MW, produced with circa 1/3 manure and 2/3 plant

material, for example, maize (on 1 ha 50 t biomass with 3.8 kg N/t and an electrical yield of 21,000 kWh can be produced). This results in a nitrogen inflow from digestates of 233,016 t N, rounded off to 250,000 t N.

- Mineral fertilizer consumption: According to the Federal Ministry of Food and Agriculture (Bundesministerium für Ernährung und Landwirtschaft—BMEL) [37], mineral fertilizer consumption in Germany dropped from 1,658,800 t in the season 2016/2017 to 1,496,600 t in the season 2017/2018.

2.1.2. Estimation of Nitrogen Removal

Nitrogen removal by crop was calculated as an average of the years 2012–2017 and for the dry year 2018, using data on yields of main crops and utilized agricultural area (UAA) per crop according to Destatis [7]. Not considered were pulses for grain production (132,000 ha), field vegetables (118,000 ha), vinyards (100,000 ha), forests and short rotation coppices (1,142,600 ha). Specifications on the average concentration of nitrogen in harvested crops were taken from the annex of the current, recently adapted German Fertilization Ordinance of 2020 [38].

2.1.3. Estimation of Nitrogen Surplus

The nitrogen surplus of a soil surface budget can be derived by subtracting the nitrogen fertilization from the nitrogen removal by plants. We calculated the mean nitrogen surplus from the data of the years 2012–2017. Owing to the climatic conditions in 2018, a certain amount of nitrogen was not transferred into plant biomass. Thus, the difference between the mean surplus in the years 2012–2017 and the year 2018 reveals this amount of nitrogen.

2.2. Water Quality Impact by Agriculture for Different Farm Types

In order to more clearly demonstrate the impact on water quality by agricultural activities during the course of the year 2018 on the farm level, we set up two basic model farms, taking into account the guidelines of the Nitrates Directive [32] and its interpretation for Germany in the Fertilization Ordinance which came into effect in 2017 [31]. On this basis, we compared the results of the nitrogen soil surface budgets of the following two virtual 100 ha model farms and assumed fertilization and budgeting according to the German Fertilization Ordinance [31]:

1. Arable farm (25 ha winter-quality wheat, 15 ha winter-barley, 15 ha summer-barley, 15 ha oats, 15 ha grain maize, 15 ha winter-oilseed rape);
2. Arable farm with pig production, producing the same crops as (1), but using the harvested crops as feedstuff for 3500 pigs per year (1250 places for fattening pigs from 28 kg to 118 kg).

3. Results

3.1. Nitrogen Supply from Manure and Other Organic Sources

The pie chart of Figure 3 shows the nitrogen supply to German agriculture by most organics—manure, compost, biogas digestate, and sewage sludge—applied to soil as organic fertilizer, organic mineral fertilizer, or soil conditioner. Approximately 75% of all organic nitrogen applied to the soil is of animal origin, deriving from solid or liquid manure from German farms and amounting to 867,463 t N per year. This net-amount does not include nitrogen emissions from stables and storage. These emissions are estimated to be 267,305 t N annually, according to calculation figures from German Fertilization Ordinance [31], on the basis of the gross nitrogen excretion of all farm animals. The gross nitrogen excretion sums up to 1,134,768 t N per year. However, this is quite particular for the German case, and almost one quarter of the N inflow is of plant origin and was processed in biogas plants. This can be explained by the support of biogas produced from energy crops through the German

Renewable Energy Law (EEG). The other types of organics—sewage sludge and compost, as well as manure imports from the neighboring Netherlands—are, on the national level, of minor importance.

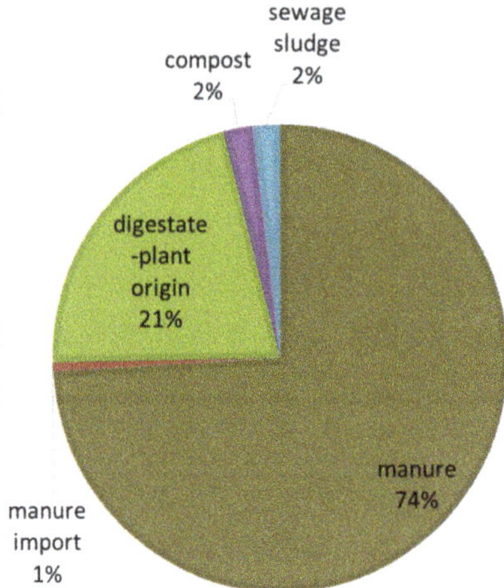

Figure 3. Nitrogen from organic fertilizers in Germany applied to soil in 2018 (source: own calculation).

3.2. Nitrogen Removals by Crops

For main crop groups, Table 1 summarizes the average yields of the years 2012 to 2017 and the yields achieved in 2018. The nitrogen demand as basis of the fertilization demand refers to the average yields in the previous years and is listed in the annex of the Fertilization Ordinance [31]. According to this ordinance, N_{min}-values in spring, N mineralized from certain crops in previous years, and N delivered from last year's organic fertilization also have to be taken into consideration for the calculation of the current fertilization need. From the organic fertilizers applied to soil, only a share is accounted for as plant available in the year of application (this figure was calculated separately for the different types of organic fertilizer). The listed level of mineral fertilization results from the nitrogen demand minus the plant available nitrogen from the different sources mentioned above.

In Table 2, total figures on production as an average for 2012 to 2017 and for the year 2018, nitrogen inflow by fertilization, and removal by harvested crops are listed with reference to the same crop groups as in Table 1. As there was less nitrogen uptake by the crops in 2018 compared with the mean of the six preceding years owing to lack of water, 464,000 t of nitrogen could not be transferred into crop biomass. Figure 4 gives a summary of the situation in 2018.

The total nitrogen excreted in livestock production and applied with other organic fertilizers in German agriculture amounts, with 1,438 kt, to the same level as the nitrogen in mineral fertilizers (1,496 kt/year). Figure 4 illustrates that, in 2018, the mineral nitrogen fertilizer applied in Germany was, with 1,496 kt, reduced by 11.5% (194 kt) in comparison with the preceding year 2017 [37]. In both cases, 267 kt nitrogen/year (18.6% of the N of organic origin) is calculated as gaseous losses in the stable or during manure storage, and 161 kt nitrogen/year (11.2% of the N of organic origin) after fertilizer application. In addition, another 464 kt of nitrogen could not be transferred to biomass in 2018.

Table 1. Calculated nitrogen supply for different crops from available organic and mineral fertilization, according to average yields (2012–2017) and rough estimates on N mineralization based on regulations of the German Fertilization Ordinance [31].

	Average Yield (2012–2017)	Yield 2018	N-Demand	N in Soil Available	N from Preceding Crop	N from Manure of Preceding Year	Fertilization Demand	N Applied with Organic Fertilizer to Soil	N Available in the First Year of Fertilization	Mineral N Fertilization
	t/ha	t/ha	kg N/ha	kg N/ha	kg N/ha	kg N/ha	kg N/ha	kg N/ha	kg N/ha	kg N/ha
W-Wheat	7.94	6.78	230	42	10	6	172	(77)	35	137
S-Wheat	5.79	4.69	180	42		6	132	(77)	35	97
Durum	5.55	4.59	165	42		6	117	(77)	35	82
Rye	5.65	4.22	148	42		6	100	(77)	35	65
W-Barley	7.21	6.07	180	42		6	132	(77)	35	97
S-Barley	6.82	4.95	140	42		6	92	(77)	35	57
Oats	4.78	4.12	115	42		6	67	(77)	35	32
S-mixed Grains	4.42	3.77	119	42		6	71	(77)	35	36
Triticale	6.39	5.42	140	42		6	92	(77)	35	57
Corn Maize/Corn Cob Mix	9.88	7.53	210	42	10	6	152	(77)	35	117
Potatoes	44.50	34.96	180	42		6	132	(77)	35	97
Sugar beet	74.20	63.10	180	42		6	132	(77)	35	97
W-Rape	3.80	3.00	195	42		6	147	(77)	35	112
Plants for green Harvest	26.29	23.04	200	42	10	6	142	(77)	35	107
Silage Maize	44.11	34.27	200	42	10	6	142	(77)	35	107
Permanent rassland	6.74	4.91	145	30		6	109	(77)	35	74

Table 2. Calculated nitrogen supply for different crops from available organic and mineral fertilization, according to average yields (2012–2017) and rough estimates on nitrogen mineralization based on regulations of German Fertilization Ordinance [31].

	Total Production (2012–2017)	Total Production (2018)	Mineral Fertilization	Organic Fertilization (net)	Total Fertilization	Removal by Crop 2012–2017	Removal by Crop 2018	Fertilizer N not Transferred into Crop Yield 2018	Fertilizer not Applied in 2018	Gaseous Losses in Stable/during Storage	Gaseous Losses during Application
	1000 t		1000 t N								
W-Wheat	24,601	19,613	425	205	630	445	355	90	37	54	33
S-Wheat	381	526	6	4	11	6	9	−2	1	1	1
Durum	98	139	1	1	3	1	2	−1	0	0	0
Rye	3621	2206	41	42	84	55	33	21	7	11	7
W-Barley	8743	7397	118	80	198	144	122	22	16	21	13
S-Barley	2660	2216	22	26	48	37	31	6	6	7	4
Oats	613	577	4	8	13	9	9	1	2	2	1
S-mixed Grains	73	43	1	1	2	1	1	0	0	0	0
Triticale	2527	1950	23	26	49	42	32	10	5	7	4
Corn Maize/Corn Cob Mix	4622	3331	55	31	86	64	46	18	6	8	5
Potatoes	10,797	8740	24	16	40	38	31	7	3	4	3
Sugar Beet	27,038	24,628	35	24	59	49	44	4	5	6	4
W-Rape	5107	3665	151	89	240	171	123	48	16	24	14
Plants for green Harvest	9359	8525	38	24	62	54	49	5	5	6	4
Silage Maize	91,661	74,229	223	137	360	348	282	66	28	36	22
Permanent Grassland	30,108	22,160	331	295	626	635	468	168	58	78	47
Sum			**1499**	**1010**	**2508**	**2100**	**1636**	**464**	**194**	**267**	**161**

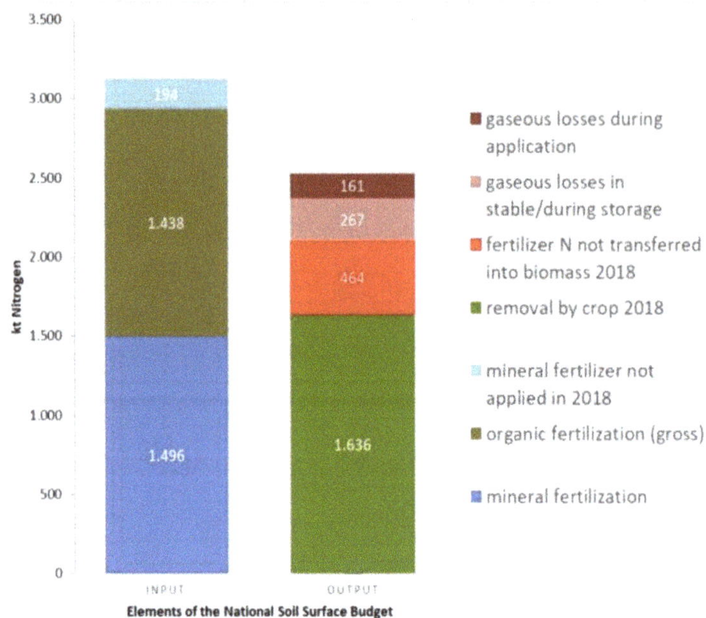

Figure 4. N-budget in 2018 (without N-deposition, legume N-fixation, and N-uptake by grazing animals in kt N/year): N-supply from organic fertilizers (left), N-removal by crop, gaseous losses in stable, storage and during fertilizer application, fertilizer not transferred into biomass (right) (source: own calculation).

3.3. Surplus of Nitrogen due to Dry Weather Conditions

From an environmental perspective, with decreasing yields due to the drought, less nitrogen was harvested from the fields. This left 464 kt of nitrogen in the soils that was not transferred to biomass. With reference to the UAA of 15,200,000 ha cultivated in Germany with arable crops and grassland, this equals an additional amount of 30 kg N/ha to the nitrogen surplus owing solely to the drought. This amount would have been even higher (up to 43 kg N/ha) if farmers had not reduced their mineral fertilizer input in 2018 by 11.5% in comparison with the preceding year 2017 [37]. As the impacts of the drought in 2018 became visible by June, most of fertilizers for arable crops had already been applied, with the exemption of late top dressing with nitrogen, especially in baking wheat. Intensive grassland receives several fertilizer inputs between spring and early autumn, allowing for adaptations during the whole vegetation period. Thus, it can be assumed that mineral N inputs have been reduced especially on wheat and on grassland.

The official national soil surface budget for 2018 has not yet been calculated. N surplus as an average of the preceding years (2015 to 2017) is reported to be 1178 kt N/year or 71 kg N/ha × year, respectively [39,40]. Thus we can estimate that the N-surplus for the dry year 2018 could amount to up to 100 kg N per ha, depending on the developments of further elements such as N-deposition, N fixation of legumes, and so on. In the past, the national soil surface budgets already indicated an increase in N surplus for dry years. In 2003, the N surplus amounted to 92 kg N/ha × year, in comparison with 74 N/ha × year as the mean value of the two preceding years, and in 2006, the surplus was 76 kg N/ha × year, compared with 66 N/ha × year as the mean of the years 2004 and 2005. As the drought 2018 was the most severe incident of this type up to now, an induced increase of the national N surplus of 30 or even 40 kg/ha × year is highly probable.

3.4. Impact of the Dry Summer 2018 on Nitrogen Flows on the Farm Level

With a soil surface budget, we show the influence of the weather conditions in 2018 on the pollution potential of nitrates inputs from agriculture for two fictive model farms according to the rules of the German Fertilization Ordinance as affective in 2018 [31]. In Table 3, N-fertilization planning and N-budgeting is outlined for an arable model-farm; the crops grown are winter quality wheat, winter barley, summer barley, oats, grain maize, and winter oilseed rape. In 2018, target yields and target qualities were calculated as three-year averages, and the corresponding nitrogen demand of the crops was drawn from the annex of the Fertilization Ordinance [31]. Plant available nitrogen in soil was determined as N_{min} per soil analysis. Elevated nitrogen delivery was expected from the catch crops preceding barley and maize. No nitrogen delivery had to be accounted for owing to high organic matter content in the soils or in connection with former organic fertilization [31]. Nitrogen fertilization need is calculated as nitrogen demand minus the share of plant available nitrogen. Nitrogen fertilization is completely realized with mineral fertilizers. Owing to the drought, realized yields were reduced, in this example, according to the average reduction as shown in Figure 2. As a consequence, less nitrogen was withdrawn from the field, resulting in a total farm-wide nitrogen surplus of 2528 kg N/100 ha and 25 kg/ha, respectively.

Table 3. Soil surface budget of a fictive arable 100 ha model farm in the dry year 2018 and the year 2019, set up according to [31].

Crop	W-Quality Wheat		W-Barley		S-Barley		Oats		Grain Maize		W-Oilseed Rape	
Year	**'18**	**'19**	**'18**	**'19**	**'18**	**'19**	**'18**	**'19**	**'18**	**'19**	**'18**	**'19**
area cultivated (ha)	25		15		15		15		15		15	
target yield(tons/ha)	8	7	7		5		5,5		9		4	
target quality (% raw protein)	16	13	13		11		11		10		23 [2]	
N demand (kg N/ha)	260	215	180	180	140	140	130	130	200	200	200	200
	plant available N in soil (kg N/ha)											
N plant available in spring (kg N/ha)	−40	−70	−40	−90	−40	−80	−40	−65	−40	−80	−40	−90
N mineralized from soil organic matter (kg N/ha)	0	0	0	0	0	0	0	0	0	0	0	0
N mineralized from former manure application (kg N/ha)	0	0	0	0	0	0	0	0	0	0	0	0
N mineralized from preceding crop (kg N/ha)	−10	−10	0	0	−40	−10	0	0	−30	−10	0	0
N fertilization need during vegetation period (kg N/ha)	210	135	140	90	60	50	90	65	130	110	160	110
mineral N fertilization (maximum, kg N/ha)	210	135	140	90	60	50	90	65	130	110	160	110
	N soil surface budget											
	N inflow											
mineral fertilizer (kg N/ha)	210	135	140	90	60	50	90	65	130	110	160	110
organic fertilizer (kg N/ha)	0	0	0	0	0	0	0	0	0	0	0	0
manure of own farm (kg N/ha)	0	0	0	0	0	0	0	0	0	0	0	0
biological N-fixation/mineralization from preceding crop (kg N/ha)	0	0	0	0	40	10	0	0	30	10	0	0
sum (kg N/ha)	210	135	140	90	100	60	90	65	160	120	160	110

Table 3. *Cont.*

Crop	W-Quality Wheat		W-Barley		S-Barley		Oats		Grain Maize		W-Oilseed Rape	
Year	'18	'19	'18	'19	'18	'19	'18	'19	'18	'19	'18	'19
	N removal (kg N/ha)											
yield realized (tons/ha)	7.5	7	6.2	6.2	6	6	5	5	9	9	3.5	3.5
N concentration in main product (kg N/tons fresh matter)	24	18.1	18	18	15	15	15	15	15	15	34	34
N removal with main product (kg N/ha)	181	127	111	111	91	91	76	76	136	136	117	117
total N removal (kg N/ha)	181	127	111	111	91	91	76	76	136	136	117	117
N surplus (kg N/ha)	29	8	29	−21	9	−31	15	−11	24	−16	43	−7
N surplus per crop (kg N/crop) [1]	731	208	435	−315	141	−288	218	−158	362	−239	641	−109

[1] 2018: N surplus on farm level 2,528 kg N/100 ha; on average, 25 kg N/ha; 2019: N surplus on farm level −900 kg N/100 ha; on average, −9 kg N/ha; [2] in dry matter (dm).

In the following year, the farmer adapted to changed climate conditions and decided to cultivate winter wheat with a minor raw protein concentration, needing less nitrogen in the late growth phase. He also reduced the target yield to 7 tons instead of 8 tons in 2018. Target yields of the other crops remained unchanged. Owing to high N_{min}-values in spring, fertilization needs were less pronounced than in 2018. Mineralization from catch crop was much lower than in 2018, as growing conditions for catch crops were not favorable and there was less plant biomass. Owing to less nitrogen input in this year, the total farm-wide nitrogen surplus was negative, with −900 kg and −9 kg/ha, respectively.

In Table 4, N-fertilization planning and N-budgeting for a pig rearing farm are outlined. The main difference to the arable model farm is that, in this farm, the produced pig manure needs to be utilized owing to limited storage capacities for liquid manure. With 1,250 pig fattening places on the farm, circa 14,250 kg nitrogen is generated per year, as calculated by standard figures from the annex of the Fertilization Ordinance [31]. The ordinance also allows the deduction of emissions from stables and storages from the excreted nitrogen (for pigs, 20% of the N-excretions). In liquid pig manure applied to soil, at least 60% of the nitrogen content is available in the first year. Further, a subsequent nitrogen delivery of 10% of the nitrogen applied in the preceding year has to be accounted for [31].

Aiming at the use of all liquid manure produced on the farm and complying with all fertilization rules, mineral fertilization in 2018 was conducted primarily for winter wheat, oilseed rape, and winter barley. This is useful for steering the plant growth via late nitrogen applications.

Owing to the yield reductions in 2018 and with manure fertilization in the fictive pig breeding model farm, a nitrogen surplus of 4774 kg N per farm and 47 kg N/ha, respectively, has been reached with a calculation as net soil surface budget. A corresponding calculation as gross farm budget, including standard gaseous nitrogen losses from stable and storage (2850 kg N per farm) and during application (1450 kg N per farm), is higher and amounts in total to 10,787 kg N and 107 kg N/ha per farm, respectively.

In 2019, owing to the elevated nitrogen concentration in soil in spring, the calculated fertilization need is lower than in the preceding year. Thus, the farm manure produced by the pig farm covers almost all plant fertilization needs, only for winter wheat there is a possibility of a mineral top dressing. This situation becomes especially aggravated when taking the elevated N_{min} concentrations into account.

These calculations show that the pressure on fertilization might be a lot higher for pig breeding farms or other types of livestock farming, as manure storage capacity is usually limited. This model farm is of course fictive and does not take into account an export of manure, which might ease the pressure.

Table 4. Soil surface budget of a fictive 100 ha pig rearing model farm in the dry year 2018 and the year 2019, set up according to [31].

Crop	W-Quality Wheat		W-Barley		S-Barley		Oats		Grain Maize		W-Oilseed Rape	
Year	**'18**	**'19**	**'18**	**'19**	**'18**	**'19**	**'18**	**'19**	**'18**	**'19**	**'18**	**'19**
area cultivated (ha)	25		15		15		15		15		15	
target yield (tons/ha)	8	7	7		5		5,5		9		4	
target quality (% raw protein)	16	13	13		11		11		10		23 [2]	
N demand (kg N/ha)	260	215	180	180	140	140	130	130	200	200	200	200
	plant available N in soil (kg N/ha)											
N plant available in spring (kg N/ha)	−40	−70	−40	−90	−40	−80	−40	−65	−40	−80	−40	−90
N mineralized from soil organic matter (kg N/ha)	0	0	0	0	0	0	0	0	0	0	0	0
N mineralized from former manure application (kg N/ha)	−6	−9	−14	−13	−9	−7	−13	−9	−17	−16	−13	−16
N mineralized from preceding crop (kg N/ha)	−10	−10	0	0	−40	−10	0	0	−30	−10	0	0
N fertilization need during vegetation period (kg N/ha)	204	126	126	77	51	43	77	56	113	94	147	94
	manure application											
N-excretion pigs (kg N/ha)	80	115	175	160	107	90	160	116	210	196	165	196
minus maximum N-emissions from stable and storage (20%)	64	92	140	128	85.6	72	128	92.8	168	156.8	132	157
minus maximum N-emissions after application to soil (10%)	56	80.5	123	112	75	63	112	81	147	137.2	115.5	137
minimum plant availability in year of application (60%)	38	55	84	77	51	43	77	56	101	94	79	94
mineral N fertilization (maximum, kg N/ha)	166	71	42	0	0	0	0	0	12	0	68	0
	N soil surface budget											
	N inflow											
mineral fertilizer (kg N/ha)	166	71	42	0	0	0	0	0	12	0	68	0
organic fertilizer (kg N/ha)	0	0	0	0	0	0	0	0	0	0	0	0
manure of own farm (kg N/ha)	56	80.5	123	112	75	63	112	81.2	147	137.2	116	137.2
biological N-fixation/mineralization from preceding crop (kg N/ha)	10	10	0	10	40	10	0	10	30	10	0	10
sum (kg N/ha)	232	161	165	122	115	73	112	92	189	147	183	147
	N removal (kg N/ha)											
yield realized (tons/ha)	7.5	7	6.2	6.2	6	6	5	5	9	9	3.5	3.5
N concentration in main product (kg N/tons fresh matter)	24	24	18	18	15	15	15	15	15	15	34	34
N removal with main product ((kg N/ha)	181	168	111	111	91	91	76	76	136	136	117	117
total N removal (kg N/ha)	181	168	111	111	91	91	76	76	136	136	117	117
N surplus (kg N/ha)	51	−7	54	11	24	−18	37	16	53	11	66	30
N surplus per crop (kg N/crop) [1]	1271	−185	803	168	359	−267	551	240	800	168	991	448

[1] 2018: N surplus on farm level 4,774 kg N/100 ha; on average, 48 kg N/ha; 2019: N surplus on farm level 573 kg N/100 ha; on average, 6 kg N/ha; [2] in dry matter (dm).

4. Discussion

Our study is limited to Germany as the case study for Central Europe in a region with temperate climate. Owing to a long-term average yearly precipitation level of circa 800 mm (regional variation from <500 mm to >1200 mm), rain-fed agriculture is dominating. According to the DIPSIR model of the European Environmental Agency EAA [41–43]—the abbreviation stands for Driving forces, Pressures, States, Impacts and Responses, we refer to nitrogen surplus as a pressure indicator for water contamination, whereby groundwater and surface water closely interfere with each other.

In Switzerland, waterbody pollution by adjacent agricultural fields is a primary water quality issue [44]. In Germany, between 2012 and 2014, around 50% of nitrogen was discharged into surface waters via groundwater [45].

Droughts affect the concentration of nitrates and other solubles in the groundwater [44]. According to a meta-analysis, higher nitrate in some agricultural catchments could be linked to reduced dilution of groundwater drainage input and increased influence of sediment nitrogen fluxes. In the immediate post-drought period, several studies found that nitrate was flushed from the catchment soils [46].

4.1. Pressure on the Environment and Groundwater Quality

In 2018, at least 464 kt N was applied across Germany as fertilizer, but not taken up by growing crops. As the severe drought in 2018 continued in many parts of Germany until late autumn [3], winter crops or cover crops sown in autumn often did not germinate. Winter crops like W-oilseed rape were often not sown at all. This had effects on the N_{min}-values in soil (after harvest, autumn, and spring). These are standardized determinations of the plant available mineral nitrogen within the root zone, usually up to 90 cm depth. Owing to the drought, N_{min}-analyses at harvest time showed values significantly above those of 2015–2017 [47]; the effect was particularly evident if crops were over-fertilized. Owing to high temperature and little precipitation after harvest and in autumn, there was often no leaching, despite high N-surpluses due to poor harvests. Often, soils below 60–70 cm depth were still completely dry in late autumn and showed a deficit of rain of 50–200 mm. Consequently, in autumn 2018 and spring 2019, high amounts of residual nitrogen could be detected in soils. Figure 5 shows an example from southern Germany, where arable cropping, partly irrigated, dominates. N_{min}-values in spring 2019 were on average 36 kg/ha higher than in spring 2018, even though, owing to irrigation, yields did not drop as far as in areas without irrigation [48].

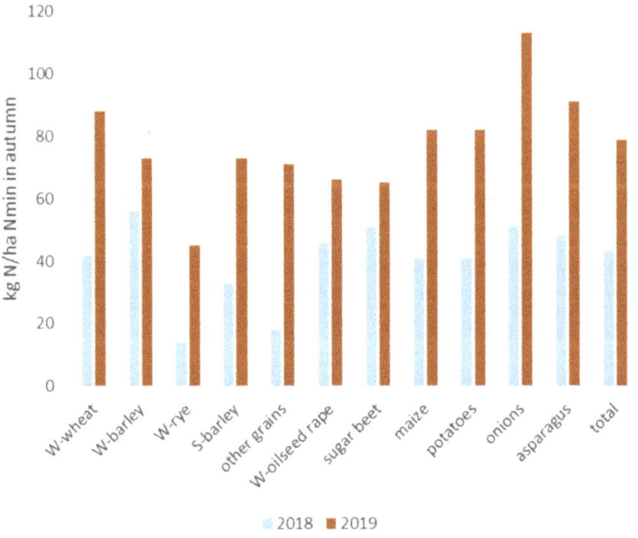

Figure 5. Spring N_{min}-values in 2018 and 2019 (0–90 cm depth) with reference to the preceding crop [47]; average increase in N_{min}-value of 36 kg/ha (source: [48], adapted and translated).

The elevated soil content of plant-available nitrogen has to be estimated precisely and crop fertilization has to be adjusted accordingly. In this way, the N surplus of the previous drought year can be counterbalanced through adapted fertilizer input in the subsequent year.

Factors explaining the elevated N_{min}-concentrations in spring are the dry weather conditions in the preceding year, which put the vegetation under water stress and reduce their nutrient uptake [49], a stimulation of mineralization and nitrification when dry soils are rewet [50] and a prolongation of the water recharge rate, in summer owing to high evaporation or extreme events connected to a large proportion of run-off, and in winter owing to less total precipitation [44].

4.2. General Economic Impact

From the farmer's perspective, a reduction in yield and/or in quality leads to less farm income. Especially dairy farms and producers of beef suffered from a decrease of harvested roughages (such as grassland or maize). One consequence of this shortage in feedstuff was that farmers had to reduce the number of livestock, for example, fattening bulls or even dairy cattle. While the prices for agricultural products of animal origin thus decreased, the prices for plant products increased (Figure 6). This was especially the case for those plant products used as feed, for example, roughage and grains, and even straw [51].

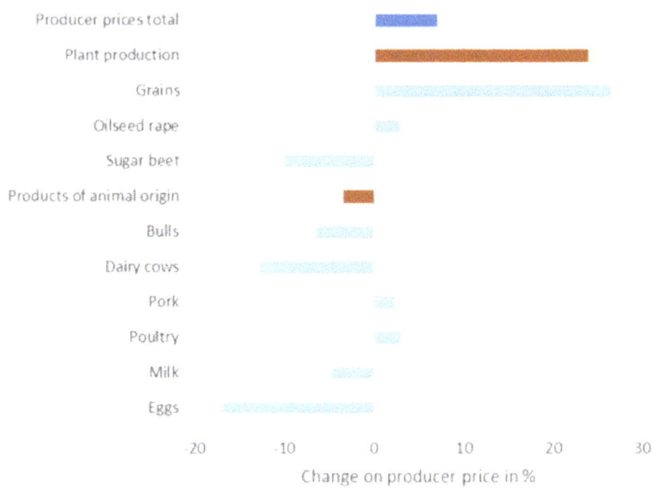

Figure 6. Development of producer prices in agriculture between January 2018 to January 2019 [52] (source: [52], adapted and translated).

As a consequence, the Federal Ministry of Food and Agriculture and the agricultural ministries of the federal states together put into effect an emergency management plan for farmers including financial support as well as some specific options, for example, on fodder provision: farmers were allowed—after having undergone a simplified application procedure—to harvest catch crop plots they had cultivated as part of the so-called "Greening" in the framework of the Common Agricultural Policy of the European Union without losing their subsidies [53].

4.3. Pressure on Specific Farm Types

Owing to the elevated concentrations on residual nitrogen in soils, the fertilization need of crops in spring 2019 was less pronounced than in preceding years. As a result, the quantity of continuously produced manure in livestock farms may have exceeded existing storage capacities. This may particularly be the case in the hot-spot region of livestock production in the north-west of Germany, where manure production, biogas plants, and—bordering the Netherlands—manure import is accumulating, while export opportunities are limited. These factors result in a particularly high local pressure on soil and ground water quality [54].

The situation for the farmers in autumn 2018 was further aggravated as the rules of the German Fertilization Ordinance, adopted in 2017 [31], largely restricted manure application on arable land in autumn after harvest. These rules only allowed fertilization in autumn for barley, W-oilseed rape, fodder crops, and catch crop, and only in cases plants were in need for nitrogen, at most with a dose of 60 kg N/ha. Owing to an infringement process on the implementation of the Nitrates Directive, new and even stricter amendments of the Fertilization Ordinance were passed in March 2020 and set into effect 1st of May 2020. Fertilization in autumn is now restricted to W-oilseed rape and W-barley. In areas with groundwater bodies in bad chemical condition, from 2021 onwards, no fertilization will be allowed in autumn [38].

Legislation, however, also produces opportunities for the application of animal manure; while according to the German Fertilization Ordinance [31], the target yield for fertilization planning had usually to be deduced from a three-year average, extreme weather events may be left out, so farmers may calculate the same long-term target yield for 2019 as in the previous year, to the detriment of the environment. However, farmers had to take the elevated spring N_{min}-values into account.

Owing to the exemption clause in autumn 2018, in order to overcome feedstuff shortages, environmentally relevant grassland and catch crop plots were regionally harvested more intensively, possibly leading to a higher nutrient export from these plots. The effect is probably limited, as—under the dry conditions in 2018—yields of grassland decreased, especially for the second and third cuts.

4.4. Measures for the Adaption to Climate Change

Subsequently, we list measures to adapt to pressures on agriculture linked to climate change.

4.4.1. Measures for Reducing Nitrogen Losses in Autumn/Winter

Late top dressing with nitrogen is always at risk with respect to droughts. However, when the situation of drought becomes obvious in May or June, these N inputs can be omitted. Therefore, crops producing high quality without high nitrogen inputs are favorable for risky cultivation sites. Baking wheat varieties exist with high baking quality performance in combination with low protein concentration in the grains. These varieties do not need high quantities of nitrogen in late development stages [55]. What hampers the introduction of these nitrogen efficient varieties is the lack of analytical methods for a quick assessment of their baking quality as criterion for the payment of the farmers replacing the raw protein content as criterion [56,57]. Newly discovered, old species like spelt, emmer, and einkorn wheat produce high quality grains with less nitrogen fertilization. Producing these kind of crops minimizes the residual nitrogen in soil after harvest [58].

A need for nitrogen fertilization in autumn is justified by the necessity to feed young crops with readily available nutrients. However, in most cases, there is sufficient nitrogen left from preceding crops, especially owing to climate change, where the vegetation period is prolonged, with warm temperatures until November or even December. In the case of sufficient precipitation or remaining soil moisture, mineralization will occur until this time. A ban of nitrogen fertilization in autumn would result in an increase of the pressure on manure storage capacities in order to store manure over the winter period (with respect to the pig model farm 450 m^3 manure storage is necessary in the case in which 60 kg/ha was applied in autumn). The new amendments of the Fertilization Ordinance include further restrictions on manure application in autumn [38].

The cultivation of catch crops is a highly efficient measure to keep nitrogen from leaching during the winter period [59–61]. In Germany, the cultivation of catch crops over the winter season was practiced in 2018 on 1.43 mio ha. If winter catch crops would be grown in all suitable crop rotations with summer crops and maize, 2.89 mio ha could be covered [62], an increase of almost 50%. With regard to water protection, winter hardy catch crops are most efficient, but seed bed preparation in spring for the following main crop might become difficult. The new amendments of the Fertilization Ordinance thus includes the cultivation of winter catch crops in areas of groundwater bodies in bad chemical conditions, in case the yearly precipitation exceeds a certain limit [38].

Owing to dry weather conditions in autumn, the germination and further establishment of a catch crop cover might be difficult. In the case of sufficient annual precipitation, the undersowing of forage or catch crops in the preceding main crop might be a solution [63].

Moreover, in the case of sufficient annual precipitation, double cropping may be the choice in order to almost permanently cover the soil and gain income from more than one harvest by year. Furthermore, N removal through a second crop offers opportunities to curb down N surplus and residual mineral N in the soil.

Grassland turnover in general leads to high mineralization rates of the soil humus, with negative effects on the concentration of the mobile soil nitrogen. Therefore, grassland renewal in autumn is problematic with respect to water protection.

4.4.2. Measures Reducing Risk of Crop Failure

The risk of crop failure can be reduced by using crop mixtures instead of the cultivation of one single breed [64].

Some crops show an adaption to changed climatic conditions, especially droughts, owing to their specific yield and quality production mechanisms [65]. Intensive efforts are made to increase plant resistance to droughts through breeding [66].

Plants composing grasslands respond differently to the combined effects of climatic change, resulting in compositional changes in plant communities [67]. Grassland with a higher percentage of herbs with deeper roots shows a better heat/drought stress tolerance, but produces lower yields. Harvesting grassland in early spring/summer would primarily ensure a certain amount of hay available for the cattle, but early cutting violates German legislation on environmental friendly production (e.g., bird protection). This implies that dairy cattle farming will probably have to increase the main forage area (thus, reducing animal density) and conserve more hay and silage to cope with feedstuff shortages [68].

4.4.3. Measures for Increasing N-Efficiency during the Season

There is a long list of measures that could be realized in order to increase nitrogen-efficiency during the whole season (for more details, see [58]):

- Soil analysis of N_{min} in autumn and spring (instead of using standard values); the autumn value gives information on the precision of the fertilization planning and the spring value informs on the current level of plant available nitrogen in soil,
- Improving prognosis of N delivery from soil (originating from organic matter, organic fertilization, and preceding crop/catch crop);
- Monitoring plant needs and adjusting fertilization accordingly (i.e., by applying of different methods of plant analysis;
- Monitoring and prognosis of water supply to plants [69] (i.e., German Weather Service (dwd) prognosis model of soil water saturation [70]);
- Using farm rapid analysis methods (i.e., near-infrared spectrocospy—NIRS) for the evaluation of the fertilizer value of organic fertilizers, especially liquid manure [71];
- Using low-emission application techniques like drip hose, trailing shoe, or injection of liquid manure and adjusting supplementary mineral fertilization accordingly;
- Using acidification of liquid manure, nitrification, or urease inhibitors in order to prevent the mineralization and evaporation of organically bound nitrogen;
- Implementation of N-efficient crops and rotations with long-term land cover (catch crop cultivation) and crop mixtures;

- Limitation of the organically bound nitrogen being applied as fertilizer to a lower value, such as 170 kg N/ha and year (as declared in the Nitrates Directive [32]); if not yet performed, extend the rule to all organic fertilizers, not only to animal manure;
- Application of precision farming techniques;
- Installation of field irrigation.

4.4.4. Tightening Legislation

After the EU court took legal action against Germany, the current Fertilization Ordinance [31] is being revised and tightened [54]. This will most likely pose further restrictions on fertilization and management options for farmers.

Unfortunately, with the recent revision of the German Fertilization Ordinance [38], nitrogen soil surface budgets on the farm level as control instruments for excessive use of manure are no longer compulsory.

On the whole, climate change poses a great challenge for the implementation of the Water Framework Directive, its "daughter directives", and the national and regional transcripts.

5. Conclusions

According to our research, extreme summer droughts like in 2018 result in less yield, less plant uptake of nitrogen, a reduced nitrogen efficiency, and an increased N surplus. This is because of reduced plant growth and thus reduced nitrogen uptake by plants, as well as owing to a prolongation of the water recharge rate and an increased mineralization of soil biomass after soil rewetting. As agri-drinking water indicator, the elevated N surplus puts more pressure on the quality of surface, ground, and drinking water.

Presuming a fertilization level defined by plant needs according to fertilization planning under regular climate conditions, the main factors influencing the N surplus are splitting and timing of fertilization and the share of organically bound nitrogen in the fertilizers applied.

Monitoring of plant needs in the framework of precision farming and the splitting and timing of fertilization accordingly may in future lead to a reduction of N surplus.

The examples of the model farms show clearly, however, that fertilization regimes with high shares of organic fertilizers produce higher nitrogen surpluses (gross and net). Taking into account the subsequent delivery of plant-available nitrogen, for example, after crop failure, the reasonable use of farm-own manure is limited quantitatively.

There are feasible options to mitigate the risks of crop failure or risk of overfertilization of crops during droughts. However, many of these options imply an extensification of current agricultural practices.

Author Contributions: Conceptualization, B.O., C.H., and S.K.; Methodology, S.K.; Investigation, S.K.; Resources, S.K.; Data curation, S.K.; Writing—Original draft preparation, S.K.; Writing—Review and editing, C.H. and B.O.; Visualization, S.K. All authors have read and agreed to the published version of the manuscript.

Funding: This project has received funding from the European Union's Horizon 2020 research and innovation program under grant agreement No. 727984.

Conflicts of Interest: The authors declare no conflict of interest.

References

1. DWD. Erste Bilanz des Deutschen Wetterdienstes zum Jahr 2018 in Deutschland—2018 ist das wärmste Jahr in Deutschland seit Messbeginn1881. In Proceedings of the Pressemitteilung, Offenbach, Germany, 20 December 2018; Available online: https://www.dwd.de/DE/presse/pressemitteilungen/DE/2018/20181220_jahr2018_rekord.pdf?__blob=publicationFile&v=2 (accessed on 28 January 2020).
2. DWD. Durchschnittlicher monatlicher Niederschlag in Deutschland von Februar 2018 bis Februar 2019 (in Liter pro Quadratmeter). 2019. Available online: http://dwd.de (accessed on 28 January 2020).

3. DWD. Auch im Herbst 2018 Riss die Trockenserie in der Landwirtschaft Nicht ab. 2019. Available online: https://www.dwd.de/DE/fachnutzer/landwirtschaft/berichte/32__rueckblicke/2018/bericht_herbst_2018.pdf?__blob=publicationFile&v=7; (accessed on 28 January 2020).
4. DWD. Durchschnittliche Sonnenscheindauer pro Monat in Deutschland von Februar 2018 bis Februar 2019 (in Stunden). 2019. Available online: http://dwd.de (accessed on 28 January 2020).
5. Destatis. *Wachstum und Ernte—Feldfrüchte—2018. Fachserie 3 Reihe 3.2.1*; Statistisches Bundesamt (Destatis): Wiesbaden, Germany, 2019.
6. Sarhadi, A.; Ausín, M.C.; Wiper, M.P.; Touma, D.; Diffenbaugh, N.S. Multidimensional risk in a nonstationary climate: Joint probability of increasingly severe warm and dry conditions. *Sci. Adv.* **2018**, *4*, 3487. [CrossRef]
7. Rosenzweig, C.; Iglesias, A.; Yang, X.B.; Epstein, P.R.; Chivian, E. Climate change and extreme weather events Implications for food production, plant diseases, and pests. *Glob. Chang. Hum. Health* **2001**, *2*, 90–104. [CrossRef]
8. OECD. *OECD Core Set of Indicators for Environmental Performance Reviews: A Synthesis Report by the Group on the State of the Environment*; Environment Monographs; No. 83; OECD Publishing: Paris, France, 1993.
9. Kremer, A.M. *Methodology and Handbook—Eurostat/OECD—Nutrient Budgets EU-27, Norway, Switzerland*; European Commission, Eurostat: Brussels, Belgium, 2007; 112p, Available online: https://ec.europa.eu/eurostat/documents/2393397/2518760/Nutrient_Budgets_Handbook_%28CPSA_AE_109%29_corrected3.pdf/4a3647de-da73-4d23-b94b-e2b23844dc31 (accessed on 28 January 2020).
10. Eurostat. Agri-Environmental Indicator—Gross Nitrogen Balance. 2018. Available online: https://ec.europa.eu/eurostat/statistics-explained/index.php/Agri-environmental_indicator_-_gross_nitrogen_balance (accessed on 15 May 2020).
11. van Beek, C.L.; Brouwer, L.; Oenema, O. The use of farm gate balances and soil surface balances as estimator for nitrogen leaching to surface water. *Nutr. Cycl. Agroecosyst.* **2003**, *67*, 233–244. [CrossRef]
12. Wick, K.; Heumesser, C.; Schmid, E. Groundwater nitrate contamination: Factors and indicators. *J. Environ. Manag.* **2012**, *111*, 178–186. [CrossRef] [PubMed]
13. Lord, I.; Antony, S.G. Agricultural nitrogen balances and water quality in the UK. *Soil Use Manag.* **2002**, *18*, 362–369. [CrossRef]
14. Sieling, K.; Kage, H. N balance as an indicator of N leaching in an oilseed rape—Winter wheat—Winter barley rotation. *Agric. Ecosyst. Environ.* **2006**, *115*, 261–269. [CrossRef]
15. Rankinen, K.; Salo, T.; Granlund, K.; Rita, H. Simulated nitrogen leaching, nitrogen mass field balanes and their correleation on four farms in south-western Finland during the peeriod 2000–2005. *Agric. Food Sci.* **2007**, *16*, 98–107.
16. Dalgaard, T.; Bienkowski, J.F.; Bleeker, A.; Dragosit, U.; Drouet, J.L.; Durand, P.; Frumau, A.; Hutchings, N.J.; Kedziora, A.; Magliulo, V.; et al. Farm nitrogen balances in six European landscapes as an indicator for nitrogen losses and basis for improved management. *Biogeosciences* **2012**, *9*, 5303–5321. [CrossRef]
17. Spiess, E. Nitrogen, phosphorus and potassium balances and cycles of Swiss agriculture from 1975 to 2008. *Nutr. Cycl. Agroecosyst.* **2011**, *91*, 351–365. [CrossRef]
18. Poisvert, C.; Curie, F.; Moatar, F. Annual agricultural N surplus in France over a 70-year period. *Nutr. Cycl. Agroecosyst.* **2017**, *107*, 63–78. [CrossRef]
19. Oenema, O.; Kros, H.; de Vries, W. Approaches and uncertainties in nutient budgets. Implications for nutrient management and environmental policies. *Eur. J. Agron.* **2003**, *20*, 3–16. [CrossRef]
20. Cameira, M.R.; Rolim, J.; Valente, F.; Faro, A.; Dragosits, U.; Cordovil, C.M.D.S. Spatial distribution and uncertainties of nitrogen budgets for agriculture in the Tagus river basin in Portugal—Implications for effectiveness of mitigation measures. *Land Use Policy* **2019**, *84*, 278–293. [CrossRef]
21. Gourley, C.J.P.; Aarons, S.R.; Powel, M. Nitrogen use efficiency and manure management practices in contrasting dairy production systems. *Agric. Ecosyst. Environ.* **2012**, *147*, 73–81. [CrossRef]
22. D'Haene, K.; Magyar, M.; Mulier, A.; De Neve, S.; Pálmai, O.; Nagy, J.; Németh, T.; Hofman, G. Comparison of N and P farm gate balances between the intensive agriculture in Flanders and the extensive agriculture in Hungary. In Proceedings of the 14th World Fertilizer Congress of the International Centre for Fertilizers, CIEC, Chiang Mai, Thailand, 22–27 January 2006.
23. Mulier, A.; Hofman, G.; Baecke, E.; Carlier, L.; De Brabander, D.; De Groote, G.; De Wilde, R.; Fiems, L.; Janssens, G.; Van Cleemput, O.; et al. A methodology for the calculation of farm level nitrogen and phosphorus balances in Flemish agriculture. *Eur. J. Agron.* **2003**, *20*, 45–51. [CrossRef]

24. Kim, H.; Surdyk, N.; Abel, H.; Møller, I.; Hansen, B. The link between agricultural impact on drinking water quality—Lessons learned in Denmark and France. *Water* **2020**. under review.
25. Klages, S.; Heidecke, C.; Osterburg, B.; Bailey, J.; Calciu, I.; Casey, C.; Dalgaard, T.; Frick, H.; Glavan, M.; D'Haene, K.; et al. Nitrogen surplus—A unified indicator for water pollution in Europe? *Water* **2020**, *12*, 1197. [CrossRef]
26. Gu, B.; Jub, X.; Chang, J.; Ge, Y.; Vitousek, P.M. Integrated reactive nitrogen budgets and future trends in China. *Proc. Natl. Acad. Sci. USA* **2015**, *112*, 8792–8797. [CrossRef] [PubMed]
27. Bouwman, L.; Klein Goldewijk, K.; Van Der Hoek, K.W.; Beusen, A.H.W.; Van Vuuren, D.P.; Willems, J.; Rufino, M.C.; Stehfest, E. Exploring global changes in nitrogen and phosphorus cycles in agriculture induced by livestock production over the 1900–2050 period. *Proc. Natl. Acad. Sci. USA* **2013**, *110*, 20882–20887. [CrossRef]
28. Dobers, E.S. Anpassungsbedarf bei der Nährstoffversorgung. In *Kühlen Kopf Bewahren—Anpassung der Landwirtschaft an den Klimawandel*; KTBL Kuratorium für Technik u. Bauwesen i. d. Landwirtschaft e.V.: Darmstadt, Germany, 2019; pp. 75–89.
29. Kolbe, H. Comparison of methods for calculation of legume N_2 fixation for use in practical agriculture. *Pflanzenbauwissenschaften* **2009**, *13*, 23–36.
30. Destatis. *Viehhaltung der Betriebe—Agrarstukturerhebung—2016. Fachserie 3 Reihe 2.1.3*; Statistisches Bundesamt (Destatis): Wiesbaden, Germany, 2017.
31. Federal Ministry of Justice and Consumer Protection. *Düngeverordnung (DüV) vom 26. Mai 2017. Verordnung über die Anwendung von Düngemitteln, Bodenhilfsstoffen, Kultursubstraten und Pflanzenhilfsmitteln nach den Grundsätzen der guten Fachlichen Praxis beim Düngen. BGBl I*; Bundesanzeiger Verlag GmbH: Köln, Germany, 2017; pp. 1305–1348. Available online: https://www.gesetze-im-internet.de/d_v_2017/D%C3%BCV.pdf (accessed on 20 April 2020).
32. Council of the European Union. *Council Directive 91/676/EEC of 12 December 1991 Concerning the Protection of Waters against Pollution Caused by Nitrates from Agricultural Sources. OJ L 375, 31.12.1991*; Council of the European Union: Brussels, Belgium, 1992; pp. 1–8. Available online: https://eur-lex.europa.eu/legal-content/EN/TXT/PDF/?uri=CELEX:31991L0676&from=EN) (accessed on 20 May 2020).
33. Haenel, H.-D.; Rösemann, C.; Dämmgen, U.; Döring, U.; Wulf, S.; Eurich-Menden, B.; Freibauer, A.; Döhler, H.; Schreiner, C.; Osterburg, B.; et al. *Calculations of Gaseous and Particulate Emissions from German Agriculture 1990–2018. Report on Methods and Data (RMD), Submission 2020. Thünen Report 57*; Johann Heinrich von Thünen-Institute: Braunschweig, Germany, 2020; p. 414.
34. Zinke, O. Wirtschaftsdünger—Weniger Gülleimporte aus den Niederlanden. *Agrarheute* **2018**. Available online: https://www.agrarheute.com/management/betriebsfuehrung/weniger-guelleimporte-niederlanden-548856 (accessed on 28 January 2020).
35. WBD Anwendung von Organischen Düngern Und Organischen Reststoffen in der Landwirtschaft. Standpunkt des Wissenschaftlichen Beirats für Düngungsfragen, 30p. 2015. Available online: https://www.bmel.de/SharedDocs/Downloads/Ministerium/Beiraete/Duengungsfragen/OrgDuengung.pdf?__blob=publicationFile (accessed on 28 January 2020).
36. German Biogas Association Figures and Facts about Biogas. Available online: https://www.biogas.org/edcom/webfvb.nsf/id/EN-Figures-and-facts-about-biogas (accessed on 28 January 2020).
37. BMEL Statistisches Jahrbuch 2018. 40. Inlandsabsatz von Handelsdünger. a. Nach Sorten und Nährstoffen. 1.000 t Nährstoff. 3060210. 2019. Available online: https://www.bmel-statistik.de/footernavigation/archiv/statistisches-%20jahrbuch/ (accessed on 28 January 2020).
38. Federal Ministry for Food and Agriculture. *Verordnung zur Änderung der Düngeverordnung und Anderer Vorschriften vom 28. BGBl. I*; Bundesanzeiger Verlag GmbH: Köln, Germany; pp. 846–861. Available online: https://www.bgbl.de/xaver/bgbl/start.xav?startbk=Bundesanzeiger_BGBl&start=%2F%2F%2A%5B%40attr_id=%27bgbl120s0846.pdf%27%5D#__bgbl__%2F%2F*%5B%40attr_id%3D%27bgbl120s0846.pdf%27%5D__1589993256631 (accessed on 15 May 2020).
39. BMEL Statistischer Monatsbericht des Bundesministeriums für Landwirtschaft und Ernährung, Kapitel, A. Landwirtschaft. Nährstoffbilanzen und Düngemittel. MBT-0111130-0000 Flächenbilanz von 1990 bis 2017—in kg N/ha, MBT-0111160-0000Flächenbilanz von 1990 bis 2017—kt N. 2020. Available online: https://www.bmel-statistik.de/landwirtschaft/statistischer-monatsbericht-des-bmel-kapitel-a-landwirtschaft/ (accessed on 31 January 2020).

40. Bach, M.; Godlinski, F.; Greef, J.-M. Handbuch Berechnung der Stickstoff-Bilanz für die Landwirtschaft in Deutschland Jahre 1990–2008. In *Berichte aus dem Julius Kühn-Institut*; Federal Research Centre for Cultivated Plants: Braunschweig, Germany, 2011; Volume 159, 28p.
41. EAA (2018): Glossary—List of Environmental Terms Used by EEA. Available online: https://www.eea.europa.eu/help/glossary#c4=10andc0=allandb_start=0andc2=dpsir (accessed on 20 February 2018).
42. Gabrielsen, P.; Bosch, P. *Internal Working Paper Environmental Indicators: Typology and Use in Reporting*; European Environment Agency: Copenhagen, Denmark, 2003; 20p.
43. Smeets, E.; Weterings, R. *Environmental Indicators: Typology and Overview. Technical Report No. 25*; European Environment Agency: Copenhagen, Denmark, 1999; 19p.
44. Benateau, S.; Gaudard, A.; Stamm, C.; Altermatt, F. *Climate Change and Freshwater Ecosystems: Impacts on Water Quality and Ecological Status. Hydro-CH2018 Project*; Federal Office for the Environment (FOEN): Bern, Switzerland, 2019; p. 110. Available online: https://www.zora.uzh.ch/id/eprint/169641/7/Climate_change_and_freshwater_ecosystems.pdf (accessed on 22 May 2020).
45. Hannappel, S.; Köpp, C.; Zühlke, S. Aufklärung der Ursachen von Tierarzneimittelfunden im Grundwasser—Untersuchung eintragsgefährdeter Standorte in Norddeutschland. *UBA Texte* **2016**, *54*, 149. Available online: https://www.umweltbundesamt.de/publikationen/aufklaerung-der-ursachen-von-tierarzneimittelfunden (accessed on 29 April 2020).
46. Mosley, L.M. Drought impacts on the water quality of freshwater systems; review and integration. *Earth Sci. Rev.* **2015**, *140*, 203–214. [CrossRef]
47. LWK Niedersachsen Vorabveröffentlichung der Ernte-N_{min}-Werte 2018—aus den Wasserschutz-und ergänzenden Untersuchungen in Pflanzenbauversuchen der LWK Niedersachsen, 26p. 2018. Available online: https://m.lwk-niedersachsen.de/?file=30940 (accessed on 28 January 2020).
48. Wbl-mr-Hessen N_{min}-Werte der WRRL-Beratung im Hessischen Ried. Stand 26.02.2018 bzw. 05.04.2019. Wasser-, Boden-und Landschaftspflegeverband Hessen, Griesheim, Germany. 2018. 2019. Available online: http://wbl-mr-hessen.de/ (accessed on 28 January 2020).
49. Prasuhn, V.; Albisser, C.V. Grundwasserqualität und Bewässerung—Eine Lysimeterstudie zur Schadstoffverfrachtung ins Grundwasser. *Aqua Gas* **2014**, *4*, 54–58.
50. Morecroft, M.D.; Burt, T.P.; Taylor, M.E.; Rowland, A.P. Effects of the 1995–1997 drought on nitrate leaching in lowland England. *Soil Use Manag.* **2000**, *16*, 117–123. [CrossRef]
51. Zinke, O. Futterversorgung—Futtermangel bedroht Existenzen. *Agrarheute* **2019**. Available online: https://www.agrarheute.com/markt/futtermittel/futtermangel-bedroht-existenzen-552437 (accessed on 15 May 2020).
52. Destatis Preise—Preisindizes für die Land-und Forstwirtschaft. Available online: https://www.destatis.de/DE/Themen/Wirtschaft/Preise/Landwirtschaftspreisindex-Forstwirtschaftspreisindex/Publikationen/Downloads-Land-und-Forstwirtschaftpreise/erzeugerpreise-land-forstwirtschaft-2170100191094.html (accessed on 1 May 2020).
53. BMEL Erweiterte Möglichkeiten der Futternutzung von ökologischen Vorrangflächen. Available online: https://www.bmel.de/SharedDocs/Pressemitteilungen/2018/136-Vorrangflaechen.html;jsessionid=EA792F11361CAE8462715428C4FB260A.2_cid385 (accessed on 28 January 2020).
54. Meergans, F.; Lenschow, A. Die Nitratbelastung in der Region Weser-Ems: Inkohärenzen in Wasser-, Energie- und Landwirtschaftspolitik. *Neues Archiv für Niedersachsen* **2018**, *2*, 105–117.
55. Böse, S. Stickstoffeffizienz von Weizensorten. *Praxisnah* **2016**, *2*, 2–4.
56. Laidig, F.; Piepho, H.-P.; Hüsken, A.; Begemann, J.; Rentel, D.; Drobek, T.; Meyer, U. Predicting loaf volume for winter wheat by linear regression models based on protein concentration and sedimentation value using samples from VCU trials and mills. *J. Cereal Sci.* **2018**, *84*, 132–141. [CrossRef]
57. Gabriel, D.; Pfitzner, C.; Haase, N.; Hüsken, A.; Prüfer, H.; Greef, J.-M.; Rühl, G. New strategies for a reliable assessment of baking quality of wheat—Rethinking the current indicator protein content. *J. Cereal Sci.* **2017**, *77*, 126–134. [CrossRef]
58. Klages, S.; Apel, B.; Feller, C.; Hofmeier, M.; Homm-Belzer, A.; Hüther, J.; Loeloff, A.; Olfs, H.-W.; Osterburg, B. *Effizient düngen—Anwendungsbeispiele zur Düngeverordnung*; Bundesanstalt für Landwirtschaft und Ernährung (BLE): Bonn, Germany, 2018; p. 68.
59. Basche, A.D.; Miguez, F.E.; Kaspar, T.C.; Castellano, M.J. Do cover crops increase or decrease nitrous oxide emissions? A meta-analysis. *J. Soil Water Conserv.* **2014**, *69*, 471–482. [CrossRef]

60. Thapa, R.; Mirsky, S.B.; Tully, K.L. Cover Crops Reduce Nitrate Leaching in Agroecosystems: A Global Meta-Analysis. *J. Environ. Qual.* **2018**, *47*, 1400–1411. [CrossRef] [PubMed]
61. Tonitto, C.; David, M.B.; Drinkwater, L.E. Replacing bare fallows with cover crops in fertilizer-intensive cropping systems: A meta-analysis of crop yield and N dynamics. *Agric. Ecosyst. Environ.* **2006**, *112*, 58–72. [CrossRef]
62. BMEL. *Referat 123 Ernteberícht 2018*; BMEL: Berlin, Germany, 2018; Available online: https://www.bmel.de/SharedDocs/Downloads/DE/_Landwirtschaft/Pflanzenbau/Ernte-Bericht/ernte-2018.pdf;jsessionid=5F69D1B154A2B0C2FBD21C5593A9C1C9.internet2851?__blob=publicationFile&v=2 (accessed on 15 May 2020).
63. Valkama, E.; Lemola, R.; Känkänen, H.; Turtola, E. Meta-analysis of the effects of undersown catch crops on nitrogen leaching loss and grain yields in the Nordic countries. *Agric. Ecosyst. Environ.* **2015**, *203*, 93–101. [CrossRef]
64. Paulsen, H.M.; Schochow, M.; Ulber, B.; Kühne, S.; Rahmann, G. Mixed cropping systems for biological control of weeds and pests in organic oilseed crops. *Asp. Appl. Biol.* **2006**, *79*, 215–220.
65. Rafiee, M.; Abdipour, F. Mixed Cropping system with Biofertilizer under Dry land condition: I. Evaluation of Drought Tolerance. *Bull. Environ. Pharm. Life Sci.* **2015**, *4*, 49–54.
66. Schimmelpfennig, S.; Heidecke, C.; Beer, H.; Bittner, F.; Klages, S.; Krengel, S.; Lange, S. *Klimaanpassung in Land- und Forstwirtschaft—Ergebnisse eines Workshops der Ressortforschungsinstitute FLI, JKI und Thünen-Institut. Thünen Working Paper 2018, 86*; Johann Heinrich von Thünen-Institute: Braunschweig, Germany, 2018; 110p.
67. Ghahramani, A.; Howden, S.M.; del Prado, A.; Thomas, D.T.; Moore, A.D.; Ji, B.; Ates, S. Climate Change Impact, Adaptation, and Mitigation in Temperate Grazing Systems: A Review. *Sustainability* **2019**, *11*, 7224. [CrossRef]
68. Elsässer, M.; Grant, K. Anpassungsstrategien zur Sicherung von Ertrag und Qualität im Grünland. In *Kühlen Kopf bewahren—Anpassungen der Landwirtschaft an den Klimawandel*; KTBL Kuratorium für Technik u. Bauwesen i. d. Landwirtschaft e.V.: Darmstadt, Germany, 2019; pp. 135–173.
69. Freebairn, D.M.; Ghahramani, A.; Robinson, J.B.; Mc Clymon, D.J. A tool for monitoring soil water using modelling, on-farm data, and mobile technology. *Environ. Model. Softw.* **2018**, *104*, 55–63. [CrossRef]
70. Zinke, O. Wetterprognose—Deutscher Wetterdienst will Dürren vorhersagen. *Agrarheute* **2019**. Available online: https://www.agrarheute.com/management/betriebsfuehrung/deutscher-wetterdienst-will-duerren-vorhersagen-552709 (accessed on 5 March 2020).
71. Severin, K.; Hoffmann, A.; Licht, F.; Olfs, H.-W.; Rest, T.; Tillmann, P. *Die Nahinfrarotspektroskopie (NIRS) zur Untersuchung von Güllen und Gärresten. Standpunkt des VDLUFA*; Verband Deutscher Landwirtschaftlicher Untersuchungs- und Forschungsanstalten e.V. (VDLUFA): Speyer, Germany, 2019.

Article

Nitrogen Surplus—A Unified Indicator for Water Pollution in Europe?

Susanne Klages [1,*], Claudia Heidecke [1,*], Bernhard Osterburg [1], John Bailey [2], Irina Calciu [3], Clare Casey [4], Tommy Dalgaard [5], Hanna Frick [6], Matjaž Glavan [7], Karoline D'Haene [8], Georges Hofman [9,10], Inês Amorim Leitão [11], Nicolas Surdyk [12], Koos Verloop [13] and Gerard Velthof [13]

1 Coordination Unit Climate, Johann Heinrich von Thünen-Institute, 38116 Braunschweig, Germany; bernhard.osterburg@thuenen.de
2 Agri-Food and Biosciences Institute (AFBI), Belfast BT9 5PX, Northern Ireland, UK; john.bailey@afbini.gov.uk
3 Department of Physics and Technology, National Research and Development Institute for Soil Science, Agrochemistry and Environment, 011464 Bucharest, Romania; irina.calciu@icpa.ro
4 Nitrates, Biodiversity and Engineering Division, Department of Agriculture, Food and the Marine, Y35PN52 Wexford, Ireland; Clare.Casey@agriculture.gov.ie
5 Department of Agroecology, Aarhus University, DK-8830 Tjele, Denmark; tommy.dalgaard@agro.au.dk
6 Research Institute of Organic Agriculture (FiBL), 5070 Frick, Switzerland; hanna.frick@fibl.org
7 Biotechnical Faculty, University of Ljubljana, 1000 Ljubljana, Slovenia; matjaz.glavan@bf.uni-lj.si
8 Research and Advisory Board on Sustainable Fertilization and ILVO Plant, 9820 Merelbeke, Belgium; karoline.dhaene@ilvo.vlaanderen.be
9 Department of Environment, Faculty of Bioscience Engineering, UGent, 9000 Gent, Belgium; Georges.Hofman@UGent.be
10 Research and Advisory Board on Sustainable Fertilization, 9820 Merelbeke, Belgium
11 Polytechnic Institute of Coimbra (IPC), Coimbra Agrarian Higher School (ESAC), Research Centre for Natural Resources, Environment and Society (CERNAS), 3045-093 Coimbra, Portugal; ines.leitao@esac.pt
12 Bureau de Recherches Geologiques et Minieres (BRGM), 45060 Orléans, France; N.Surdyk@brgm.fr
13 Environmental Research, Wageningen University and Research, 6700 AA Wageningen, The Netherlands; koos.verloop@wur.nl (K.V.); gerard.velthof@wur.nl (G.V.)
* Correspondence: susanne.klages@thuenen.de (S.K.); claudia.heidecke@thuenen.de (C.H.); Tel.: +49-531-596-1111 (C.H.)

Received: 10 March 2020; Accepted: 15 April 2020; Published: 22 April 2020

Abstract: Pollution of ground-and surface waters with nitrates from agricultural sources poses a risk to drinking water quality and has negative impacts on the environment. At the national scale, the gross nitrogen budget (GNB) is accepted as an indicator of pollution caused by nitrates. There is, however, little common EU-wide knowledge on the budget application and its comparability at the farm level for the detection of ground-and surface water pollution caused by nitrates and the monitoring of mitigation measures. Therefore, a survey was carried out among experts of various European countries in order to assess the practice and application of fertilization planning and nitrogen budgeting at the farm level and the differences between countries within Europe. While fertilization planning is practiced in all of the fourteen countries analyzed in this paper, according to current legislation, nitrogen budgets have to be calculated only in Switzerland, Germany and Romania. The survey revealed that methods of fertilization planning and nitrogen budgeting at the farm level are not unified throughout Europe. In most of the cases where budgets are used regularly (Germany, Romania, Switzerland), standard values for the chemical composition of feed, organic fertilizers, animal and plant products are used. The example of the Dutch Annual Nutrient Cycling Assessment (ANCA) tool (and partly of the Suisse Balance) shows that it is only by using farm-specific "real" data that budgeting can be successfully applied to optimize nutrient flows and increase N efficiencies at the farm level. However, this approach is more elaborate and requires centralized data processing under consideration of data protection concerns. This paper concludes that there is no unified indicator for nutrient management and water quality at the farm level. A comparison of regionally calculated nitrogen budgets across European countries needs to be interpreted carefully, as methods as well as data and emission factors vary across countries. For the implementation of EU nitrogen-related policies—notably, the Nitrates Directive—nutrient budgeting is currently ruled out as an entry point for legal requirements. In contrast, nutrient budgets are highlighted as an environment indicator by the OECD and EU institutions.

Keywords: nitrogen budget; nitrogen balance; water pollution; nitrates; agriculture; drinking water

---

1. Introduction

1.1. Nitrogen Budgets and Relation to Water Quality

Nitrogen (N) surplus is used as an indicator to compare the environmental status with regard to nitrogen between the EU and OECD member states. The OECD [1,2] suggested the gross nitrogen budget (GNB) as an appropriate method to calculate comparable indicators at the regional and the national scale. The OECD approach was later adopted by Eurostat [3]. Budgets calculating the nitrogen surplus at the farm level are particularly used to determine the leaching potential from arable land [4]. According to Eurostat/OECD [3], the term "nitrogen budget," as introduced by [5], is more comprehensive and appropriate than the term "nitrogen balance," as it includes a summary of all major N flows between the major compartments of agriculture and the environment. One aim of the Horizon2020 European Union-funded project FAIRWAY (farm systems management and governance for producing good water quality for drinking water supplies), running from 2017 to 2021, is to identify "agri-drinking water indicators (ADWIs)": harmonized indicators—preferably at the farm level—for detecting nitrate pollution of ground and surface waters and for monitoring implemented mitigation measures. These ADWIs should be "ready to use" and produce comparable results across Europe. Therefore, we conducted a survey among the experts within the project and selected experts of further countries in order to evaluate the current state of applying nitrogen budgets at the farm level in Europe and the possibility of establishing nitrogen budgets as a unified ADWI at the farm level. In this paper, we thus focus on calculation methods and implementation set ups for nitrogen budgets at the farm scale in different European countries. Looking at nitrogen budgets from a comparability and applicability perspective, we discuss the differences and decisive parameters and so contribute to the current state of research.

In Austria, Wick et al. [6] used the nitrogen land budget to compare nationwide regional agricultural budgets with the concentration of nitrates in groundwater of corresponding catchments, finding a good statistical correlation.

Wick et al. [6] reported that a couple of authors cast doubt upon the applicability of the soil surface budget to assess the actual nitrate leaching. The budget is a theoretical concept describing only the potential for groundwater contamination [7–9]. Wick et al. [8] further explain that some authors find only a poor statistical relationship between the soil surface budget result and nitrate leaching, using correlation analysis [10], analysis of covariance [8,10] and regression analysis [9–11]. According to [8], these statistical evaluations are geographically and temporarily limited. This can be explained by the fact that the soil surface budget is an indicator for potential N loss through nitrate leaching, subject to denitrification and shifts of soil organic matter (SOM) content as interfering elements. The rate of denitrification as well as the share of N input fixed in SOM differs between soils, climatic zones and types of farm management.

Hansen et al. [12] found significant correspondence between developments in N surplus and nitrate concentrations in upper groundwater for four subsequent development periods for Danish agriculture in the period 1946–2012, and similarly for developments in the nitrogen use efficiency.

Dalgaard et al. [13] calculated gross farm budgets for six European landscapes in Poland, The Netherlands, France, Italy, Scotland and Denmark as an indicator for N losses: the highest surpluses were found in livestock-intensive landscapes. The authors found significant correlations of N surplus to both nitrate concentrations in soils (A-horizons) and groundwater, measured during the period of N management data collection in the landscapes (2007–2009).

Andelov et al. [14], went a step further, assessing nitrogen reduction levels necessary to reach groundwater quality targets in Slovenia. For this purpose, the hydrological model GROWA–DENUZ (Großräumiges Wasserhaushaltsmodell-Denitrifikation in der ungesättigten Zone) was coupled with agricultural N budgets and applied consistently to the whole territory of Slovenia. Model results indicated that additional specific N reduction measures should be implemented in priority areas rather than across a whole country. This would allow the development of specific measures and spatial allocation of financial funds.

1.2. Nitrogen Surpluses as a Result of Budgets and as Indicators for Measures and Policies

For Switzerland, Spiess [15] calculated nitrogen (as well as phosphorus and potassium) budgets in the period 1975–2008. Intensification of Swiss agriculture after 1950 was characterized by increasing plant nutrient inputs into the agricultural production system. The farm gate budget for 2008 showed a surplus of 108 kg N. Nutrient surpluses rose between 1975 and 1980 and then decreased significantly until 2008, with a higher percentage of relative reductions for P (80%) and K (54%) than for N (27%). A pronounced decrease in nutrient surpluses was achieved through the introduction of direct payments for ecological programs in 1993. Reductions in mineral fertilizer use and N deposition primarily caused lower surpluses. Uncertainties in the budget calculation were mostly due to biological N fixation and atmospheric deposition. Recent figures show that N surplus has stagnated on a high level for the last two decades.

In France, Poisvert et al. [16] calculated N soil surface budgets according to [17] between 1940 and 2010 for metropolitan France. The authors showed that there was high variation for the N budgets between regions and time and performed a sensitivity analysis, revealing that N removal by crop production and N inflow by organic fertilizer use contributed most to total uncertainties in the N budgets. The average of the departmental imprecision in N surpluses ranged from 6 to 45 kg N/ha per utilized agricultural area (UAA) per year for the whole period, with an average of 21 kg N/ha per UAA per year, calculated by a Monte Carlo simulation analysis. Imprecisions were mainly linked to N export and organic fertilization, but also to symbiotic N fixation. Between 1940 and 1991, there was an upward trend in N surplus for 82% of the studied area. Between 1991 and 2010, there was a downward or stable trend for more than 90% of the area, probably as a consequence of the implementation of the Nitrates Directive.

In Portugal, Cameira et al. [18] quantified the gross soil nitrogen budget between 1989 and 2016, according to [17], for representative groups of crops in a catchment located in a nitrate vulnerable zone (NVZ). The N surplus varied over the 27 years, showing a general decrease of 1.8 kg N/ha/year, according to the type of crop. A higher decrease (26%) in N surplus occurred between 1989 and 1999 due to the reduction in fertilizer sales at the country level. After 2005, although at a lower rate, N surplus continued to decrease due to the restrictions imposed by the national action program linked to the Nitrates Directive.

1.3. Nitrogen Budgets for Investigating Farm Performance and as Tool for Farm Advice

Gourley et al. [19] calculated farm budgets for N, P and K for 41 representative Australian dairy farms. N surpluses ranged from 47 to 601 kg*ha^{-1} and nitrogen use efficiencies ranged from 14% to 50%. Imports and exports were measured individually. The authors detected a wide within-farm variation in the nutrient; content of forages, whereas variation in nutrient concentrations, except for sulfur, of the produced milk was small (CV < 15%). In organic farms, nutrient concentrations of forages were generally lower than in conventional farms. The authors conclude that an increase in milk production per ha is associated with rising nutrient surpluses at the farm scale, with potential for growing environmental impact.

In Flanders, a farm gate advice system for livestock farms, including deposition but not fixation, was developed in the first years of the new millennium [20]. In order to validate the advice system in practice, farms of the different livestock sectors were followed closely. The farm gate advice system was tested in Hungary at a limited number of farms, revealing that that agriculture is still clearly less intensive in Hungary. Four major restrictions to the accurate calculation of farm-level nutrient budgets were identified: (1) the wide variability that is allowed between the actual and reported nutrient composition of concentrated feed; (2) the estimates of the amount and the composition of manure; (3) the assessment of changes in standing stock on the farm between the beginning and end of the reporting period; (4) the accuracy of the data supplied by the farmers [21]. At present, in Flanders, nutrient budgeting at the farm level is compulsory only in certain cases. In order to restrict the nitrogen and phosphorus fertilization and still meet the crop nutrient needs, now yield and nitrogen dose response curves are developed by a re-evaluation of field experiments, also calculating soil surface budgets [22]. The N_{min} (0–90 cm) is measured regularly in some farms between October 1 and November 15, as the main leaching period is during the winter. The measurement of N_{min} also gives a good evaluation of the N fertilization rate [22]. However, the nitrate leaching risk also depends on the crop residue and presence of catch crop [23].

In Switzerland, in its annual environmental report, the federal office for agriculture publishes N budgets for farms in different regions (valley, hills, and mountains). These aggregated data are deduced from 300 farms whose owners voluntarily calculated N budgets according to the OECD soil surface budget [24].

In France, the CORPEN (Comité d'ORientation pour des Pratiques agricoles respectueuses de l'ENvironnement) soil surface budget [25] was introduced on a voluntary basis to measure nitrogen surplus at the farm level. Updates are regularly available. The budget is mainly used by animal breeding farms as, in France, this type of farm has to prove that there is no structural over-fertilization with breeding effluents.

In Germany, due to the legislation, a range of tools are in use to calculate N budgets according to the national legislation and specifications of the federal States.

In the Netherlands, permissible manure and N fertilizer use rates in arable and dairy farming systems were calculated using N budgets based on standard farms on sandy soils. Nitrate leaching was estimated from the N surplus and leaching fractions (the fraction of the N surplus that leaches a nitrate from the rooting zone) that depend on land use and soil type. Through calculations based on experimental data from various sources, the limits on the use of cattle slurry and mineral fertilizer in grass and silage maize production on sandy soils were calculated [26].

2. Materials and Methods

As our research is focused on the use of nitrogen budgets at the farm level, we define the framework within which we carried out our survey.

2.1. Theoretical Framework

Budgets are applied on various levels and can be calculated differently, according to the system boundary (farm, soil or land) they refer to [3].

2.1.1. Budget Type

The (net) nitrogen soil (surface) budget takes the soil surface as the boundary. Only nutrient inflows to the soil and nutrient removal from the soil are taken into account. The soil budget, therefore, requires data on manure and fertilizer applications to the soil. The term "net" refers to the fact that the result of the soil surface budget approach results in N surpluses excluding N gaseous emissions occurring before the application of manure and other fertilizers to the soil [3] (Figure 1a).

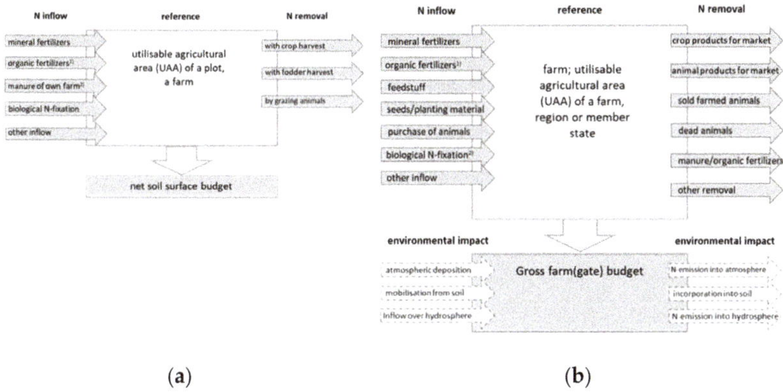

Figure 1. (**a**) Net (soil) surface budget [3,27]: (1) manure from other farms, organic and organic-mineral fertilizers: reductions are made for nitrogen losses due to a volatilization during fertilizer application to the land; (2) N in manure production by livestock, calculated by N excretion (using standard coefficients), reductions are made for nitrogen losses due to volatilization in stables, storages and with application to the land. **Source**: [3,27], translated. (**b**) Gross farm (gate) budget [27,28]: (1) all kinds of organic fertilizers including manure. **Source**: [27,28], translated.

The (gross) farm (gate) budget refers to the farm boundaries and records the nutrients in all products that enter and leave the farm gate. N emissions from stables and manure storages or during the application of organic fertilizers to land are not explicitly accounted for and neither are changes in soil pools and storages. However, all these emissions are part of the nitrogen losses and thus contribute to the total nitrogen surplus in the gross budget (Figure 1b). Farm budgets can also be calculated at larger scales, e.g., at the country level—in that case, the whole farming sector in a country is considered as a single farm [3].

The (gross) land budget approach aims to estimate the total nutrient loads at risk of pollution (air, soil and water). The land budget, therefore, requires data on excretion. The term "gross" refers to the fact that the result of the land budget, the gross nitrogen budget (GNB), includes all N emissions to the air [3] (Figure 2).

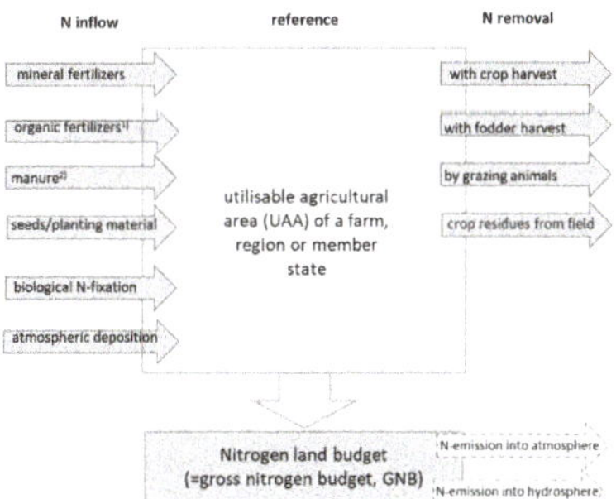

Figure 2. The gross nitrogen budget (GNB) [3]: (1) according to the current methodology, no reductions are made for nitrogen losses due to volatilization in stables, storages and with application to the land; (2) "manure": total manure production by livestock (calculated by N excretion) minus manure withdrawals, plus manure import, plus change in manure stocks. **Source**: own design, according to [3].

In summary, nitrogen surplus data thus may be derived from different calculation methods, including the above-mentioned total N budget assessment [5] and system boundaries may vary between methods.

2.1.2. Level of Application of Budget

Nitrogen surplus is used as an indicator on different decision levels—at the national, regional, farm and even field levels—as a parameter in legislation, for consulting farmers, benchmarking, and monitoring, but also for taking political decisions and is thus quite a common indicator for land and water managers. Nitrogen surpluses are calculated from individual or standard data. From the field level or the farm level towards higher levels, the degree of data aggregation increases and the share of individual farm data and direct measurements decreases (Figure 3).

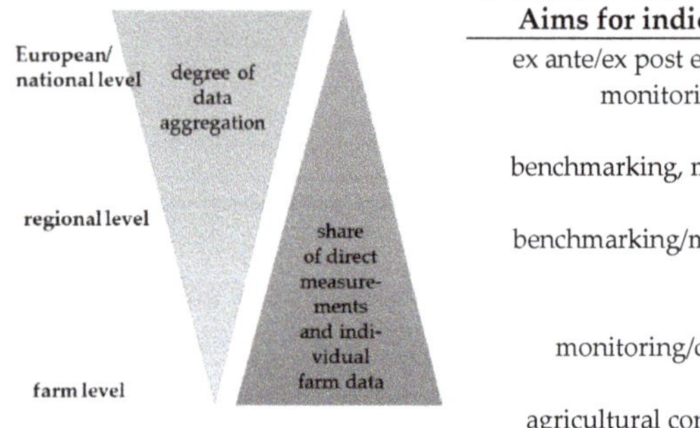

Aims for indicator use	Examples
ex ante/ex post evaluation, monitoring	conceptual or mechanistic simulation model, e.g., AGRUM[1], MERLIN[2], …
benchmarking, monitoring	national gross N budget [kg N/ha and year]
benchmarking/monitoring	livestock density; gross N budget at the regional level [kg N/ha and year]
monitoring/control	N budget at the farm level [kg N/ha and year]
agricultural consultancy	N_{min} in soil in autumn

Figure 3. Degree of aggregation and individualization in relation to the operational level of agri-environmental indicators (with selected examples related to N assessment); **Source:** [29], with modifications. [1] Agricultural and Environmental Measures for Agricultural Water Protection [30], [2] Méthode de hiErarchisation du Risque de LIxiviation du Nitrate [31].

At the level of EU member states, the GNB is applied as the only agri-environmental indicator (AEI) for nitrogen efficiency and, consequently, as an indicator for the pollution of ground- and surface water with nitrogen [28]. Data from all EU member states, as well as Switzerland and Norway, are reported to the European Commission regularly, such as in the four-yearly report of the Nitrates Directive. However, Eurostat [3] points out that the current national budgets quoted are not comparable between different member states due to differences in the definitions, methodologies and data sources used. Budgets at the national level are also calculated in order to monitor the effect of the implementation of policies and measures [15,32].

At the regional level (for administrative districts or (sub) catchments), nitrogen budgets are calculated in order to indicate the regional nitrate leaching potential [6,14,16].

At the farm level, budgets are used to investigate farm performance [19,20,22,26,33] and are used as a tool for farm advice [21,25]. One aim of the survey conducted was to determine the prevalence of fertilization planning and budgeting as nutrient management tools at the farm level.

2.2. *Survey on Nitrogen Budgets at the Farm Level*

Fertilization planning and nutrient budgeting represent two different perspectives: while fertilization planning is performed ex ante in order to limit input of nitrogen and other plant nutrients both timely and according to plant needs, budgeting the inflow and removal of N or other nutrients from the system is applied ex post and reveals the efficiency of the process. Consequently, in this paper, we cover both perspectives.

We investigated to what extent and at which levels nitrogen budgets are applied in the different member states. We contacted experts of the following member states (contacts within the framework of the Horizon2020 project FAIRWAY are marked with *): Belgium (Flanders)-B(Flan), Denmark-DK*, England*, France*-FR, Germany*-GE, Greece*, the Netherlands*-NL, Northern Ireland*-UK-NI,

Portugal*-PT, Romania*-RO, Republic of Ireland-IE, Slovenia*-Sl and from outside the European Union Switzerland-CH and Norway*.

Using an expert survey, we conducted a questionnaire to collect information on:

- The type and elements of fertilization plans and budgets used at the farm and plot levels,
- The type and elements of budgets used at the regional or national levels,
- The legally binding procedures in connection with the use of fertilization plans and budgets,
- The informative value of budgets, according to budget type, and
- The barriers in the use/implementation of budgets according to the individual experience and the experience in the FAIRWAY case studies.

The results of the questionnaire were analyzed and compared, and then discussed in detail with the experts.

3. Results

3.1. Prevalence of Fertilization Planning and Budgeting at the Farm Level

In all countries participating in this survey, farmers regularly set up a fertilization plan ex ante fertilization (Table 1). In the Netherlands and Flanders (Belgium), the fertilization plan is compulsory for farms with derogation (higher amount of N from manure). In Flanders, the fertilization plan is also compulsory for farms with high soil mineral nitrogen (N_{min}) values in autumn and after a bad audit. Fertilization between 1 September and 30 October is only allowed for horticultural crops and based on fertilization advice. In Regions 1–3, which have bad water quality, fertilization advice is needed for a portion of strawberry, vegetable and floricultural fields.

Table 1. Fertilization planning and budgeting at the farm level in EU member states and Switzerland in 2019.

Country	Fertilization Plan (y/n)	Legally Bound (y/n)	N Budgets (y/n)	Legally Bound (y/n)
B (Flan)	(y)	(y)[(1)]	(y)	(y)[(2)]
CH	(y)	(y)[(3)]	y	y
DK	y	y	N[(4)]	-
FR	y	(y)[(5)]	n	-
GE	y	y	y	y
NL	y	(y)[(6)]	y	n
PT	(y)	(y)[(5)]	n	
IE	y	(y)[(7)]	n	-
RO	y	y	y	y
Sl	y	y	(y)[(8)]	-
UK-NI	y	y	n	-

[(1)] B (Flanders), fertilization plan compulsory only for farms with derogation (higher amount of N from manure) and with high soil N_{min} values in autumn and after a bad audit. Fertilization between 1 September and 30 October is only allowed for horticultural crops and based on fertilization advice. In Regions 1–3, which have bad water quality, fertilization advice is needed for a portion of strawberry, vegetable and floricultural fields. [(2)] B (Flanders), 6th action program: nutrient and mass balances of manure processing and anaerobic digestion facilities as risk analysis. [(3)] CH: fertilization plans are not mandatory, but can be requested upon control of the Suisse Balance; in general, fertilization has be performed according to Swiss fertilization guidelines. [(4)] Danish farmers are legally bound to make an ex post fertilizer account, with updated information on the fertilizer plan. [(5)] PT, FR: fertilization plans are mandatory only in NVZ and for certified agriculture. [(6)] NL: fertilization plans are mandatory only for dairy farms with derogation. [(7)] IE: fertilization plans are mandatory only for farms with derogation. [(8)] Sl: in the framework of farm advice.

Within the 6th action program in Flanders, nutrient and mass balances are suggested for manure processing and anaerobic digestion facilities as risk analysis of the manure export of farms. In France and in Portugal, fertilization plans have to be calculated only on farms situated in NVZs. Farmers in France may deviate from the fertilization plan in cases where (1) they use a planning tool (2), their yields exceed the standards or (3) they had a problem on the particular plot. Fertilization plans are not generally legally binding in Switzerland and the Republic of Ireland. In the Republic of Ireland, farms with derogation, however, have to set up a fertilization plan based on soil analysis results and stocking rates. In Slovenia, fertilization plans have to refer to the exact field in case set target values for total N use on certain crops are exceeded.

Nitrogen budgets ex post fertilization are obligatory only in Switzerland, Germany and Romania. In the Netherlands, nitrogen budgets are calculated in the dairy sector at the farm level, but they are not compulsory in a legislative sense. In Slovenia, a simple N budget is calculated at the administrative level for the use of organic fertilizers based on livestock unit (LU) and land area (LU/ha). In order to receive subsidy payments, farmers need to prove that they comply with these cross-compliance standards linked to the Nitrates Directive.

3.2. Results on Fertilization Planning

3.2.1. Factors Considered in Fertilization Planning

In different countries, the complexity of fertilization planning varies considerably (Figure 4). In most cases, basic information is drawn from databases. These figures are subsequently modified according to individually expected yields, qualities, plant varieties or whether grassland is grazed or cut. The fertilizing demand is subsequently further reduced according to the amount of N_{min} available at the beginning of the vegetation period (N_{min}), other residual effects (of preceding crop or previous organic fertilization) or the soil type. In Romania, certain basic information is aggregated to indices.

3.2.2. Impact of Technical Progress on Nitrogen Requirement of Crop

Technical progress may also increase nitrogen efficiency and thus reduce initial nitrogen requirements for the fertilization planning (Figure S1). Split application was quoted by five respondents of this survey, and low-emission application techniques particularly for liquid manure and fermentation residue (drag hose, trailing shoe, strip application and injection) by three respondents as a reason for reduced nitrogen emissions and, consequently, increased plant availability of nitrogen in fertilizers to crops. Precision farming techniques have not yet reached a wider practice level and, therefore, are not referred to in official guidelines for fertilization planning. However, in Denmark, there is a special demonstration scheme on this issue, which may develop into a permanent scheme. The Swiss fertilization guidelines (GRUD) [34] address fertilizer application according to "good agricultural practice", meaning that fertilizer amounts per application are limited to certain thresholds and application should be performed in such a way that losses are minimized. Thus, N availabilities listed in the guidelines indirectly imply low-emission application techniques and/or rapid incorporation of organic fertilizers in order to minimize emissions.

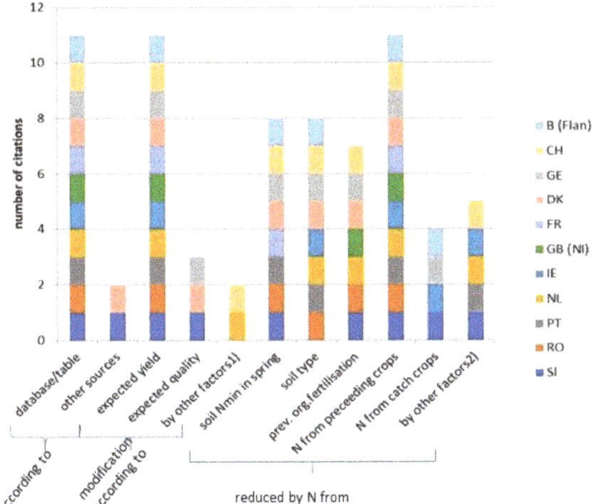

Figure 4. Factors of fertilization planning at the farm level in different member states and Switzerland; (1) NL: grazing or cutting; CH: variety (e.g., for potatoes); DK: soil type and irrigation; (2) PT: N available in irrigation water; IE: nitrogen index. **Source:** expert survey results.

3.2.3. Factors Used to Estimate Plant-Available Nitrogen in Manure and Other Organic Fertilizers

For the estimation of plant-available nitrogen, a range of variables is referred to in the different countries. In some member states, standard figures for gaseous nitrogen losses from stables and storages are deduced from standard excretion figures for the different animal categories. In other member states, only the available nitrogen as a percentage of the applied nitrogen to soil with the organic fertilizer (manure, biogas digestate or compost) is referred to.

The scales of reference for the calculation figures vary as well: in Denmark and in Slovenia, for example, the amount of manure, its dry matter content, nutrient concentration and corresponding N emissions are listed for different animal categories with reference to the head of animal. In Germany, nutrient excretions are listed for different animal categories and nutritional regimes, with reference to one animal place (i.e., head of animal per year).

Aerial nitrogen losses occurring during manure application in some member states are considered, with reference either to the excreted or to the applied N (Table 2). In Germany, ammonia emissions after fertilizer application are not taken into account for the calculation of the plant-available N, but for setting up the soil surface budget ex post fertilization.

Swiss fertilizer guidelines provide information on N excretion, the amount of manure and different nutrient concentrations with reference to different animal categories. Whether figures refer to animal head or animal places depends on the animal category. The data consider unavoidable losses in stables and storages. Emissions during/after application are not taken into account as application technique and timing should be chosen in such a way that NH_3 losses are minimized.

Similarly, in Flanders, the measured or average N concentration in manure (after losses from stables) is considered as plant-available N. As manure has to be applied with low-emission techniques, it has to be worked into the soil within 2 h, and no further ammonia losses are taken into account, as they have to be minimized. In Slovenia and Portugal, there is also no deduction of gaseous emissions during manure application.

Table 2. Methods applied for the calculation of plant-available nitrogen in organic fertilizers.

	N emissions From Stables and Storages	Availability of Applied N in Soil for Crops	N Emissions during/after Organic Fertilizer Application	
	% of N excretion	% of applied N	% of N excretion	% of applied N
B (Flan)[(1)]	y	y	n	n
CH	y	y	n	n
DK	y[(4)]	y[(3,4)]	n	n
FR	n	n.a.[(2)]	n	n
GE	y	y	y[(5)]	y[(6)]
IE	n	y	n	n
NL	y	y	n[(7)]	y[(8)]
PT	n	y	n	y
RO	y	y	y	n
Sl	y	y	n	y
UK-NI	n	y	n	n

[(1)] B: no regulation on determination of N concentration in manure; [(2)] FR: manure N availability is expressed qualitatively with respect to N concentration and the C/N ratio in the manure [35]; [(3)] DK: [35]; [(4)] DK: [36]; [(5)] GE: farm-own manure; [(6)] GE: manure or other organic fertilizers imported to farm; [(7)] NL: excretion not calculated; [(8)] NL: depending on crop, method and TAN concentration.

3.2.4. Emission Factors for Nitrogen

Standard emission factors for nitrogen (predominantly as ammonia) after organic fertilizer application were reported by five countries. The emission factors vary broadly between 5% and 30% of applied nitrogen for Germany and Romania, indicated as the % of excreted nitrogen. There is a slight tendency for solid organic fertilizers towards a higher standard excretion factor than for liquid organic fertilizers. However, based on the length of time the animals spend outside on the pasture, higher emissions can be considered in some countries, e.g., in Germany (Figure S2).

3.2.5. The Plant Availability of Nitrogen in Manure

Table 3 lists the reported values for the crop availability of nitrogen in manure according to Webb et al. [35]. The values of the current survey are inserted as red, bold figures. Nitrogen availabilities refer to the applied nitrogen on or into the soil. There are slight differences in the values reported by some of the countries in the present survey, but generally a higher nitrogen availability is assigned to liquid manure than to solid manure. The nitrogen availability in cattle manure is lower than in pig manure while the figures vary a lot for poultry manure. Anaerobic digestion results in an increased nitrogen availability of the digestate in comparison to the initial substrate (manure, renewable raw materials). Aerobic treatment, on the other hand, leads to a decreased and prolonged nitrogen availability of the compost in soil (figures on N availability in compost are not included in Table 3).

Slovenia recently updated the values on crop availability. Availabilities are defined for three consecutive years, there is no distinction between animal type, but between solid and liquid manure and between arable and grassland: the crop availability of nitrogen in solid manure on grassland is comparably low, at 50% [37]. For the Suisse Balance, farmers who receive subsidies for manure application by drag hose have to add 3 kg $N_{available}$/ha to their Suisse Balance. By this, increased efficiency upon drag hose application is considered at least for the N budget, but not for the fertilization planning.

Countries which do not declare specific emission factors for nitrogen losses prior or during manure application (e.g., Flanders, Northern Ireland and the Republic of Ireland) do not stipulate lower nitrogen availabilities—the effect is a slight improvement in the legal requirements. However, the overall impact also depends on the underlying N excretion coefficients per head of animal.

Table 3. Reported values of crop available % of total nitrogen (=manure N efficiency) [35], supplemented by data from the current survey (**red, bold figures**).

	Crop Available % of Total Nitrogen (=Manure N Efficiency)							
	Cattle		Pigs		Layer		Broiler	Sheep
Country	Slurry	Solid	Slurry	Solid	Slurry	Solid		
AT	50	5/15	65	5/15	60	30	30	
B(Flan)	60	60	60	30	60	30	30	30
	60	**30**	**60**	**30**	**60**	**30**	**30**	**30**
BG	20–35	20	40–45	20	40–50	40–50	40–50	
CH [(1)]	**45**	**20**	**50**	**35**		**35**	**35**	**30**
CZ	60	40	60	40	60	40	40	40
DK [(2)]	70	65	75	65	70	65	65	65
EE	50	25	50	25	50	25	25	25
GE [(2)]	50	25	60	30/40		60	60	25
	60	**35**	**70**	**40**		**70**	**70**	**35**
FR	45	10/15	60	20/30	45	45	35	10
GR	20–35	10	20–45	10	20–30	20–30	20–30	10
IE	40	30	50	50	50	50	50	30
		40		**50**		**50**	**50**	**40**
IT [(3)]	24–62	24–62	28–73	28–73	32–84	32–84	32–84	
LV	50	25	50	25	30	25	25	
LT		35		35		35		
LU	25–50	30–50	30–60	30–50		50	50	
NL	60/40	40/25	60–70	55	60/70	55	55	
PL	50–60	30	50–60	30	50–60	30	30	30
PT	55–75	30–60	50–80	40–60	50–70	40–60	40–60	40–60
	60	**20**	**80**			**90**	**90**	
RO	50	30	50	30		30	50	
SE	40–50	36–41	57	47		48	47/57	
SI	50	30	50	30	30	50	50/	
	75–85	**50–70**	**75–85**	**50–70**	**75–85**	**50–70**	**50–70**	
SK	50	30	50	30	30	50	50	
UK	20/35	10	25/50	10		20/35	20/30	10
UK-NI		30	50			30		
	38	**30**	**50**	**30**		**30**	**30**	

(1) Values referred to are reported as N availability in the year of application for arable farming; for forage production, 5%–10% higher availabilities are assumed. (2) Also includes residual N effects in the following years after application. (3) Availabilities are presented as a matrix according to soil type and time of application.

3.3. Elements of Soil Surface Budgets

Five countries gave further information on the elements they include in their soil surface budgets (Germany, the Netherlands, Romania, Switzerland and Slovenia, although in Slovenia, the information required by law is limited to the relation of organic N production to land area (LU/ha). Nitrogen emissions from organic fertilizers and manure are considered in all above-listed practitioners of soil surface budgets, so the net budget is the more widely used budget version. The Suisse Balance, however, does not strictly follow a soil surface approach, but rather a combination between soil surface and farm gate balance. However, as it is more related to a soil surface budget, it was considered as such.

On the inflow side of the farm budget, the chemical composition of mineral fertilizers and traded organic fertilizers (other than manure) is mostly taken from product declarations or databases. For the determination of the nitrogen content in manure imported into the farm, various options are used, including chemical lab analysis and rapid on-farm analysis, declarations and databases on nitrogen concentration in manure and on nitrogen excretion of farm animals. Manure produced on the own farm is analyzed less frequently. Instead, in this case, databases are used to determine nitrogen concentration (Figure 5).

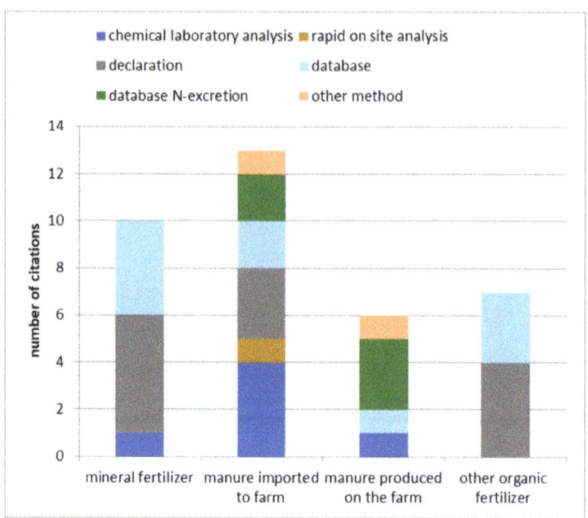

Figure 5. Qualitative determination of elements of nitrogen inflows of soil surface budgets; multiple answers are possible (Source: expert survey results).

The quantity of fertilizers is mostly weighed, less frequently estimated or taken from the product declarations. Quantitative information on seeds and planting material mostly derives from product declarations. Quantification of leguminous nitrogen fixation and nitrogen deposition is mostly deduced from tables, or the amounts of nitrogen entering the system are estimated (Figure S3).

Nitrogen removal take places through various kinds of harvested crops. For arable crops which are regularly analyzed post-harvest as part of a quality assessment (e.g., for a payment according to their raw protein concentration), these data are used for the budgets—otherwise, the nitrogen concentration is deduced from databases. The same may be the case for harvested fodder plants. The largest uncertainty is related to the amount of nitrogen removed by grazing animals (Figure 6).

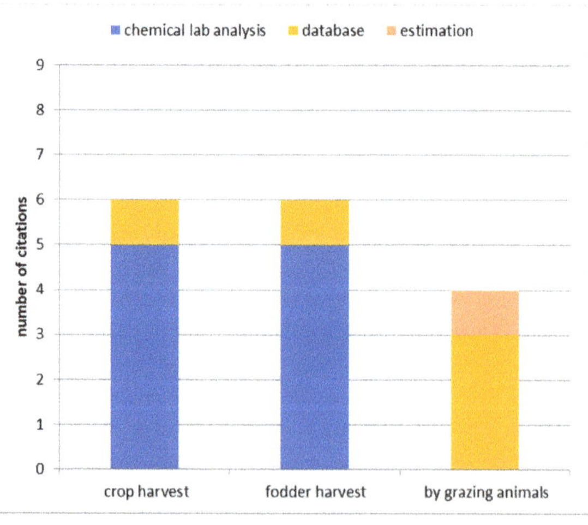

Figure 6. Qualitative determination of elements of nitrogen removal of soil surface budgets; multiple answers are possible (Source: expert survey results).

Quantities of cash crops and fodder harvests are weighed, but also estimated. The quantity of forage consumption by grazing animals is estimated, partly on the basis of tables and with reference to the particular animal performance (Figure S4).

3.4. Elements of Farm Budgets

Germany, Romania, Northern Ireland and the Netherlands reported on the elements used for farm budgets in their countries. On the inflow side of the farm budget, the sources of information on the chemical composition of the different elements of the budget are databases (and tables, respectively) and chemical laboratory analyses of the product, and the latter is also used for product declaration. The nitrogen concentration of manure imported to farms is analyzed with various methods. This is due to national legislation on the placement on the market of fertilizers. The actual nutrient concentration of manure imported to a specific farm determines its fertilizing value and its price; therefore, both seller and buyer are interested in reliable information on fertilizer value. Manure produced on the farm usually has to be taken into account if it is exported from the farm (Figure 7).

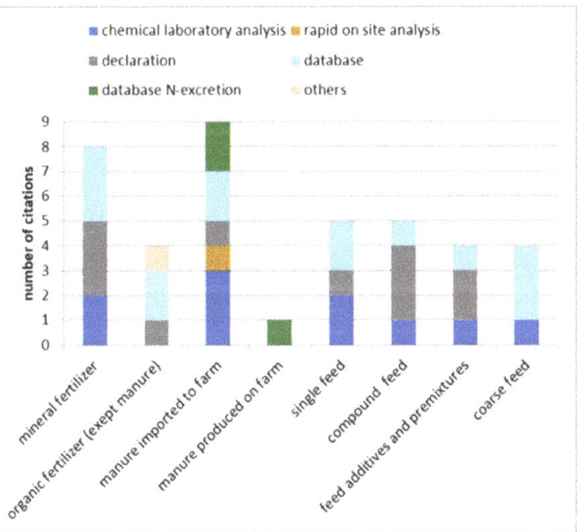

Figure 7. Qualitative determination of elements of nitrogen inflows of farm budgets; multiple answers are possible (Source: expert survey results).

Concerning the quantitative registration of inflow elements of the farm budgets for those goods which are traded regularly with an accompanying goods receipt or a product declaration on the package, this declaration is used for budgeting. For elements that are difficult to estimate, e.g., animals entering the farm, listings of an estimation table are used (Figure S5).

Concerning nitrogen removal from the farm, data on the nitrogen concentration of the harvested crops mostly derive from databases or tables (Figure 8), in contrast to the reports on soil surface budgets (Figure 6). Nitrogen removal by farm animals, sold or dead, leaving the farm is usually taken from databases. As an exemption, manure to be exported from the farm is also analyzed using rapid on-site analysis, e.g., near-infrared spectroscopy (NIRS) technology. So, in the Netherlands and Flanders, manure that is transported between farms has to be weighed and analyzed for N and P. Data of this origin are found in [38].

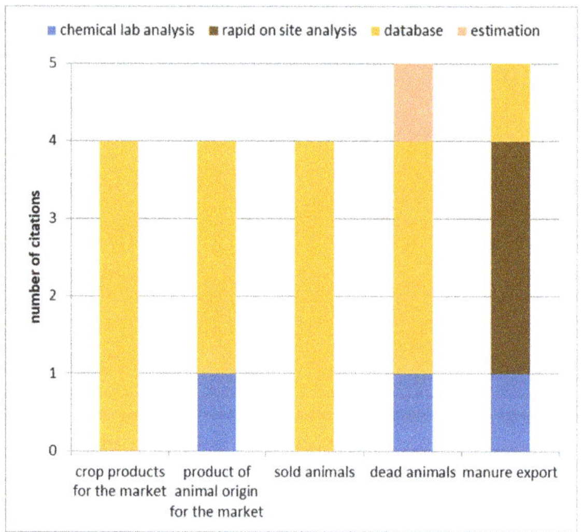

Figure 8. Qualitative determination of elements of nitrogen removal of farm budgets; multiple answers are possible (Source: expert survey results).

For the quantitative assessment of nitrogen removal, goods are often weighed or counted (Figure S6).

3.5. Comparison of Budgets in Different Countries

Below, five selected budget examples for the assessment of nitrogen flows at the farm level are presented and their characteristics are compared. The supplementary table gives an in-depth overview over the different budgets (Table S1).

3.5.1. Germany

The German fertilization ordinance [4] functions as the main legal act for the national implementation of the Nitrates Directive. Before the recent revision of the fertilization ordinance [18,39], all farmers were obliged to calculate a soil surface budget ex post fertilization (net budget, referring to total nitrogen inflow of applied fertilizer). After the EU court took legal action against Germany, the German fertilization ordinance [18] was revised and improved in March 2020 [39]. The nitrogen soil surface budget used as a control instrument against excessive use of N fertilizer is no longer compulsory—instead, a plotwise record of fertilizer use has now been introduced.

In addition, since 2018, intensive animal breeding farms and biogas production facilities using farm inputs are obliged to calculate a farm budget ex post [40]. By 2023, it is planned to include more farms into the obligation to calculate a farm budget. The aim of the farm budget is to transparently and verifiably monitor nutrient flows in breeding farms [41].

Standard values (e.g., excretion factors, nutrient concentrations in harvested crops and in animal products) have to be used for the farm gate budget and were also compulsory for the soil surface budget. Farmers also have to estimate individual calculation data, taking into account set minimum values (e.g., for nutrient concentration in roughages). For some data, nutrient declarations have to be used (e.g., for animal feed and for commercial fertilizers). For the soil surface budget, N removal of harvested roughage crops was estimated on the basis of animal categories and numbers. However, considerable losses could be added to the calculated N intake through roughage. The target value of the now abandoned net soil surface budget was a nitrogen surplus of less than 50 kg*ha^{-1}, the so called control value, as the farm average over a three-year period [4]. The target value of the gross farm budget [40] is a surplus of less than 175 kg*ha^{-1}, which is equivalent to the target value of the net soil surface budget of the former German fertilization ordinance for animal breeding farms [27]. In the course of the latest infringement procedure, the European Commission criticized the mandatory nitrogen soil (surface) budget in the German fertilization ordinance [4] and the uniform control value above zero, which furthermore is not adapted to different soil and climatic zones. The federal German agricultural ministry, therefore, in turn, replaced the soil surface budget by a precise, plot-wise record of all applied fertilizers [39].

3.5.2. Switzerland

A prerequisite for receiving subsidies for farmers in Switzerland is to present the Suisse Balance, an individual-farm nutrient budget for nitrogen and phosphorus as 'proof of ecological performance'. It is a combination of the soil surface and farm gate budget in which manure produced within the farm plus fertilizer imports are balanced against the "net nutrient requirements" of the crops [42]. Approximately 90% of all Swiss farmers participate in this scheme. Farmers who do not use fertilizers containing nitrogen and phosphorus are exempted from this obligation in cases where in addition the animal densities on their farms are below certain thresholds, depending on defined land zones. The Suisse Balance is a net budget, since standard nitrogen emissions from stables and storages and from grazing animals are considered. Furthermore, the nitrogen availability of organic fertilizers is calculated as the farm-specific nitrogen use efficiency, also taking into account ammonia emissions and other losses upon application. The resulting nutrient supply from fertilizers (farm-own and farm-external) is compared to the standard nutrient requirement of crops. This means that plant residues which remain on the field, as well as biological nitrogen fixation, deposition and nitrogen release from soil, are indirectly taken into account. Nutrient removal from grassland is calculated indirectly by estimating feedstuff consumption of farm animals [43]. The target is that the nitrogen and phosphorus supply may exceed standard plant requirements for N and P by at maximum 10%, respectively. Although the Suisse cantons are responsible for the enforcement of the national legislation, there is no guideline on a unified implementation and execution at the canton level [44]. An evaluation of the Suisse Balance in 2011/2012 revealed the need for improvements with regard to method and administration [43].

3.5.3. The Netherlands

In the Netherlands, in 1998, the Mineral Accounting System (MINAS)—a gross farm budget—was fully introduced as a policy instrument. It was based on the principles that the nutrient surplus is an indicator for nutrient efficiency and that a certain nitrogen surplus is unavoidable. The remaining surplus was taxed. Although this approach was successful in the reduction of nitrogen and phosphorus surpluses, in 2003, the EU Court of Justice banned the Dutch farm-level budget system MINAS as, according to the Nitrates Directive, application standards are required while acceptable loss standards are rejected. Accordingly, at the legislative level, MINAS was replaced by a system of nitrogen and phosphorus application standards [45].

The Dutch Annual Nutrient Cycling Assessment (ANCA), a management tool for dairy farmers, was introduced parallel to legislation. Since 2016, the use of ANCA is, according to an agreement among representatives of the dairy processing industry and the farmers union, mandatory for all dairy farms. Farmers who do not join the ANCA scheme are not permitted to deliver milk to the milk-processing industries. As ANCA is a web-based tool, approximately 90% of the data can be directly supplied by the industry (milk processing, feedstuff, laboratories, etc.). It is a tool for benchmarking purposes. Data can be evaluated in different directions. The aim is to increase nitrogen use efficiencies (NUEs):

- A farm with a high individual herd NUE produces less N in manure, and thus reduces the need for N-export;
- A farm with a high individual herd NUE consumes less feedstuff;

- A high NUE with respect to soil indicates a low farm N surplus;
- A high NUE of grassland indicates a high N yield;
- A high N input goes along with a lower NUE;
- A high NUE, in turn, results in substantial savings for the farmer (expenses for fertilizer, feedstuff and for manure export) [46].

3.5.4. Romania

In order to receive subsidies, farmers in Romania have to comply with on-going legislation that transposes the EU Nitrates Directive. As in all EU member states, the legal Code of Good Agricultural Practices for water protection against nitrate pollution from agricultural sources and the corresponding action program have to be revised every four years. The rules of this program are mandatory for all farmers who request subsidies. Payment agencies for interventions in agriculture monitor a sample of farmers once a year (5% of all farms) for compliance with the applied standards related to maximum quantities of fertilizers or with the accomplished fertilization plans. Sanctions are applied if the farm does not comply with these requirements.

Farms which apply higher fertilizer quantities than the maximum standards have to complete an N budget and a fertilization plan. For the calculation of economic optimum fertilizer rates, the measured agrochemical soil indicators, costs of fertilizers, expected crop yields and the limits imposed by legislation are taken into account.

Each farm has to retain documents with evidence related to nutrient management at the farm/parcel level for the control authorities.

4. Discussion

4.1. Prevalence of N Fertilization Planning and N Budgeting at the Farm Level

Budgeting as ex post evaluation of the nitrogen use efficiency (NUE) at the farm level by determination of N surplus is practiced only in a minority of the participating countries in this survey (Table 1). At present, it is legally binding only in Germany, Romania and Switzerland (not an EU member state). In the Netherlands, the statutory anchoring of the farm budget MINAS was banned by the EU Court of Justice in 2003. In Germany, as a reaction to the pressure from the EU Commission to improve the implementation of the Nitrates Directive, the soil surface budgeting system, in effect since 2007 and amended in 2017 [4], has now been replaced by an improved mandatory fertilizer planning, with the duty to keep plotwise records on the applied fertilizers [40,41].

The limited legal anchoring of nitrogen (and phosphorus) farm-level budgets might be due to a lack of appreciation of the Commission for this indicator for evaluation and control of farm nutrient management. It is in any case surprising, as at the national, regional and even landscape level, the use of nutrient budgeting is well established and highly appreciated [3,5,6,13,15,16,19,26,28]. In particular, the OECD and EU institutions have highlighted the value of nutrient budgeting as an agri-environmental indicator [3,47].

Fertilization planning, on the other hand, "arrived" at the farm level in accordance with the requirements of the Nitrates Directive. It is no surprise that it is practiced in all participating countries of this survey, and it is a legally binding procedure for the farmer in the majority of countries (Table 1), although the complexity of the system varies a lot between the countries (Figure 4).

4.2. Need for the Standardization of Data Collection

This survey revealed that the methods used for nutrient budgeting at the farm level are far from harmonized (Figures 5–8; Figures S3–S6). This refers to the budget type, the elements of the budgets considered as well as the data sources. There is a clear discrepancy between the precision of data needed to exactly interpret nutrient flows at the farm level (Figure 1) and the availability of data at this level. Generally, databases and tables play a dominant role in data generation. There seems to be no standard method for the assessment of chemical manure composition, both in soil surface and farm budgets. Quantitative estimations and weighing of N removal (yield) prevail for soil surface budgets, while counting and weighing is often used with N removal (products of animal and plant origin) in farm budgets. These findings are confirmed by other authors [21]. As an example, the assessment of the concentration of nitrogen in manure is discussed below.

4.2.1. Nitrogen Concentration in Different Manure Types

Nitrogen concentration in manure is specific for the type of animal and related to its concentration in the feedstuff used: matching the composition of the livestock diet to the nutritional demand thus reduces nitrogen excretion [35]. The composition of the liquid and solid manure types also differs according to housing system and manure management practices. Standard values for the different manure types in the member states often take into account the varying straw–water–dung relation and, therefore, differ in their nutrient concentrations [36,48]. In the framework of the Interreg/Baltic Sea Region project "Manure Standards," the different mass-balance approaches to calculate manure composition in Denmark, Estonia, Finland, Sweden, Germany, Russia and Poland were investigated [49]. Luostarinen and Kaasinen [50] detected differences in the level of detail, in the algorithms applied, and in the background data used in the national calculation tools. Standards for manure quantity and composition are used, e.g., as the basis for fertilization planning or the calculation of manure storage capacity.

Nitrogen excretion is calculated in order to standardize nitrogen flows, budgeting the quantity of nitrogen fed (=input) and the amount of nitrogen in the products of animal origin, e.g., milk, meat or eggs (=output). However, N excretion factors for dairy cattle, other cattle, pigs, laying hens, broilers, sheep and goats differ significantly between policy reports (IPCC guidelines, the EMEP/EEA inventory guidebook, the EU Nitrates Directive, OECD/EUROSTAT, the GAINS model and the CAPRI model) and countries [38]. These differences can be related to disparities in animal production between the member states, but also to differences in the aggregation of livestock categories and in estimation procedures [38].

4.2.2. Analyses and Measurements versus Standard Data and Estimations

For the establishment of nitrogen budgets on farm level, apart from standard values as described previously in 4.2.1., sampling of in and outputs and direct (=on farm) measurements or analysis respectively laboratory analysis are possible. On-farm analysis possesses the advantage that the information on, e.g., nitrogen concentration in manure, is directly available and can be used to adjust supplementary mineral fertilization. Nevertheless, producing representative samples from heterogeneous substances is difficult and the accuracy of on-farm analytical methods is generally not very high [51]. Moreover, manipulative estimations by the farmer could result in voluntary unbalanced budgets (e.g., a higher than realistic yield, lower N concentration in manure) that are even more difficult to detect for control officers [44].

Near-infrared spectroscopy (NIRS technology) achieves satisfying results for the analysis of liquid manure. However, the technology is not yet widespread [51]. A new approach to calculate nutrient excretion more precisely was chosen for the Dutch Annual Nutrient Cycling Assessment (ANCA) tool. The online tool is fed in "real time" with data necessary to calculate farm-specific nutrient excretions. Therefore, with this tool, optimization of the milk production process is possible, increasing the farm-specific NUE [52] and reducing pollution potential and costs [53].

4.3. Detection of "Hidden" Surpluses Using N Budgets

While fertilization planning refers to the plant-available N in fertilizers, budgets take the total N amount into account. There are a range of factors influencing the level of "hidden" surplus in fertilizer use.

4.3.1. Nitrogen Losses during Housing and Storage and during/after Manure Application

Nitrogen losses from livestock housing and from manure storages are mainly in the form of NH_3 emissions, but high N_2 and N_2O losses are also reported from solid manures [54]. For the calculation of the share of nitrogen emitted, emission factors are applied, referring to different livestock categories, manure types and diets. Emission factors are either based on total N excretion or N concentration, or on the total ammonia nitrogen (TAN) [55–57], resulting in considerable variation in emission factors for the same livestock categories among member states. Moreover, abatement measures in an "upstream part" of the manure management system (e.g., manure storage) affect downstream losses (e.g., during manure application) by increasing the TAN concentration of the manure, and thus have an impact on the emission factors of application [55].

In the countries participating in this study, the share of nitrogen in manure and other organic fertilizers actually incorporated into soil is calculated in different ways (Table 2). Nitrogen emitted as standardized gaseous losses from stables and storages is deduced from the N amount excreted by farm animals, as in B, CH, GE, NL, RO and Sl. Gaseous emissions after manure application are considered in GE, NL, PT, RO and Sl. The N surplus of gross budgets includes the share of nitrogen emitted via these paths.

4.3.2. The Plant Availability of Nitrogen in Organic Fertilizers and Soil Conditioners

In the framework of fertilization planning, the plant availability of nitrogen is a necessary variable for the calculation of the amount of organic fertilizer and soil conditioners to be applied and of supplementary mineral fertilization. Webb et al. [35] carried out a survey among all 27 member states on these availability factors for the liquid and solid manure of cattle, pigs and layers and for the solid manure of broilers and sheep. The availability factor is used throughout the member states, not only within the NVZs. It describes the manure N efficiency and is expressed as the percent of total nitrogen (Table 3, updated). In some member states, the availability factors include residual effects for one or several followings years. It is not only that these availability factors do not exist in every member state for all livestock categories, they also vary considerably between them.

What is listed in Table 3 is in fact the actual plant-available share of the fertilized nitrogen which can be considered for fertilization planning purposes. This share increases with a decrease in the C/N ratios in manure, respectively, in organic fertilizers or soil conditioners. Therefore, liquid manure and digestates which are characterized by comparably narrow C/N ratios show higher plant availabilities than, e.g., solid manure. The share of N unavailable to plants—together with organic carbon—accumulates in soils with repeated application of the organic fertilizers. The effect increases with the application rates (temporal and quantitative). In the long term, elevated organic matter concentrations in soils leads to an increase in the mineralization of soil organic matter and, consequently, higher N_{min} concentrations in soil and soil water, particularly if the C/N ratio is—due to high application rates of liquid manure—narrow [58]. This mineralization effect of soil organic matter is expected to even increase with a prolongation of the cultivation season due to climate change [59].

Another effect of changing climate conditions is the unpredictability of growing conditions. Droughts, for example, can reduce yields considerably, and high rates of precipitation in the winter season can reduce mineral nitrogen in the root zone from the start of the growing season—an issue which may be corrected for by an annual prognosis and monitoring program, as implemented in Denmark. The gap between predicted and accomplished yields results in a gap between fertilized nitrogen and plant uptake of nitrogen. At best, nutrients that are not transferred into plant biomass remain in soil and are available in the following growing period—at worst, they are leached below the root zone and contaminate groundwater [60].

In soil surface budgets, gaseous nitrogen losses are excluded. This budget type is, therefore, particularly suitable for detecting the "hidden" surpluses of nitrogen, which increase the N stock in soil and might threaten the environment by leaching, erosion or denitrification.

4.4. Evaluation of the Presented Budget Types

The Suisse Balance compares the nitrogen supply by fertilizers with the standard nutrient requirement of crops. The soil surface budget of the German fertilization ordinance [4] compares nitrogen inflow by fertilization calculated from standard N animal excretion and fertilizer rates and concentrations with current nitrogen removal by crop, determined from yields and standard N concentrations. The farm gate budget of the German material flow ordinance (StoffBilV) [40] compares nitrogen inflow and output at the farm level, based on standards as well as product declarations. ANCA (the Netherlands) calculates more realistically, with up-to-date analysis data supplied from laboratories and the industry. While the ANCA evaluation is variable, referring to the farm level or in detail to the herd, manure, soil or crop, the budgets used in Romania, Germany and Switzerland have a fixed set up.

While nitrogen budgets in Germany [4,39], Romania and Switzerland (Suisse Balance) are statutory requirements, with *de minimis* limits for smaller and/or less intensive farms, the ANCA tool in the Netherlands is an agreement beyond the legal level, between dairy farmers and the dairy industry. This overall arrangement can be explained as a reaction to the decision of the EU Court of Justice stating that the previous tool, MINAS, is incompatible with the Nitrates Directive.

ANCA is a benchmarking tool for dairy farmers, aiming to increase N efficiency. The German StoffBilV [40] was installed to improve transparency on nutrient flows on intensive animal breeding farms. The German fertilization ordinance [4] and the Suisse Balance both have to be attained in order to receive subsidies, and both aim at the reduction of water pollution due to nitrates. In the federal structures of both Germany and Switzerland, control mechanisms at the federal state level or the canton level are not unified. In Romania and Germany, yearly increasing sanctions related to the rules of Cross Compliance (CC) according to the Common Agricultural Policy of the EU may be imposed. Additionally, legal fines for the violation of the rules of the German fertilization ordinance could amount up to 50,000 €. Sanctions in Switzerland are comparable to the CC sanctions in the EU member states—in the case of repeated violations, subsidies might even be cut completely.

In the German system, no third parties—such as farm advisors or control bodies—have unwanted direct access to farm data: farmers have to present the results and some of the basis of their calculations to the control authority on request. In Switzerland, at least some data on manure and organic fertilizer flow between farms or from/to biogas plants are stored centrally. In contrast, in the Netherlands, data for ANCA from different sources are collected centrally, evaluated by a farm advisor and communicated to and discussed with the farmers. In Denmark, the fertilizer accounts are publicly accessible.

Table S1 also lists the elements of the different budgets. As for ANCA, a range of data are processed, different budget types with different system boundaries can be calculated and internal data flows between the systems can be extracted and evaluated.

All systems except the farm budget StoffBilV use an indirect method to determine the uptake of roughages, particularly of grass, by the number of animals and the grazing hours. Although this is not a precise method, the alternative, estimating or measuring the grass growth and grass quality (protein concentration), is not yet standard practice among farmers.

The Suisse Balance and ANCA offer the possibility of using farm-specific excretion rates. This enables the possibility of increasing NUE by using selected N-reduced feeds.

The German budgets have to be interpreted versus the target values: 50 kg*ha^{-1}*year^{-1} in 2020 for the soil surface budget [4] and 175 kg*ha^{-1}*year^{-1} for the gross farm budget [40]. The difference between the two values can be explained by the sum of nitrogen losses from stables and storages, losses from roughages and ammonia deposition [27]. The ANCA benchmarks are calculated as the annual mean of all farms and amounted to 218 kg*ha^{-1}*year^{-1} as the farm budget and to 158 kg*ha^{-1}*year^{-1} as the soil surface budget in 2018 [47]. From these figures, it can be concluded that, in Dutch farms, the gap between gross and net value (60 kg*ha^{-1}*year^{-1}) is lower than between farm and soil surface budget in Germany (125 kg*ha^{-1}*year^{-1}). The target value of 175 kg*ha^{-1}*year^{-1} of the StoffBilV in the German farm budget seems even more elevated with respect to the underlying N application standards. They amount to 250/230 kg*ha^{-1}*year^{-1} with reference to manure in the Netherlands in comparison to 170 kg*ha^{-1}*year^{-1} with reference to manure and other organic fertilizers, like compost and biogas effluent, in Germany. The target of the Swiss budget is to have less than 10% surplus in comparison to the standard nutrient requirement of the crop.

5. Conclusions

The harmonization of the methodology of nitrogen budgets as agri-environmental indicators at the farm level is limited due to European legislation. This was seen in the examples of the Netherlands (MINAS) and Germany (fertilization ordinance). European legislation does not accept N surplus as an indicator for the success of the implementation of the Nitrates Directive into national legislation. Our expert survey in several European member states and Switzerland reveals that, on the other hand, fertilization planning, defined as "good agricultural practice" in the Nitrates Directive, has clearly reached the farm level.

In most of the cases where budgets are in regular use at farm level (Germany, Romania, Switzerland), standard values for the chemical composition of feedstuff, organic fertilizers, and animal and plant products are applied. However, standard values cannot depict individual, farm-specific conditions. Therefore, further joint research activities and development of methods and/or products are necessary in order to carry out plausibility checks of the underlying nutrient flows at the farm level.

Consequently, nitrogen budgets are not yet fit to be applied as standard "ready to use" "agri-drinking water indicators (ADWIs)" for the pollution of waters with nitrates, which would produce comparable results across Europe. Further, the comparability of soil surface and farm nutrient budgets used in the different countries is rather limited. However, nitrogen budgets at the farm level can serve as a comparably simple benchmark instrument, next to autumn N_{min} values of the soil, in order to identify best management practices under defined environmental and farming conditions.

The example of the Dutch Annual Nutrient Cycling Assessment (ANCA) tool (and partly of the Suisse Balance) shows that it is only by using farm-specific "real" data that budgeting can be successfully applied to optimize nutrient flows and increase N efficiencies at the farm level. However, this approach is more elaborate and requires centralized data processing where data protection concerns are considered. The detailed data analysis does not correspond with governmental control and sanctions.

The Dutch example also shows that in intensive dairy production, even when comparatively high N use efficiencies of slightly below 40% are attained at the farm level, large amounts of nitrogen, approximately 200 kg N per ha and year, are still released into the environment, questioning the livestock density and intensity level of dairy production.

Supplementary Materials: The following are available online at http://www.mdpi.com/2073-4441/12/4/1197/s1: Figure S1. Impact of technical progress on N-requirement of the crop according to national fertilization planning systems; Figure S2. N emissions during and after organic fertilizer application in % of applied N in % of excreted N; Figure S3. Quantitative determination of elements of nitrogen inflows of soil surface budgets; Figure S4. Quantitative determination of elements of nitrogen removal of soil surface budgets; Figure S5. Quantitative determination of elements of nitrogen inflows of farm budgets; Figure S6. Quantitative determination of elements of nitrogen removal of farm budgets; Table S1. Characteristics of five budgets from four countries practiced on farm level.

Author Contributions: Conceptualization: S.K. and B.O.; methodology: S.K. and C.H.; formal analysis: S.K.; investigation and data curation: J.B., I.C., C.C., T.D., H.F., M.G., K.D., G.H., I.A.L., N.S. and K.V.; writing—original draft preparation: S.K.; writing—review and editing: G.V.; visualization: S.K.; supervision: B.O.; project administration: C.H. All authors have read and agreed to the published version of the manuscript.

Funding: This project was funded from the European Union's Horizon 2020 project FAIRWAY (https://www.fairway-project.eu/) under grant agreement No. 727984.

Conflicts of Interest: The authors declare no conflict of interest.

References

1. OECD. *OECD Core Set of Indicators for Environmental Performance Reviews. A Synthesis Report by the Group on the State of the Environment*; Environment Monographs No 83; OECD Publishing: Paris, France, 1993.
2. Kremer, A.M.; Methodology and Handbook—Eurostat/OECD—Nutrient Budgets EU-27, Norway, Switzerland. European Commission, Eurostat. 2017, 112p. Available online: https://ec.europa.eu/eurostat/documents/2393397/2518760/Nutrient_Budgets_Handbook_%28CPSA_AE_109%29_corrected3.pdf/4a3647de-da73-4d23-b94b-e2b23844dc31 (accessed on 28 January 2020).
3. *Eurostat Nutrient Budgets—Methodology and Handbook*; Version 1.02.; Eurostat and OECD: Luxembourg, Luxembourg, 2013.
4. Düngeverordnung (DüV) vom 26. Mai 2017. Verordnung über die Anwendung von Düngemitteln, Bodenhilfsstoffen, Kultursubstraten und Pflanzenhilfsmitteln nach den Grundsätzen der guten fachlichen Praxis beim Düngen. *BGBl. I*, 1305 pp. Available online: https://www.gesetze-im-internet.de/d_v_2017/D%C3%BCV.pdf (accessed on 20 April 2020).
5. Leip, A.; Britz, W.; Weiss, F.; de Vries, W. Farm, land, and soil budgets for agriculture in Europe calculated with CAPRI. *Environ. Pollut.* **2011**, *159*, 3243–3253. [CrossRef] [PubMed]
6. Wick, K.; Heumesser, C.; Schmid, E. Groundwater nitrate contamination: Factors and indicators. *J. Environ. Manag.* **2012**, *111*, 178–186. [CrossRef] [PubMed]
7. de Ruijter, F.J.; Boumans, L.j.M.; Smit, A.L.; von den Berg, M. Nitrate in upper groundwater on farms under tillage as affected by fertilizer use, soil type and groundwater table. *Nutr. Cycl. Agroecosyst.* **2007**, *77*, 155–167. [CrossRef]
8. Lord, I.; Antony, S.G. Agricultural nitrogen balances and water quality in the UK. *Soil Use Manag.* **2002**, *18*, 362–369. [CrossRef]
9. Sieling, K.; Kage, H. N balance as an indicator of N leaching in an oilseed rape—Winter wheat—Winter barley rotation. *Agric. Ecosyst. Environ.* **2006**, *115*, 261–269. [CrossRef]
10. Buczko, U.; Kuchenbuch, R.O. Environmental indicators to assess the risk of diffuse nitrogen losses from agriculture. *Environ. Manag.* **2010**, *45*, 201–1222. [CrossRef]
11. Rankinen, K.; Salo, T.; Granlund, K.; Rita, H. Simulated nitrogen leaching, nitrogen mass field balanes and their correleation on four farms in south-western Finland during the peeriod 2000–2005. *Agric. Food Sci.* **2007**, *16*, 98–107.
12. Hansen, B.; Thorling, L.; Schullehner, J.; Termansen, M.; Dalgaard, T. Groundwater nitrate response to sustainable nitrogen management. *Sci. Rep.* **2017**, *7*, 1–12. [CrossRef]
13. Dalgaard, T.; Bienkowski, J.F.; Bleeker, A.; Dragosit, U.; Drouet, J.L.; Durand, P.; Frumau, A.; Hutchings, N.J.; Kedziora, A.; Magliulo, V.; et al. Farm nitrogen balances in six European landscapes as an indicator for nitrogen losses and basis for improved management. *Biogeosciences* **2012**, *9*, 5303–5321. [CrossRef]
14. Andelov, M.; Kunkel, R.; Uhan, J.; Wendland, F. Determination of nitrogen reduction levels necessary to reach groundwater quality targets in Slovenia. *J. Environ. Sci.* **2014**, *26*, 1806–1817. Available online: https://www.sciencedirect.com/science/article/pii/S1001074214000734 (accessed on 16 April 2020). [CrossRef]
15. Spiess, E. Nitrogen, phosphorus and potassium balances and cycles of Swiss agriculture from 1975 to 2008. *Nutr. Cycl. Agroecosyst.* **2011**, *91*, 351–365. [CrossRef]
16. Poisvert, C.; Curie, F.; Moatar, F. Annual agricultural N surplus in France over a 70-year period. *Nutr. Cycl. Agroecosyst.* **2017**, *107*, 63–78. [CrossRef]
17. Oenema, O.; Kros, H.; de Vries, W. Approaches and uncertainties in nutient budgets. Implications for nutrient management and environmental policies. *Eur. J. Agron.* **2003**, *20*, 3–16. [CrossRef]
18. Cameira, M.R.; Rolim, J.; Valente, F.; Faro, A.; Dragosits, U.; Cordovil, C.M.d.S. Spatial distribution and uncertainties of nitrogen budgets for agriculture in the Tagus river basin in Portugal—Implications for effectiveness of mitigation measures. *Land Use Policy* **2019**, *84*, 278–293. [CrossRef]
19. Gourley, C.J.P.; Aarons, S.R.; Powel, M. Nitrogen use efficiency and manure management practices in contrasting dairy production systems. *Agric. Ecosyst. Environ.* **2012**, *147*, 73–81. [CrossRef]
20. D'Haene, K.; Magyar, M.; Mulier, A.; De Neve, S.; Pálmai, O.; Nagy, J.; Németh, T.; Hofman, G. Comparison of N and P farm gate balances between the intensive agriculture in Flanders and the extensive agriculture in Hungary. In Proceedings of the 14th World Fertilizer Congress of the International Centre for Fertilizers, CIEC Chiang Mai, Thailand, 22–27 January 2006.

21. Mulier, A.; Hofman, G.; Baecke, E.; Carlier, L.; De Brabander, D.; De Groote, G.; De Wilde, R.; Fiems, L.; Janssens, G.; Van Cleemput, O.; et al. A methodology for the calculation of farm level nitrogen and phosphorus balances in Flemish agriculture. *Eur. J. Agron.* **2003**, *20*, 45–51. [CrossRef]
22. D'Haene, K.; Salomez, J.; De Neve, S.; De Waele, J.; Hofman, G. Environmental performance of the nitrogen fertiliser limits imposed by the EU Nitrates Directive. *Agric. Ecosyst. Environ.* **2014**, *192*, 67–79. [CrossRef]
23. De Waele, J.; D'Haene, K.; Salomez, J.; Hofman, G.; De Neve, S. Simulating the environmental performance of post-harvest management measures to cope with the Nitrates Directive. *J. Environ. Manag.* **2017**, *187*, 513–526. [CrossRef]
24. Frei, J. ZA-AUI: Jüngste Entwicklungen und Resultate. BLW, Fachbereich Agrarumweltsysteme und Nährstoffe, 2017. Available online: https://2017.agrarbericht.ch/de/umwelt/agrarumweltmonitoring/agrarumweltindikatorenaui (accessed on 16 April 2020).
25. CORPEN Des Indicateursd'azotepourgérer des Actions de Maîtrise des Pollutions à L'échelle de la Parcelle, de L'exploitation et du Territoire. Ministère de l'Écologieet du Développement Durable, Paris, France, 2006. Available online: http://www.developpement-durable.gouv.fr/IMG/pdf/DGALN_2006_09_azote_indicateur.pdf (accessed on 25 July 2018).
26. Schröder, J.J.; Aarts, H.F.M.; van Middelkoop, J.C.; de Haan, M.H.A.; Schils, R.L.M.; Velthof, G.L.; Fraters, B.; Willems, W.J. Permissible manure and fertilizer use in dairy farming systems on sandy soils in The Netherlands to comply with the Nitrates Directive target. *Eur. J. Agron.* **2007**, *27*, 102–114. [CrossRef]
27. Klages, S.; Osterburg, B.; Hansen, H.; Betriebliche Stoffstrombilanzen für Stickstoff und Phosphor—Berechnung und Bewertung. Dokumentation der Ergebnisse der Bund-Länder-Arbeitsgruppe "Betriebliche Stoffstrombilanzen" und der Begleitenden Analysen des Thünen-Instituts 2017. Johann Heinrich von Thünen-Institut, Braunschweig, Germany, 108p. Available online: https://literatur.thuenen.de/digbib_extern/dn059490.pdf. (accessed on 28 January 2020).
28. Eurostat—Agri-Environmental Indicator—Gross Nitrogen Balance. 2018. Available online: https://ec.europa.eu/eurostat/statistics-explained/index.php/Agri-environmental_indicator_-_gross_nitrogen_balance#Key_messages (accessed on 9 March 2020).
29. Brandt, J. *Review Report of Agri-Drinking Water Quality Indicators and IT/Sensor Techniques, on Farm Level, Study Site and Drinking Water Source*; Brandt, J. FAIRWAY Project Deliverable 3.1; 2018; 180p, Available online: https://fairway-is.eu/index.php/farm-management/workpackages/agri-drinking-water-quality-indicators (accessed on 16 April 2020).
30. Hirt, U.; Mahnkopf, J.; Venohr, M.; Kreins, P.; Heidecke, C.; Schernewski, G. How can German river basins contribute to reach the nutrient emission targets of the Baltic Sea Action Plan. In *Environmental Science and Technology: Proceedings from the Fifth International Conference on Environmental Science and Technology, Houston, TX, USA, 25–29 June 2012*; Sorial, G.A., Hong, J., Eds.; American Science Press: Houston, TX, USA, 2012.
31. Aimon-Marie, F.; Angevin, F.; Guichard, L. *MERLIN: Une Méthode Agronomique Pour Apprécier les Risques de Pollution Diffuse par les Nitrates D'origine Agricole*; Agrotransfert: Lusignan, France, 2001; 27p.
32. Häußermann, U.; Bach, M.; Klement, L.; Breuer, L. Stickstoff-Flächenbilanzen für Deutschland mit Regionalgliederung Bundesländer und Kreise—Jahre 1995 bis 2017; Methodik, Ergebnisse und Minderungsmaßnahmen. *Abschlussbericht TEXTE 131/2019*. 2019. Available online: https://www.umweltbundesamt.de/publikationen/stickstoff-flaechenbilanzen-fuer-deutschland (accessed on 20 April 2020).
33. BLW Weisungen und Erläuterungen 2019. *Verordnung über die Direktzahlungen an die Landwirtschaft (Direktzahlungsverordnung, DZV*; SR 910.13; Bundesamt für Landwirtschaft: Bern, Switzerland, 2018.
34. Richner, W.; Sinaj, S. Grundlagen für die Düngung landwirtschaftlicher Kulturen in der Schweiz (GRUD 2017). *Agrarforschung Schweiz* **2017**, *8*, 276.
35. Webb, J.; Soerensen, P.; Velthof, G.; Amon, B.; Pinto, M.; Rodhe, L.; Salomon, E.; Hutchings, N.; Burczyk, P.; Reid, J. An assessment of the variation of manure nitrogen efficiency throughout Europe and an appraisal of means to increase manure-N efficiency. *Adv. Agron.* **2013**, *119*, 371–441.
36. Lund, P.; Frydendahl Hellwing, A.L.; Børsting, Ch.F. (Eds.) Normtal for husdyrgødning 2019, 38.Institut for husdyrvidenskab, Aarhus Universitet, Tjele, Denmark. Available online: http://anis.au.dk/normtal/ (accessed on 28 January 2020).

37. Anonymous Decree on the Protection of Waters against Pollution Caused by Nitrates from Agricultural Sources. 2009. Available online: http://www.pisrs.si/Pis.web/pregledPredpisa?id=URED5124 (accessed on 9 March 2020).
38. Velthof, G.L.; Hou, Y.; Oenema, O. Nitrogen excretion factors of livestock in the European Union: A review. *J. Sci. Food Agric.* **2015**, *95*, 3004–3014. [CrossRef] [PubMed]
39. Verordnung zur Änderung der Düngeverordnung und Anderer Vorschriften. Bundesrat Drucksache 98/20,20.02.20. Available online: https://www.bundesrat.de/SharedDocs/drucksachen/2020/0001-0100/98-20.pdf?__blob=publicationFile&v=2 (accessed on 8 April 2020).
40. Stoffstrombilanzverordnung (StoffBilV) vom 14. Dezember 2017. Verordnung über den Umgang mit Nährstoffen im Betrieb und betrieblichen Stoffstrombilanzen. *BGBl. I*, pp. 3942; 2018 I pp. 360. Available online: https://www.gesetze-im-internet.de/stoffbilv/StoffBilV.pdf (accessed on 20 April 2020).
41. BMEL Stoffstrombilanz: Mehr Transparenz über Nährstoffe in Landwirtschaftlichen Betrieben. Available online: https://www.bmel.de/DE/Landwirtschaft/Pflanzenbau/Ackerbau/_Texte/Stoffstrombilanz.html (accessed on 27 January 2020).
42. Avater-Esper, S.; Neue Düngeverordnung soll ab April 2020 Gelten. Top Agrar. Available online: https://www.topagrar.com/acker/news/neue-duengeverordnung-soll-ab-april-2020-gelten-11837988.html (accessed on 27 January 2020).
43. Agrida, B.L.W. *Wegleitung Suisse-Bilanz*; Auflage 1.13; Agridea und Bundesamt für Landwirtschaft: Bern, Switzerland, 2016.
44. Bosshard, C.; Spiess, E.; Richner, W. *Überprüfung der Methode SuisseBilanz*; Schlussbericht. Forschungsanstalt Agroscope, Reckenholz-Tänikon ART: Zurich, Switzerland, 2012.
45. Gassner, A. Gewässerschutzbestimmungen in der Landwirtschaft. Ein internationaler Vergleich. In *Umwelt-Wissen 2006, Nr. 0618*; Bundesamt für Umwelt: Bern, Switzerland, 2006; 76p.
46. Poppe, K.J. MINAS—The Dutch Mineral Accounting System For the California Department of Food and Agriculture. LEI Wageningen, 2013. Available online: https://www.cdfa.ca.gov/is/ffldrs/frep/pdfs/6_Poppe.pdf (accessed on 28 January 2020).
47. EEA (2005): Agriculture and Environment in EU-15—The IRENA Indicator Report. Agriculture and Environment. p. 128. Available online: https://www.eea.europa.eu/publications/eea_report_2005_6 (accessed on 9 March 2020).
48. Menzi, H. Manure management in Europe: Results of a recent survey. In Proceedings of the 10th International Conference of the RAMIRAN Network, High Tatras, Strbské Pleso, Slovak Republic, 14–16 May 2002; pp. 93–102.
49. Kaasik, A.; Lund, P.; Damgaard Poulsen, H.; Kuka, K.; Lehn, F.; Kuoppala, K.; Rinne, M.; Nousiainen, J.; Perttilä, S.; Koivunen, E.; et al. Overview of Calculation Methods for the Quantity and Composition of Livestock Manure in the Baltic Sea Region. Report on Current National Manure Calculation Systems Produced in Manure Standards Work Package 3: Guidelines for Calculated Manure Systems. 2019. Available online: https://www.luke.fi/manurestandards/wp-content/uploads/sites/25/2019/06/WP3-report_ManureStandards_Final2.pdf (accessed on 28 January 2020).
50. Luostarinen, S.; Kaasinen, S. *Manure Nutrient Content in the Baltic Sea Countries*; Natural Resource Institute: Helsinki, Finland, 2016; 45p, Available online: http://urn.fi/URN:ISBN:978-952-326-272-0 (accessed on 28 January 2020).
51. Snauwaert, E.; Forrestal, P.; Bonmati, A.; Riiko, K.; Klages, S.; Brandsma, J.; Provolo, G.; Bernard, J.-P.; Mini-Paper—On Farm Tools for Accurate Fertilization. EIP-AGRI Focus Group—Nutrient Recycling. 2016. Available online: https://ec.europa.eu/eip/agriculture/sites/agri-eip/files/2_mp_on_farm_tools_final.pdf (accessed on 28 January 2020).
52. Daatselaar, C.H.; Reijs, J.R.; Oenema, J.; Doornewaard, G.J.; Aarts, F.M. Variation in nitrogen use efficiencies on Dutch dairy farms. *J. Sci. Food Agric.* **2015**, *95*, 3055–3058. [CrossRef] [PubMed]
53. Oenema, J. ANCA: The Dutch way. Wageningen University & Research Denmark cattle congress, Herning, Denmark, 26.02. 2019. Available online: https://www.landbrugsinfo.dk/Kvaeg/Dansk-Kvaeg-kongres/Filer/kk19_57_Jouke_Oenema.pptx (accessed on 28 January 2020).
54. Shah, G.M.; Shah, G.A.; Groot, J.C.J.; Oenema, O.; Raza, A.S.; Lantingal, E.A. Effect of storage conditions on losses and crop utilization of nitrogen from solid cattle manure. *J. Agric. Sci.* **2016**, *154*, 58–71. [CrossRef]

55. Velthof, G.L.; Lesschen, J.P.; Webb, J.; Pietrzak, S.; Miatkowski, Z.; Kros, J.; Pinto, M.; Oenema, O. *The Impact of the Nitrates Directive on Gaseous N Emissions—Effects of Measures in Nitrates Action Programme on Gaseous N Emissions*; Final Report 2010. Contract ENV.B.1/ETU/2010/0009; Alterra: Wageningen, The Netherlands, 2011; 58p.
56. Reidy, B.; Dämmgen, U.; Döhler, H.; Eurich-Menden, B.; van Evert, F.K.; Hutchings, N.J.; Luesink, H.H.; Menzi, H.; Misselbrook, T.H.; Monteny, G.-J.; et al. Comparison of models used for national agricultural ammonia emission inventories in Europe: Liquid manure systems. *Atmos. Environ.* **2007**, *42*, 3452–3464. [CrossRef]
57. Reidy, B.; Webb, J.; Monteny, G.-J.; Misselbrook, T.H.; Menzi, H.; Luesink, H.H.; Hutchings, N.J.; Eurich-Menden, B.; Döhler, H.; Dämmgen, U. Comparison of models used for national agricultural ammonia emission inventories in Europe: Litter-based manure systems. *Atmos. Environ.* **2009**, *43*, 1632–1640. [CrossRef]
58. Gebauer, W.-G.; Pfister, P.; Schaaf, H.; Homm-Belzer, A.; Fischer, S. Humus—Chancen und Risiken für den Grundwasserschutz (Ergebnisse aus der Praxis). In *Verbraucherschutz als Herausforderung für die landwirtschaftliche Produktion*; VDLUFA Kongress, Kurzfassungen; Verband Deutscher Landwirtschaftlicher Untersuchungs- und Forschungsanstalten e.V.: Darmstadt, Germany, 2019; pp. 13–14.
59. Schimmelpfennig, S.; Heidecke, C.; Beer, H.; Bittner, F.; Klages, S.; Krengel, S.; Lange, S. Klimaanpassung in Land- und Forstwirtschaft—Ergebnisse eines Workshops der Ressortforschungsinstitute FLI, JKI und Thünen-Institut. In *Thünen Working Paper 2018, 86*; Johann Heinrich von Thünen-Institut: Braunschweig, Germany, 2018; 110p.
60. Klages, S.; Heidecke, C.; Osterburg, B. The impact of agricultural production and policy on water quality during the dry year 2018. *Water* **2020**, under review.

Article

Model-Based Analysis of Nitrate Concentration in the Leachate—The North Rhine-Westfalia Case Study, Germany

Frank Wendland [1,*], Sabine Bergmann [2], Michael Eisele [2], Horst Gömann [3], Frank Herrmann [1], Peter Kreins [4] and Ralf Kunkel [1]

1 Forschungszentrum Juelich, IBG-3, 52425 Juelich, Germany; f.herrmann@fz-juelich.de (F.H.); r.kunkel@fz-juelich.de (R.K.)
2 LANUV NRW, 47051 Duisburg, Germany; Sabine.Bergmann@lanuv.nrw.de (S.B.); michael.eisele@lanuv.nrw.de (M.E.)
3 LWK NRW, 50765 Köln-Auweiler, Germany; Horst.Goemann@LWK.NRW.DE
4 Thünen-Institute, 38116 Braunschweig, Germany; peter.kreins@thuenen.de
* Correspondence: f.wendland@fz-juelich.de

Received: 20 December 2019; Accepted: 11 February 2020; Published: 15 February 2020

Abstract: Reaching the EU quality standard for nitrate (50 mg NO_3/L) in all groundwater bodies is a challenge in the Federal State of North Rhine-Westfalia (Germany). In the research project GROWA+ NRW 2021 initiated by the Federal States' Ministry for Environment, Agriculture, Nature and Consumer Protection, amongst other aspects, a model-based analysis of agricultural nitrogen inputs into groundwater and nitrate concentration in the leachate was carried out. For this purpose, the water balance model mGROWA, the agro-economic model RAUMIS, and the reactive *N* transport model DENUZ were coupled and applied consistently across the whole territory of North Rhine-Westfalia with a spatial resolution of 100 m × 100 m. Besides agricultural *N* emissions, *N* emissions from small sewage plants, urban systems, and *NOx* deposition were also included in the model analysis. The comparisons of the modelled nitrate concentrations in the leachate of different land use influences with observed nitrate concentrations in groundwater were shown to have a good correspondence with regard to the concentration levels across all regions and different land-uses in North Rhine-Westphalia. On the level of ground water bodies (according to EU ground water directive) *N* emissions exclusively from agriculture led to failure of the good chemical state. This result will support the selection and the adequate dimensioning of regionally adapted agricultural *N* reduction measures.

Keywords: nitrate; groundwater; source apportionment; modelling; nitrogen sources; leachate rate

1. Introduction

Excessive nitrate inputs into groundwater have been recognized as a main reason for failing drinking water standards for decades [1–6]. Agricultural *N* emissions originating from mineral or organic fertilizers are regarded as the most relevant source of nitrate in groundwater worldwide [7]. Accordingly, strategies to cope with the nitrate pollution of groundwater are focused on controlling the agricultural sources of nitrate [8,9]. In Europe, this is reflected in the water legislation at the EU level, that is, the EU Water Framework Directive (EU-WFD) [10], the EU Marine Strategy Framework Directive [11], and the EU Nitrates Directive [12], obliging the polluter to implement measures to reduce the nitrogen impact on groundwater.

The EU-WFD [10] requires the assessment of the groundwater status at the level of groundwater bodies for both quantitative and qualitative status. The latter is assessed by comparing measured concentrations and trends at groundwater observation wells to groundwater quality standards or threshold values [13] and by extrapolating the findings at the monitoring sites to the area represented by these monitoring sites. As suggested by the European Commission in a guidance document on groundwater status and trend assessment [14] groundwater bodies in the EU are classified as being not in good status in cases of 20% or more of the total groundwater body area being polluted [15–17]. Consequently, a groundwater body is classified as failing in terms of good status due to nitrate in cases where the area represented by the monitoring station(s) shows nitrate concentrations above 50 mg NO_3/L and/or rising trends exceed 20% of the total groundwater body area.

In the groundwater bodies failing in terms of good status due to nitrate, measures reducing the nitrogen input into the aquifer have to be implemented [10,11]. Taking nitrate concentrations in groundwater above 50 mg NO_3/L as a reference, the implementation of measures is limited to oxidized aquifers. Significant denitrification capacities are lacking in oxidized aquifers, and thus nitrate concentrations in the upper groundwater are directly correlated to the nitrate input [18,19].

In reduced aquifers, such a relationship is missing due to denitrification processes [20–23], and thus the related groundwater bodies are in good status with regard to nitrate in spite of high nitrogen inputs. Consequently, no measures to reduce *N* input have to be implemented in such cases. In North Rhine-Westfalia, such aquifers occur in the Northern parts of the Lower Rhine Embayment and parts of the Westphalian lowland. In this context, it must be remembered that the fossil pyrite and/or lignite particles involved in the denitrification process in aquifers are exhaustible resources [24,25]. Once this inventory of an aquifer is consumed, a dramatic nitrate concentration rise in groundwater can be observed [26,27]. Therefore, it is in generally not expedient to allow higher nitrogen inputs for reduced aquifers as is done for oxidized aquifers [28].

In order to ensure that the nitrate concentrations in oxidized aquifers will not exceed 50 mg/L and to conserve the natural denitrification capacity of reduced aquifers, the German Working Group on water issues of the Federal States and the Federal Government (LAWA) requires that the nitrate concentration in the leachate should not exceed 50 mg NO_3/L [28]. Similar considerations are reflected in the groundwater strategies of other countries [29,30]. Against this background, the nitrate concentration in the leachate can be considered as the decisive starting-point to determine the *N* reduction required [31].

In order to implement measures efficiently, a stepwise procedure can be used [18,32]. At first, the source areas contributing to measured nitrogen concentrations in groundwater beyond quality standards are identified within the groundwater bodies failing good chemical status. This is achieved by using modeled land use specific nitrate concentrations in the leachate to indicate nitrogen emission into groundwater. Subsequently, the absolute *N* reduction required to reach the groundwater quality target in the areas concerned is determined. Finally, regionally adapted and accurately dimensioned *N* reduction measures are selected and implemented.

Thus far, the model approach developed by the authors has focused on *N* emissions from agriculture, assuming that these *N* sources mainly have to be considered for determining the *N* reduction need [33–35]. The predominance of agricultural *N* sources in rural areas is quite obvious. In densely populated and industrialized regions, however, there are clear indications that also non-agricultural *N* sources may contribute to high nitrate concentrations [36,37].

With an average population density of 525 inhabitants/km^2 [38], the Federal State of North Rhine-Westfalia represents a prime example for a densely populated region in Germany. In light of the above, there is a debate as to what extent non-agricultural *N* emissions contribute to high nitrate concentrations. In cases where the nitrate concentrations in the leachate above 50 mg/L can be attributed to *N* sources other than agricultural *N* sources, agriculture alone cannot be obliged to implement *N* reduction measures according to EU-WFD requirements [10]. In the regions where this is the case, *N* reduction measures to counteract non-agricultural *N* sources would have to be implemented.

Against this background, a simultaneous consideration of the *N* emissions from all relevant *N* sources is indispensable for assessing the entire *N* loading of the leachate as well as the *N* inputs into groundwater and surface waters. In practical terms, this means that in addition to the agriculturally produced *N* emissions (agricultural *N* balance surplus, atmospheric *NHx* deposition), the *N* emissions from urban systems; small sewage treatment plants seeping into groundwater; and the atmospheric *NOx* deposition from industry, household, and traffic should also be determined, doing so enables the identification of the "hot-spot areas", wherein the nitrate concentration in the leachate exceeds 50 mg/L. Additionally, the proportion of the individual *N* sources with regard to the modelled nitrate concentration in the leachate in a certain region can be assessed objectively. The latter is an important prerequisite for addressing the regionally relevant *N* emission source(s) and for deriving correctly dimensioned *N* reduction measures.

2. Materials and Methods

The co-operation project GROWA+ NRW 2021 initiated by the Federal States' Ministry for Environment, Agriculture, Nature and Consumer Protection started at the end of 2015. In the context of the GROWA+ NRW 2021 project, the coupled agro-economic-hydrologic model network RAUMIS-mGROWA-DENUZ-WEKU-MONERIS [31] is applied and further developed in order to support the implementation process of the EU-Water Framework Directive, the EU Nitrates Directive, and the EU Marine Strategy Framework Directive in the Federal State of North Rhine-Westphalia. The questions to be answered in the GROWA+ NRW 2021 project include;

- The determination of the actual *N* input into groundwater and surface waters;
- The identification of actual "hot spot" areas of *N* pollution;
- The assessment of the necessary reduction of *N* emissions to reach the EU quality standards;
- The prediction of the time lags until *N* reduction measures will show effect (target achievement).

Nitrate concentration in the leachate is determined within the framework of an area-differentiated modeling of the diffuse and point source *N* inputs into groundwater and surface waters. For this purpose, the model system [31,32], which had already proven its suitability in macro-scale applications on the level of States and river basins in Germany, was applied in NRW on the basis of statewide available input data. The individual sub-models and their interplay are documented in the literature [32–36] and will only be briefly described here. The interplay of the individual sub-models is summarized in Figure 1.

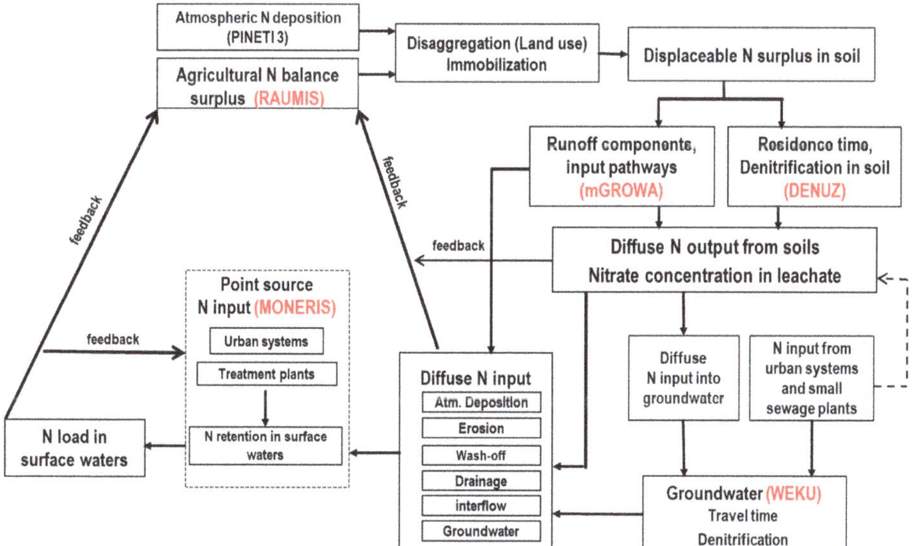

Figure 1. Simulation of nitrate concentration in the leachate in the framework of the coupled model system RAUMIS-mGROWA-DENUZ-WEKU-MONERIS.

In the agricultural sector model RAUMIS [39], a set of agro-environmental indicators is linked to agricultural production. Currently, the model comprises indicators such as fertilizer surplus (nitrogen, phosphorus, and potassium), pesticides expenditures, a biodiversity index, and corrosive gas emissions. These indicators help to evaluate direct and indirect environmental impacts of policy-driven changes in agricultural production. Regarding diffuse water pollution, the indicator "nitrogen surplus" is of particular importance. Agricultural statistics with data, for instance on crop yields, livestock farming, and land use, were used to balance the nitrogen supplies and extractions for the agricultural area. The long-term nitrogen balance averaged over several vegetation periods is calculated by considering the organic nitrogen fertilization, the mineral nitrogen fertilization, the symbiotic *N* fixation, the atmospheric *N* inputs, the compost and sewage sludge spreading, and the *N* removal with the crop substance [40,41]. As a rule, the difference between nitrogen supplies, primarily by mineral fertilizers and farm manure, and nitrogen extractions, primarily by field crops, leads to a positive *N* balance.

A certain amount of the mineral *N* surpluses in soils is denitrified to molecular nitrogen. Denitrification losses in the soil are calculated using the DENUZ model as a function of the diffuse *N* surpluses, denitrification conditions, and the residence time of percolation water in the soil according to Michaelis–Menten kinetics [31,34]. The kinetical parameters are assessed on the basis of observed denitrification rates in soils [42,43] according to the geological substrate, the influence of groundwater, and perching water of the soils and the average residence time of perching water in the soil.

In order to determine the diffuse *N* inputs into groundwater and surface waters according to individual pathways, the *N* outputs from soil were subsequently coupled to runoff components. The latter were quantified with the water balance model mGROWA [44,45]. The mGROWA modelling approach comprises several sequential parts. Within the basic part, the hydrologic processes at the earth's surface and in the root zone of soil are simulated. In particular, soil moisture dynamics including percolation water movement, capillary rise from groundwater to the root zone, actual evapotranspiration, and total runoff generation are calculated on the basis of the multilayer soil water balance model BOWAB [46]. Total runoff values are used to derive runoff components including the leachate rate and input pathways for nutrients into groundwater and surface water.

Nitrate inputs into an aquifer may be reduced in groundwater by denitrification processes in a reduced (oxygen-free) groundwater milieu. This reactive nitrate transport in reduced groundwater is modelled using the WEKU model [47,48] assuming a first-order denitrification kinetics depending on the nitrate inputs and the residence time in groundwater with a reaction constant of 0.17–0.56 a-1 [49–51].

N inputs from point sources and urban areas were taken into account using the procedures elucidated in the MONERIS model [52]. The nitrogen load in surface waters within a catchment is calculated by integration of the diffuse nitrogen inputs into the surface waters via the considered pathways calculated for each grid cell over the catchment area of the considered station plus the sum of the *N* inputs from point sources. The observed *N* loads, however, are influenced by nitrogen retention processes in the river itself, which may reduce the nitrogen loads of rivers. Nitrogen conversion in rivers is considered using the size of the catchment area, total runoff, the size of open water areas and temperature as inputs [53].

The summation of the modelled *N* inputs from diffuse and point sources minus the riverine *N* retention were compared to the measured *N* loads from monitoring stations. The observed discharge and *N* total concentrations were transformed into N_{tot}-loads according to the OSPAR method [54].

The results of three sub-models were used for the simulation of the nitrate concentration in the leachate, namely, the mean long-term leachate rate determined with the water balance model mGROWA [44], the agricultural *N* balance surplus assessed on the basis of the agro-economic model RAUMIS [39], and the reactive *N* transport in soils determined on the basis of the DENUZ model [31].

Atmospheric *N* deposition was taken into account, separated for the agricultural-related *NHx* emission and the atmospheric *NOx* deposition originating from households, *industry*, and transport [55].

Both atmospheric N sources contribute to the diffuse N output from soils. Additionally, the N emissions from urban systems and small sewage treatment plants were considered.

Equation (1) shows the general relation between the N sources and the leachate rate in order to determine the nitrate concentration in the leachate. For an area covering analysis on the Federal State level, the individual terms of Equation (1) were determined:

$$C_{NO3} = \frac{4.43 \times (Nsoil + KKA + KS)}{0.01 \times Q_{sw}} \quad (1)$$

where

C_{NO3}: nitrate concentration in the leachate (mg/L);
Q_{sw}: leachate rate (mm/year);
KKA: N emissions from small sewage treatment plants (kg N/(KKA and year));
KS: N emissions from urban systems (kg N/(community and year));
N_{soil}: diffuse N output from soils (kg N/(ha and year));
4.43: factor to convert nitrate-N (mg/L) to nitrate-NO_3 (mg/L)(-);
0.01: factor to convert mm in liter (-).

The leachate rate (Q_{SW}) was determined on the basis of the mGROWA model [39], which simulates in the basic part the hydrologic processes at the earth's surface and in the root zone of soils. In particular, soil moisture dynamics including the movement of the leachate in soils, capillary rise from groundwater to the root zone, actual evapotranspiration, and total runoff generation were calculated in daily time steps on the basis of grass reference evapotranspiration, land use-specific crop coefficients, and a topography correction function. Additionally, mGROWA calculated the water balance for impervious surfaces (e.g., sealed surfaces in urban areas) and open water surfaces using approaches adapted to specific storage characteristics. The second part in mGROWA separated total runoff into the direct runoff components (natural interflow, drainage runoff, surface runoff, direct runoff from urban areas) and groundwater recharge from the total runoff by using base flow indices [56]. The leachate rate results from the difference between surface runoff including runoff from urban areas and total runoff and designates the water volume leaving the root zone downwards. It is in this regard not identical to groundwater recharge, which designates the water volume infiltrating into the aquifer.

The diffuse N output from soil (*Nsoil*) included N originating from two N sources, on one hand the N output from soil originating from agriculture (*Nsoil*(agri)), and on the other hand the N output from soil originating from NOx deposition. The diffuse N output from soil originating from agriculture (*Nsoil*(agri)) was derived according to Equation (2) considering the agricultural-related N loads. The corresponding diffuse N output from soil originating from NOx (*Nsoil* (NOx)) was derived according to Equation (3). In both cases, two important processes controlling the N losses from soils, these being N immobilization (NI) and denitrification (ND), needed to be considered.

The diffuse N output from soil originating from agriculture (Nsoil(agri)) included the agricultural N balance surplus (N_u) on a community level averaged over several vegetation periods provided by the regionally differentiating agricultural sector model RAUMIS [40]. As a rule, the difference between nitrogen supplies, primarily by organic and mineral fertilizers, and the nitrogen extractions, primarily by field crops, leads to a positive N balance for the agriculturally used areas (N_u).

$$Nsoil(agri) = N_U + NHx - NI - ND \quad (2)$$

$$Nsoil(NOx) = NOx - NI - ND \quad (3)$$

where

Nsoil (agri): N output from soil originating from agriculture (kg N/ha·year);
Nsoil (NOx): N output from soil originating from NOx (kg N/ha·year);

N_U: mean agricultural N balance surplus (kg N/ha·year);
NH_x: mean atmospheric NHx deposition (kg N/ha·year);
NI: mean N immobilization in soil (kg N/ha·year);
ND: mean denitrification in soil (kg N/ha·year).

The NHx emission (NH_x), particularly from livestock, is the second most important agricultural N source that contributes to the diffuse N output from soil originating from agriculture (see Equation (2)). It can be assumed that the entire agricultural share in atmospheric deposition was represented by the NHx deposition. Equivalent to the NHx deposition, it was assumed that the entire NOx deposition originating from households, traffic, and industry represented the diffuse N output from soil originating from non-agriculture N sources ($Nsoil(NOx)$).

Both NHx and NOx were derived from an official German-wide dataset on atmospheric Ndeposition determined in the framework of the PINETI project series (PINETI-3, Pollutant INputan EcosysTemImpact) [55] for the time series 2013-2015. In order to adapt the spatial resolution of the PINETTI-3 datasets to the spatial resolution of the modelling on NRW scale, the land-use specific NHx and NOx values available as 1 × 1km grids were disaggregated according to the land use grids of 100 × 100 m. Apart from its impact on agriculturally used land, the atmospheric NHx and NOx deposition also affected forests and settlement areas, and thus it had to be considered as area-covering data in the modelling of the nitrate concentration in the leachate.

The NOx deposition, as well as the sum of agricultural N balance surpluses and atmospheric NHx deposition, do not correspond to the diffuse N output from soils as two important processes controlling the N losses from soils, that is, N immobilization (N_I) and denitrification (N_D) have to be considered. For pasture and forested areas, it is assumed that a certain part of the N input into soils is immobilized in the soil organic matter [57] and contributes to the development of biomass. The immobilization rates depend on the interplay of certain site conditions, for instance, leachate rates, soil texture, and topography. They can consequently be regarded as regionally variable and are determined in the framework of the model calibration for pasture and forested areas. For arable land, however, a significant N immobilization in soil can be neglected [33].

The N load resulting from the summation of agricultural N balance surplus, atmospheric NHx and NOx deposition, and N immobilization in soil is defined as the displaceable N surplus in soil (see Figure 1). From this intermediate result, the N output from soil is derived by subtracting the N losses in soil due to denitrification. In contrast to N immobilization, denitrification in soil takes place independent from the type of land cover. Equation (3) specifies the DENUZ approach to determine denitrification rates in soil.

$$\frac{dN_s t_{soil}}{dt_{soil}} + D_{max} \cdot \frac{N_s t_{soil}}{k + N_s t_{soil}} = 0 \quad (4)$$

where

$N_s t_{soil}$: N output from soil from agricultural N sources and NOx deposition after residence time (t_{soil}) (kg N/(ha·year));
t_{soil}: residence times in soil (year);
D_{max}: maximum yearly denitrification rate in soil (kg N/(ha·year));
k: Michaelis-constant (kg N/(ha·year));
0: displaceable N surplus in soils (kg N/(ha·year)).

Using observed denitrification rates as a reference, the maximum denitrification rate (D_{max}) was assessed [18]. This was performed on the basis of a ranking of the soils predominantly according to their groundwater and perching water influence, as well as their organic carbon content [34,35]. Denitrification rates in soils were determined as a relationship between the realizable maximum annual denitrification rates assigned by the denitrification conditions and the related soil-specific reaction rates of denitrification (k) under consideration of the effective residence times of the leachate in the root zone (*t*). In this way, it is assumed that, for example, in a case were the residence time in a certain soil amounts to 6 months, the extent of denitrification is half of the maximum denitrification rate per year for the given soil.

The final result of the DENUZ model, that is, the diffuse *N* output from soils from agricultural *N* sources and *NOx* deposition, was subsequently used to calculate the nitrate concentration in the leachate according to Equation (1). Additionally, the *N* outputs from urban systems (KS) and small sewage treatment plants (KKA) (see Equation (1)) were derived from directories about inhabitants and data about water management facilities (for the North Rhine-Westfalia-specific numbers, see Section 3.2.1).

3. Results

All input parameters and model results to calculate the nitrate concentration in the leachate were processed as area-covering data for the entire state of Northrhine-Westfalia with a spatial resolution of 100 m × 100 m. This resulted in a subdivision of the area of the Federal State into around 3.4 Mio individual grids.

The leachate rate (see Figure 2) diluted the *N* emissions from agricultural and non-agricultural *N* sources, as specified in Equation (1). It was calculated with the mGROWA model [39] for the hydrological reference period of 1981–2010. Using mean long-term leachate rates as a reference has two advantages with regard to the modelled nitrate concentrations in the leachate. On one hand, as a short-term climate-induced blurring of leachate rates was excluded, the modelled nitrate concentrations in the leachate were representative of the regional long-term hydrologic conditions. On the other hand, the regional varying *N* emissions were related to a uniform leachate rate, which was of importance when the impact of *N* reduction measures was assessed.

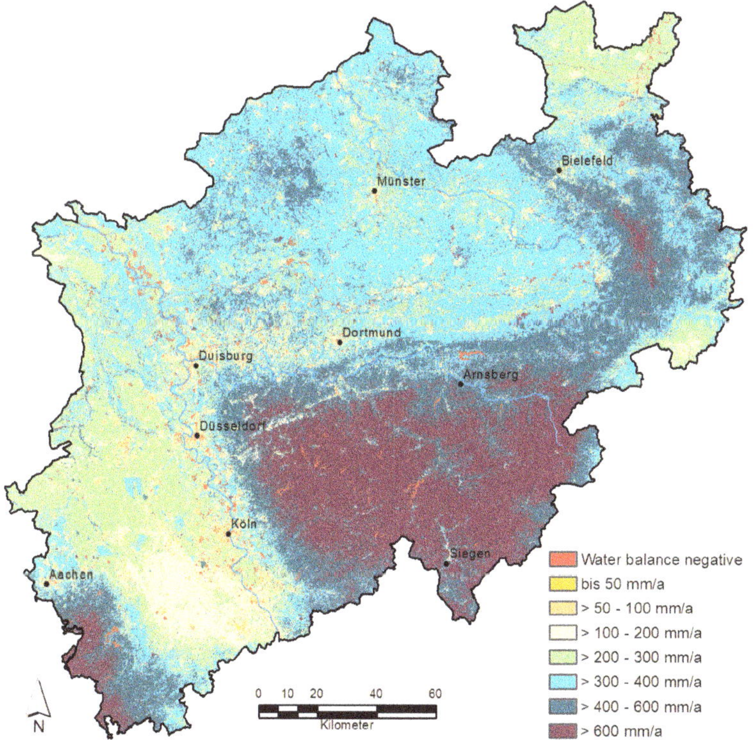

Figure 2. Mean long-term leachate rate of the hydrologic period of 1981–2010.

The mean long-term leachate rates shown in Figure 2 reveal wide disparities between the regions. Whereas leachate rates of 600 mm/year or more frequently occurred in all consolidated rock regions (e.g., Rhenish Massif), leachate rates of less than 200 mm/year may have occurred in the southern part of the Lower Rhine embayment. There, the dilution of *N* emissions from the different *N* sources was up to three times higher in the consolidated rock regions.

3.1. Nitrogen Emissions from Agricultural Sources and Resulting Nitrate Concentrations in the Leachate

3.1.1. Agricultural *N* Sources

The *N* sources taken into account for determining the *N* output from soil originating from agriculture, that is, the agricultural *N* balance surplus and the atmospheric *NHx* deposition, are shown in Figure 3a,b, respectively.

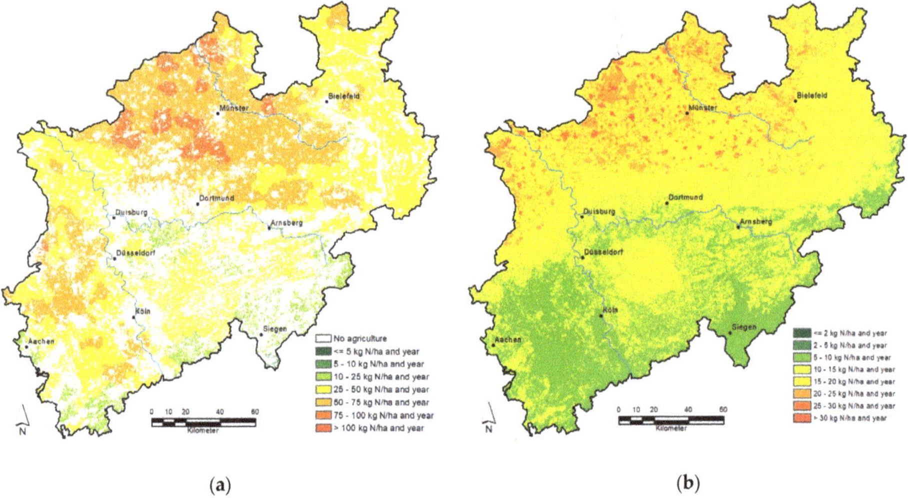

Figure 3. (**a**) Agricultural *N* balance surplus and (**b**) atmospheric *NHx* deposition.

In order to guarantee that crop- and withdrawal-related fluctuations were biased out, the agricultural *N* balance surplus on a community level provided by the RAUMIS model [40] was determined as an averaged value for the period of 2014–2016. In total, the agricultural *N* balance surplus in North Rhine-Westphalia summed up to around 82.400 t N/a. Major differences became evident in cases where this sum was considered regionally differentiated (see Figure 3a). Agricultural *N* balance surpluses of less than 25 kg N/ha and year were found in regions where extensive grassland dominates (e.g., Rhenish Massif). In the food crop cultivation regions (e.g., Lower Rhine Embayment), where the *N* demand was predominantly covered by well-controllable mineral fertilizer applications, the agricultural *N* balance surpluses ranged between 25 and 50 kg N/ha per year. In contrast, nitrogen surpluses above 50 kg N/ha per year were determined for regions where livestock density was so high that the amount of organic fertilizers (and additionally applied mineral fertilizers) exceed the *N* demand of the agriculturally used land. Areas in white indicate the non-agriculturally used forest and settlement areas.

In the agricultural regions with intensive livestock production in the north, the atmospheric *NHx* deposition provided by the PINETI model summed up to more than 20-25 kg N/ha per year and locally to more than 30 kg N/ha per year [58] (see Figure 3b). To the south, that is, in greater distance from the regions with intensive livestock production, *NHx* deposition rates between 10 and 20 kg N/ha per year predominated. This region extended from the northern part of the Lower Rhine embayment in the west to the Weserbergland in the east. In the southern part of the Lower Rhine Embayment, crop farms predominate. There, the atmospheric *NHx* deposition rarely exceeded 10 kg N/ha per year. The same *NHx* deposition rates were determined for the most southern parts of the Rhenish Massif situated in the lee of the Sauerland Mountains. Averaged over the entire Federal State of North Rhine-Westfalia, the atmospheric *NHx* deposition corresponded to a total of around 44,000 t N/a.

3.1.2. *N* Output from Soils Originating from Agricultural *N* Emission Sources

In order to determine the *N* output from soils, *N* immobilization and denitrification (in soil) was considered. *N* immobilization is related to the sum of the agricultural *N* balance surpluses and atmospheric *NHx* deposition. Whereas the determined *N* immobilization rates comprised around 50% for the land-use category pasture, the *N* immobilization rates for the land-use categories of coniferous forests and deciduous forests were in the range of 10% and 20%, respectively. Averaged over the entire Federal State of North Rhine-Westfalia, *N* immobilization retained around 18.500 tons N/a. Consequently, the displaceable *N* surplus in soil, that is, the sum of agricultural *N* balance surpluses plus atmospheric *NHx* deposition minus *N* immobilization was in the range of around 108,000 t N/a.

The displaceable *N* surplus in soil minus the *N* losses in soil due to denitrification modeled with the DENUZ model (Figure 4a) determined the *N* output from soils (see Figure 4b). In the shallow brown earth soils of the Rhenish Massif, denitrification reduced the displaceable *N* surplus in soil to less than 25%. The low denitrification rates were due to the interaction of small residence times in the soil (often less than 3 months) and poor denitrification conditions in the soil.

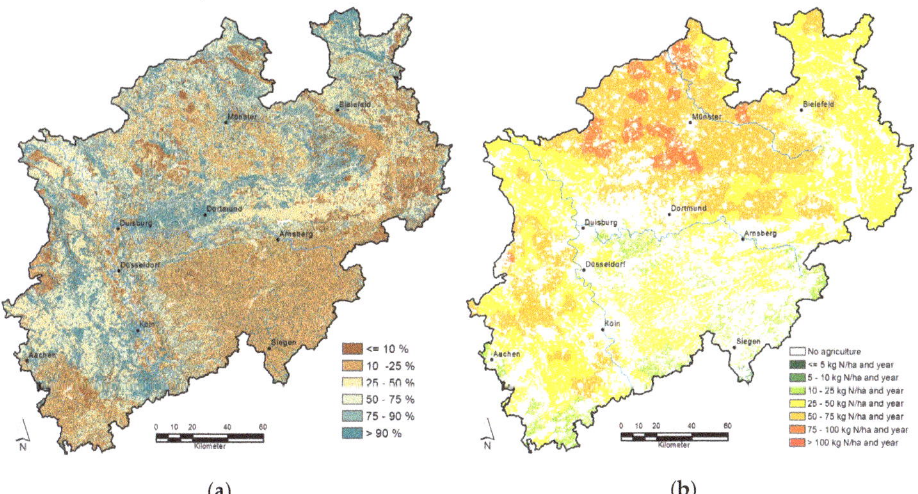

Figure 4. (**a**) Denitrification losses in soil relative to the displaceable *N* surplus in soil, and (**b**) *N* output from soils originating from agricultural sources.

In the fertile loess regions of North Rhine-Westfalia (Lower Rhine embayment, Soester Börde), up to 50% of the agricultural *N* input into soils was denitrified. Although the denitrification capacity of the predominantly occurring luvisol was quite unfavorable, these soils displayed a high water storage capacity and relatively low leachate rates (see Figure 2). Consequently, residence times in soil was often in the range of 1 year (or more), and thus the maximum possible denitrification rates in soil could nearly be reached or even be exceeded. As shown in Figure 4b, the resulting *N* output from soils in these regions ranged between 25 and 50 kg N/ha per year.

Generally, high displaceable *N* surpluses in soil occurred in the Münsterland basin in the Northern part of North Rhine-Westfalia. There, sandy podzol soils showing poor denitrification capacities alternated with loamy lowland soils (e.g., Gleyic soils) showing high denitrification capacities. Consequently, denitrification losses of more than 75% may be achieved in the lowland soils, whereas denitrification rates in the pozol soils hardly exceeded 25% (see Figure 4a). Accordingly, soils with *N* outputs of less than 25 kg N/ha per year and soils with *N* losses of more than 75 kg N/ha per year occurred closely adjacent to one another (see Figure 4b). In the core of the Münsterland, vast areas with *N* outputs from the soil in the range of 75 kg N/ha per year were identified. There, soils with moderate denitrification capacities coincided with very high displaceable *N* surpluses in soil. Clearly recognizable in Figure 4b is the metropolitan area Rhein-Ruhr. As there were no other agricultural *N* inputs other than the atmospheric *NHx* deposition, the *N* output from soil that can be assigned to agricultural *N* emissions was below 5 kg N/ha per year.

Averaged over the entire Federal State of North Rhine-Westfalia, around 45% of the displaceable *N* surplus in soil (around 108,000 t N/a) was denitrified, and thus the *N* output from soil originating from agricultural *N* emissions was reduced to approximately 69,000 t N/a.

3.1.3. Nitrate Concentration in the Leachate Originating from Agricultural Sources

The *N* output from soil attributed to agricultural *N* emissions (Figure 4b) were combined with the leachate rates (Figure 2) to determine the nitrate concentration in the leachate originating from agricultural sources (see Figure 5a). In Figure 5b, the areas where a nitrate concentration in the leachate exceeded 50 mg NO_3/L are highlighted.

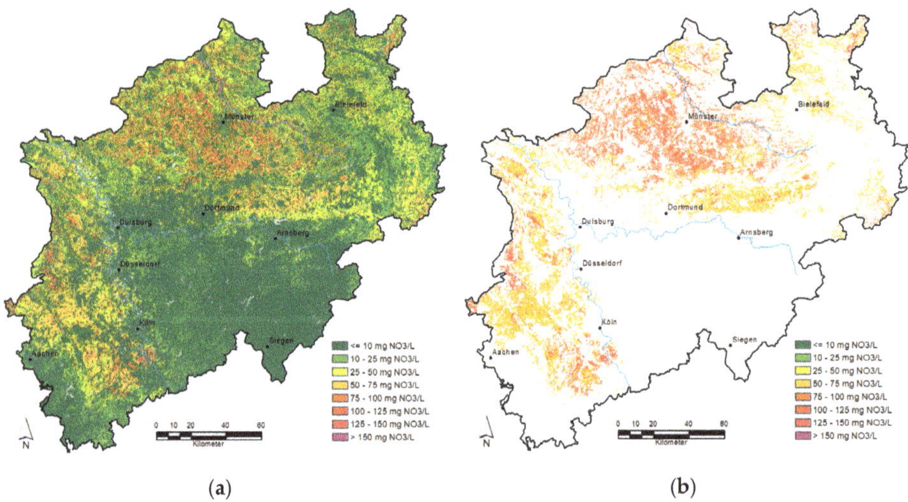

Figure 5. (**a**) Nitrate concentration in the leachate originating from agricultural sources, and (**b**) areas showing nitrate concentrations in the leachate above 50 mg NO_3/L due to agricultural *N* emissions.

For many areas in the Münsterland, particularly high values arose due to the high livestock density, which led to an accordingly high nitrogen balance surplus. In the Lower Rhine Embayment, nitrate concentrations in the leachate between 50 and 75 mg/L predominated. There, moderate nitrogen balance surpluses arising from cash crop farms were combined with low leachate rates. Nitrate concentrations in the leachate below 25 mg/L could be found in the urbanized areas of North Rhine-Westfalia, as well as in the Rhenish Massif, where high leachate rates coincided with low *N* emissions. Figure 5b illustrates in this regard how yet again there was a remarkable extent of areas with nitrate concentrations in the leachate above 50 mg/L in the Lower Rhine Embayment and the Münsterland.

3.2. Nitrogen Emissions from Non-Agricultural Sources and Resulting Nitrate Concentrations in the Leachate

The calculation of the nitrate concentration in the leachate due to non-agricultural *N* sources included on one hand the *N* emission from small sewage treatment plants and urban systems. On the other hand, it included the *N* outputs from soil attributed to the *NOx* deposition from industry, traffic, and households (see Equation (3)).

3.2.1. Non-Agricultural *N* Sources

The *N* output from urban systems (Figure 6a) was quantified on the basis of data provided by an actual management report about the development and level of sewage disposal in North Rhine-Westphalia [59]. According to this report, the total amount on nitrogen emitted from municipal waste water disposals (including indirect discharges) divided by all inhabitants resulted in 11 g *N* per day. Under the assumption that 15% of this disposal was released into groundwater due to leakage from the waste water systems, a uniform nitrogen release from urban systems into groundwater of 1.65 g *N* per inhabitant per day was estimated. The *N* losses from urban systems determined in this way ranged between 10 kg and 50 kg N/ha per year. The highest values occurred, as expected, in densely populated regions along the river Rhine and the Ruhr area. On Federal State scale, the total *N* losses from urban systems accumulated to approximately 11.200 t N/a.

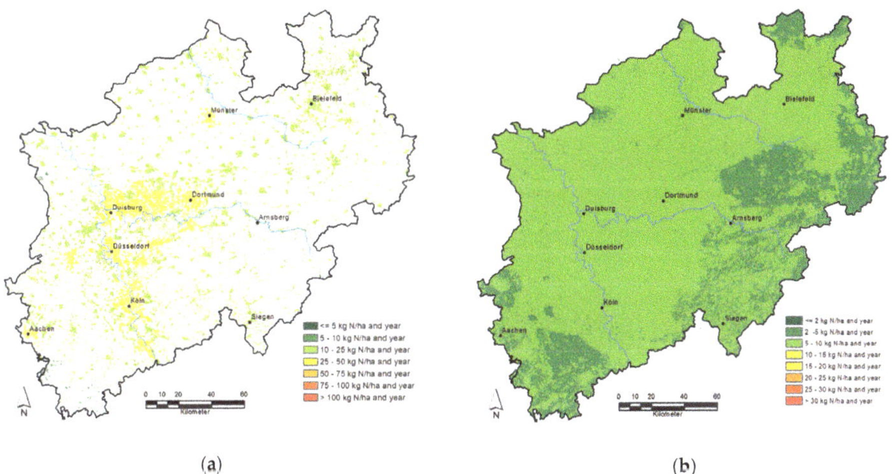

Figure 6. (**a**) Nitrogen output from urban systems and small sewage treatment plants, and (**b**) atmospheric *NOx* deposition (2013–2015).

For considering the *N* output from small sewage treatment plants, the recorded annual *N* emissions of around 20,000 individual facilities emitting into groundwater were taken into account [60]. For this purpose, the recorded *N* loads were assigned to the corresponding 100 × 100 m grids. In most of all grids concerned, the *N* outputs from small sewage treatment plants showed values below 25 kg N/ha per year. Higher *N* outputs occurred locally only, but could reach 100 kg N/ha per year and more. At the Federal State level, the total *N* loss of the small sewage treatment plants summed up to around 500 t N/a. As the visibility of the individual 100m grids was limited, the *N* outputs from small sewage treatment plants are shown in Figure 6a, together with the *N* output from urban systems.

The atmospheric *NOx* deposition originating from households, industry, and transport was provided from a Germany-wide dataset on atmospheric *N* deposition [58]. Depending on the region, the *N* output from soils from *NOx* deposition ranged between <5 to >10 kg N/ha per year (see Figure 6b). For the total area of NRW, this relatively low average annual atmospheric *NOx* deposition summed up to around 20,000 tons of N/a.

3.2.2. Nitrate Concentration in the Leachate Originating from Non-Agricultural *N* Sources

The *N* output attributed to urban systems and small sewage treatment plants (Figure 6a), as well as the *N* output from soil resulting from *NOx* deposition (Figure 6b) was combined with the leachate rates (Figure 2) to determine the nitrate concentration in the leachate originating from non-agricultural sources (Figure 7).

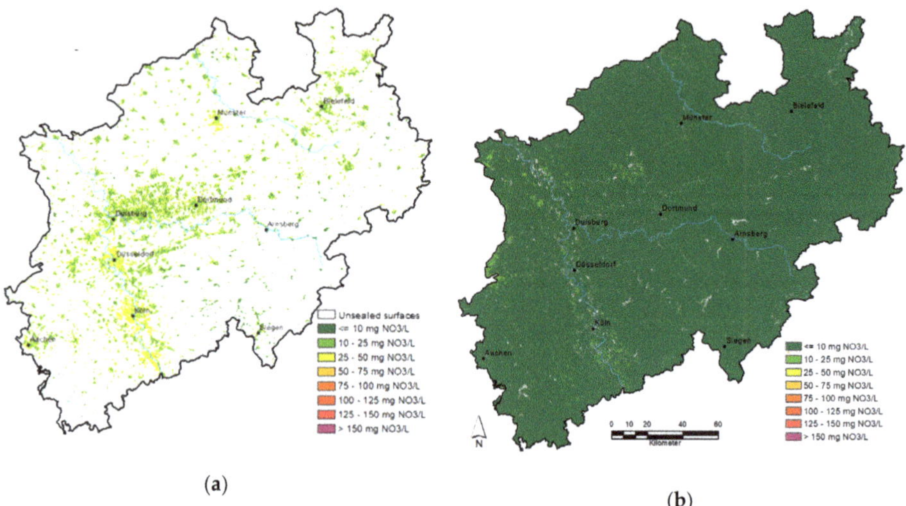

Figure 7. (**a**) Nitrate concentration in the leachate originating from urban systems and small sewage treatment plants, and (**b**) nitrate concentration in the leachate originating from the *N* output from soil due to *NOx* deposition.

Figure 7a shows that the nitrate concentration in the leachate originating from urban systems and small sewage treatment plants ranged in most areas between 10 and 25 mg/L (e.g., in the entire Ruhr area). It was evident that the nitrate concentration in the leachate originating from these sources only to a minor extent and locally exceeds 50 mg/L. As the areas concerned did not exceed 20% of the area of a groundwater body, no measures to reduce the *N* emissions from urban systems and small sewage treatment plants had to be included in the program of measures according to EU-WFD requirements [14]. This, however, does not detract from the fact that measures to reduce *N* output from urban systems may be necessary to improve surface water quality.

Figure 7b shows the nitrate concentration in the leachate originating from *NOx* deposition. For the latter, the denitrification in soil was accounted for as described in Equation (3). In most areas of North Rhine-Westfalia, the nitrate concentration in the leachate originating from *NOx* deposition was less than 10 mg/L. In the southern part of the Lower Rhine Embayment, values of up to 25 mg/L occurred due to the low leachate rates (Figure 2). In conclusion, *NOx* deposition did not at all lead to NO_3 concentrations in the leachate above 50 mg/L.

3.3. *Plausibility Check of Modelled Nitrate Concentration in the Leachate*

For comparing (modeled) nitrate concentrations in the leachate with (observed) nitrate concentrations in groundwater, it was necessary to merge the *N* emissions from the individual *N* sources. The corresponding nitrate concentrations in the leachate are represented in Figure 8 as the underlying colored areas.

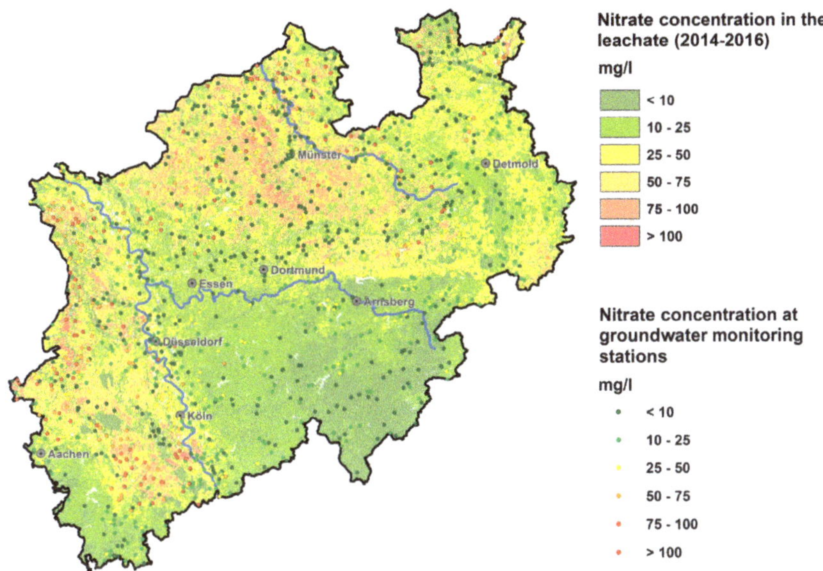

Figure 8. Plausibility check of the mean annual nitrate concentration in the leachate. The dots show observed values at groundwater monitoring sites—the underlying colored areas are the modelled nitrate concentrations in the leachate per grid.

The modelled nitrate concentration in the leachate was compared to nitrate concentrations observed at around 1500 groundwater monitoring stations from the upper aquifer for the period 2014 to 2017. In Figure 8, the mean values of the observed nitrate concentrations at the groundwater monitoring stations are represented as dots. For the comparison, the same class widths and the same color gradation were selected for both the underlying areas and the dots. The proportion of the main land use types (e.g., arable land, pasture, forest, settlement) occurring in North Rhine-Westfalia was approximately reflected by the distribution of the monitoring stations within these land use types, thus guaranteeing representability.

As can be seen from Figure 8, the modeled values in the Lower Rhine Embayment, the Weserbergland, and the Rhenish Massif corresponded spatially and with regard to their concentration levels very well to the observed nitrate concentrations in groundwater.

In the Münsterland, however, the modeled nitrate concentration in the leachate systematically exceeded the observed nitrate concentrations in groundwater (see Figure 8). There, due to the occurrence of denitrification capacities in the aquifers, the nitrate concentration in the leachate was decreased once the leachate entered into the groundwater. Consequently, the apparent disagreement between (high) modeled nitrate concentrations in the leachate and (low) observed nitrate concentrations in groundwater was attributable to the considerable denitrification capacity of reduced aquifers and did not indicate wrong model results [24].

The plausibility check described above was accompanied by a statistical evaluation of the modelled nitrate concentrations in leachate and the observed nitrate concentrations at monitoring stations for different land uses (Figure 9).

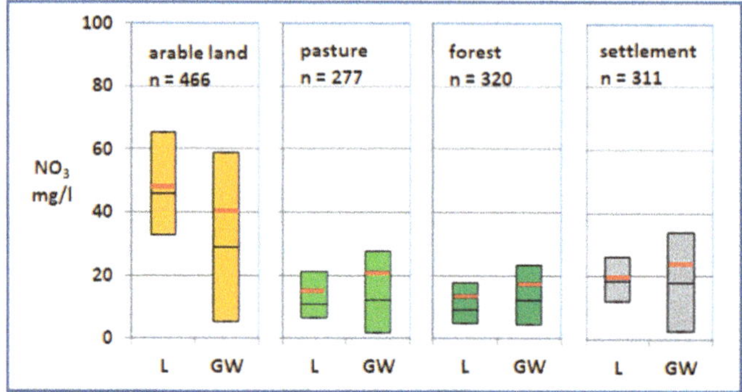

Figure 9. Comparison of modelled nitrate concentrations in the leachate (L) and observed nitrate concentrations in groundwater at monitoring stations (GW) for different land use types. The lower and upper end of the bars indicate the statistical distribution with the 25th percentile and 75th percentile, respectively, and the lines inside the bars mark the median (black line) and mean (red line).

Figure 9 shows that the 25th percentile and 75th percentile values, as well as the median and mean values of the modelled nitrate concentrations in the leachate and the observed nitrate concentrations in groundwater corresponded quite well for the land use classes of pasture, forests, and urban areas. For the land use class of arable land, however, the observed nitrate concentrations in groundwater showed lower values than the modelled nitrate concentrations in the leachate.

A more detailed analysis showed that the absence of denitrification capacities in aquifers explained the good agreement for the land use classes of pasture, forests, and settlements. Groundwater samples from these land use types showed oxidized groundwater, and thus the nitrate concentration in the leachate was preserved after infiltration into the upper aquifer [18,31].

A considerable number of the groundwater monitoring stations from the land use class of arable land were, however, located in aquifers displaying high denitrification capacities, mainly in Münsterland (see Figure 8). As denitrification in groundwater was not accounted for in the modelled nitrate concentrations in the leachate, it was obvious that the modelled nitrate concentration in the leachate showed higher vales than the overserved nitrate concentrations in groundwater.

4. Discussion

The application of Equation (1) at the Federal State level was indispensable for assessing the entire *N* loading of the leachate as well as the *N* inputs into groundwater and surface waters. The grids showing nitrate concentrations in the leachate above 50 mg NO_3/L represent the hot-spot areas of nitrate pollution of groundwater. The good correspondence of the modelled nitrate concentrations in the leachate with the observed nitrate concentrations in groundwater detected in the course of the plausibility check demonstrated that the model results were appropriate in indicating the nitrate pollution of groundwater at the Federal State level. Consequently, the modelled nitrate concentration in the leachate can be regarded as a reliable starting point for scenario analyses, for instance as an adequate reference value for assessing the extent to which a nitrate reduction is necessary to reach the EU-WFD quality target for groundwater.

Inside the groundwater bodies failing good quality status due to nitrate, the areas showing nitrate concentrations in the leachate above 50 mg/L were predestined for implementing *N* reduction measures. Modelling nitrate concentration in the leachate apportioned between the individual agricultural and non-agricultural *N* sources facilitated the identification of the main polluter in a certain region. It was obvious that only the latter had to implement measures to reduce the nitrogen impact on groundwater according to EU-WFD requirements.

Results of the model analysis in North Rhine-Westfalia proved that *N* emissions from small sewage plants, urban systems, and *NOx* deposition only locally caused nitrate concentrations in the leachate above 50 mg NO_3/L, in spite of the high population density (525 inhabitants/km^2). Consequently, as the areas concerned did not exceed 20% of the area of a groundwater body, the focus relied on agricultural measures to reduce nitrogen losses. This, however, did not detract from the fact that the implementation of measures to reduce the *N* output from urban systems may have been necessary, but not within the program of measures according to EU-WFD.

Model analysis confirmed instead that *N* emissions from agriculture exclusively led to extended areas of nitrate concentrations > 50 mg NO_3/L, especially in the north (Münsterland) and the west (Lower Rhine Embayment). As in general more than 20% of the areas in the respective groundwater bodies were concerned, the implementation of measures to reduce agricultural *N* emissions in the context of the WFD program of measures was necessary.

Results presented for the Federal State of Northrhine-Westfalia clearly illustrated the importance of systematically considering not only agricultural *N* emissions in modelling nitrate concentrations in the leachate. Especially in densely populated regions, the consideration of the *N* inputs from all *N* sources (as shown in the maps) was the only way to objectively assess the relevant polluter(s) in a region and to evaluate the relative contribution of each polluter to the overall *N* pollution. The latter was once again important to dimension appropriate polluter-specific *N* reduction measures.

A comparison of modelled nitrate concentrations in the leachate and observed nitrate concentrations in groundwater at monitoring stations for different land use types showed that the modelled nitrate concentration in the leachate showed higher vales than the overserved nitrate concentrations in groundwater due to denitrification processes in groundwater. The modelled high nitrate concentrations in the leachate were an indication that the *N* emissions in groundwater have to be reduced. Even in cases where the observed nitrate concentrations in groundwater were found to be considerably below 50 mg NO_3/L, a reduction of the *N* emissions in groundwater is the only way to conserve the natural denitrification capacity of reduced aquifers for as long as possible [28].

Although the model can be recommended for large-scale assessments on the level of states or entire river basins, the use of the model results for local issues is subjected to several constraints. Most important are limitations in the local representativeness of the model input parameters derived from statewide available databases. It is evident that, for instance, on-site occurring soil properties may not be represented in a statewide available soil map at a scale of 1:50,000. Consequently, local obviously deviating model results may have been due to databases being insufficiently accurate for local issues. Therefore, an effective monitoring network and local expert knowledge will remain indispensable for implementing local nitrogen strategies. In this regard, it must also be remembered, that the modelled values represent reference values characterizing mean long-term conditions rather than fixed values tracing specific nitrate concentrations at certain sites and at certain times.

Author Contributions: Conceptualization: F.W., S.B., M.E., H.G., F.H., P.K., R.K.; Initial manuscript writing: F.W.; Review: S.B., M.E., H.G., F.H., P.K., R.K.; Methodology and investigation: F.W., S.B., M.E., H.G., F.H., P.K., R.K.; Visualization: M.E., R.K., F.W. All authors have read and agree to the published version of the manuscript.

Funding: This research was funded by the Ministry for Environment, Agriculture, Conservation and Consumer Protection of the State of North Rhine-Westphalia in the framework of the cooperation project GROWA+ NRW 2021.

Conflicts of Interest: The authors declare no conflict of interest.

References

1. Strebel, O.; Duynisveld, W.H.M.; Böttcher, J. Nitrate pollution of groundwater in Western Europe. *Agric. Ecosyst. Environ.* **1989**, *26*, 189–214. [CrossRef]
2. Sutton, M.A.; Howard, C.M.; Erisman, J.W.; Billen, G.; Bleeker, A.; Grennfelt, P.; van Grinsven, H.; Grizzetti, B. *The European Nitrogen Assessment: Sources, Effects and Policy Perspectives*; Cambridge University Press: Cambridge, UK, 2011; p. 612.

3. Power, J.F.; Schepers, J.S. Nitrate contamination of groundwater in North America. *Agric. Ecosyst. Environ.* **1989**, *26*, 165–187. [CrossRef]
4. Rosenstock, T.S.; Liptzin, D.; Dzurella, K.; Fryjoff-Hung, A.; Hollander, A.; Jensen, V.; King, A.; Kourakos, G.; McNally, A.; Stuart Pettygrove, G.; et al. Agriculture's contribution to nitrate contamination of Californian groundwater (1945–2005). *J. Environ. Qual.* **2014**, *43*, 895–907. [CrossRef] [PubMed]
5. Thorburn, P.J.; Biggs, J.S.; Weier, K.L.; Keating, B.A. Nitrate in groundwaters of intensive agricultural areas in coastal Northeastern Australia. *Agric. Ecosyst. Environ.* **2003**, *94*, 49–58. [CrossRef]
6. Wu, M.; Wu, J.F.; Liu, J.; Wu, J.C.; Zheng, C.M. Effect of groundwater quality on sustainability of groundwater resource: A case study in the North China plain. *J. Contam. Hydrol.* **2015**, *179*, 132–147. [CrossRef] [PubMed]
7. Haller, L.; McCarthy, P.; O'Brien, T.; Riehle, J.; Stuhldreher, T. *Nitrate Pollution of Groundwater*; Alpha Water Systems Inc.: Paramount, CA, USA, 2013; Available online: http://www.reopure.com/nitratinfo.html (accessed on 10 July 2019).
8. Bechmann, M.; Blicher-Mathiesen, G.; Kyllmar, K.; Iital, A.; Lagzdins, A.; Salo, T. Nitrogen application, balances and the effect on nitrogen concentrations in runoff from small catchments in the Nordic-Baltic countries. *Agric. Ecosyst. Environ.* **2014**, *198*, 104–113. [CrossRef]
9. Billen, G.; Garnier, J.; Lassaletta, L. The nitrogen cascade from agricultural soils to the sea: Modeling nitrogen transfers at regional watershed and global scales. *Philos. Trans. R. Soc. B* **2013**, *368*, 1–13. [CrossRef]
10. EU-WFD. Directive 2000/60/EC of the European Parliament and the Council of 23 October 2000 Establishing a Framework for Community Action in the Field of Water Policy. *Off. J. Eur. Communities* **2000**, *L 327*, 1–73.
11. EU-MSFD. Directive 2008/56/EC of the European Parliament and the Council of 17 June 2008 Establishing a Framework for Community Action in the Field of Marine Environmental Policy. *Off. J. Eur. Communities* **2008**, *L 164/19*, 1–40.
12. European Parliament and Council of the European Union. Council Directive 91/676/EEC of 12 December 1991 Concerning the Protection of Waters against Pollution Caused by Nitrates from Agricultural Sources. *Off. J. Eur. Communities* **1991**, *L 375/1*, 1–8.
13. Müller, D.; Blum, A.; Hart, A.; Hookey, J.; Kunkel, R.; Scheidleder, A.; Tomlin, F.; Wendland, F. *Final Report for a Methodology to Set Up Groundwater Threshold Values in Europe*; Report D18; Specific Targeted EU—Research Project Bridge: Vienna, Austria, 2006; pp. 1–63.
14. European Commission. *Guidance on Groundwater Status and Trend Assessments, CIS Guidance Document No. 18, Luxembourg: Office for Official Publications of the European Communities*; European Commission: Brussels, Belgium, 2009; p. 84.
15. Bikšel, J.; Retikel, I. An approach to delineate groundwater bodies at risk: Seawater intrusion in Liepāja (Latvia). *E3S Web Conf.* **2018**, *54*. [CrossRef]
16. Garrido, T.; Iglesias, M.; Fraile, J.; Munne, A. Chemical status assessment of groundwater bodies and measures proposed within the framework of the Catalan River basin district management plan. In Proceedings of the European Groundwater Conference, Madrid, Spain, 20–21 May 2010.
17. LAWA. *Fachliche Umsetzung der Richtlinie zum Schutz des Grundwassers vor Verschmutzung und Verschlechterung 2006/118/EG*; Sachstandsbericht vom 31.01.2008; LAWA-Unterausschuss, Fachliche Umsetzung der Grundwassertochterrichtlinie: Berlin, Germany, 2008.
18. Kuhr, P.; Haider, J.; Kreins, P.; Kunkel, R.; Tetzlaff, B.; Vereecken, H.; Wendland, F. Model based assessment of nitrate pollution of water resources on a federal state level for the dimensioning of agro-environmental reduction strategies: The North Rhine-Westphalia (Germany) case study. *Water Resour. Manag.* **2013**, *27*, 885–909. [CrossRef]
19. Wendland, F.; Bogena, H.; Goemann, H.; Hake, J.F.; Kreins, P.; Kunkel, R. Impact of nitrogen reduction measures on the nitrogen loads of the river Ems and Rhine (Germany). *Phys. Chem. Earth* **2009**, *30*, 527–541. [CrossRef]
20. van Berk, W.; Fu, Y. Redox roll-front mobilization of geogenic uranium by nitrate input into aquifers: Risks for groundwater resources. *Environ. Sci. Technol.* **2017**, *51*, 337–345. [CrossRef] [PubMed]
21. Rivett, M.O.; Buss, S.R.; Morgan, P.; Smith, J.W.N.; Bemment, C.D. Nitrate attenuation in groundwater: A review of biogeochemical controlling processes. *Water Res.* **2008**, *42*, 4215–4232. [CrossRef] [PubMed]

22. Højberg, A.L.; Lausten Hansen, A.; Wachniew, P.; Żurek, A.J.; Virtanen, S.; Arustiene, J.; Strömqvist, J.; Rankinen, K.; Refsgaard, J.C. Review and assessment of nitrate reduction in groundwater in the Baltic Sea Basin. *J. Hydrol. Reg. Stud.* **2017**, *12*, 50–68. [CrossRef]
23. Kolbe, T.; de Dreuzy, J.R.; Abbott, B.W.; Aquilina, L.; Babey, T.; Green, C.T.; Fleckenstein, J.H.; Labasque, T.; Laverman, A.M.; Marçais, J.; et al. Stratification of reactivity determines nitrate removal in groundwater. *Proc. Natl. Acad. Sci. USA* **2019**, *116*, 2494–2499. [CrossRef]
24. Kunkel, R.; Bach, M.; Behrendt, H.; Wendland, F. Groundwater-borne nitrate intakes into surface waters in Germany. *Water Sci. Technol.* **2004**, *49*, 11–19. [CrossRef]
25. Kunkel, R.; Wendland, F.; Albert, H. Zum Nitratabbau in den grundwasserführenden Gesteinsschichten des Elbeeinzugsgebietes. *Wasser Boden* **1999**, *51*, 16–19.
26. Cremer, N.; Schindler, R.; Greven, K. Nitrateintrag ins Grundwasser und Abbaumechanismen an verschiedenen Fallbeispielen. *Korresp. Wasserwirtsch.* **2018**, *11–16*, 352–360.
27. Rohmann, U.; Sontheimer, H. *Nitrat im Grundwasser: Ursachen, Bedeutung, Lösungswege*; DVGW-Forschungsstelle am Engler-Bunte-Institut der Universität Karlsruhe: Karlsruhe, Germany, 1985; p. 468.
28. LAWA. Empfehlungen für eine harmonisierte Vorgehensweise zum Nährstoffmanagement (Defizitanalyse, Nährstoffbilanzen, Wirksamkeit landwirtschaftlicher Maßnahmen) in Flussgebietseinheiten. *Produktdatenblätter* **2017**, *35–37*, 42.
29. Fraters, D.; van Leeuwen, T.; Boumans, L.; Reijs, J. Use of long-term monitoring data to derive a relationship between nitrogen surplus and nitrate leaching for grassland and arable land on well-drained sandy soils in the Netherlands. *Acta Agric. Scand. B Soil Plant* **2005**, *65*, 144–154. [CrossRef]
30. Dalgaard, T.; Hansen, B.; Hasler, B.; Hertel, O.; Hutchings, N.J.; Jacobsen, B.H.; Stoumann Jensen, L.; Kronvang, B.; Olesen, J.E.; Schjørring, J.K.; et al. Policies for agricultural nitrogen management—Trends, challenges and prospects for improved efficiency in Denmark. *Environ. Res. Lett.* **2014**, *9*, 115002. [CrossRef]
31. Wendland, F.; Behrendt, H.; Gömann, H.; Hirt, U.; Kreins, P.; Kuhn, U.; Kunkel, R.; Tetzlaff, B. Determination of nitrogen reduction levels necessary to reach groundwater quality targets in large river basins: The Weser basin case study. *Nutr. Cycl. Agroecosyst.* **2009**, *85*, 63–78. [CrossRef]
32. Hirt, U.; Kreins, P.; Kuhn, U.; Mahnkopf, J.; Venohr, M.; Wendland, F. Management options to reduce future nitrogen emissions into rivers: A case study of the Weser river basin, Germany. *Agric. Water Manag.* **2012**, *115*, 118–131. [CrossRef]
33. Andelov, M.; Kunkel, R.; Uhan, J.; Wendland, F. Determination of nitrogen re-duction levels necessary to reach groundwater quality targets in Slovenia. *Int. J. Environ. Sci.* **2014**, *29*, 1806–1818.
34. Kunkel, R.; Herrmann, F.; Kape, H.E.; Keller, L.; Koch, F.; Tetzlaff, B.; Wendland, F. Simulation of terrestrial nitrogen fluxes in Mecklenburg-Vorpommern and scenario analyses how to reach N-quality targets for groundwater and the coastal waters. *Environ. Earth Sci.* **2017**, *76*, 146. [CrossRef]
35. Wendland, F.; Kunkel, R.; Gömann, H.; Kreins, P. Water fluxes and diffuse nitrate pollution at the river basin scale: Interfaces for the coupling of agroeconomical models with hydrological approaches. *Water Sci. Technol.* **2007**, *55*, 133–142. [CrossRef]
36. Wakidaa, F.T.; Lerner, D.N. Non-agricultural sources of groundwater nitrate: A review and case study. *Water Res.* **2005**, *39*, 3–16. [CrossRef] [PubMed]
37. Zhang, Q.; Sun, J.; Liu, J.; Huang, G.; Lu, C.; Zhang, Y. Driving mechanism and sources of groundwater nitratecontamination in the rapidly urbanized region of south China. *J. Contam. Hydrol.* **2015**, *182*, 221–230. [CrossRef] [PubMed]
38. Wikipedia. Available online: https://de.wikipedia.org/wiki/Metropolregion_Rhein-Ruhr (accessed on 20 November 2019).
39. Henrichsmeyer, W.; Cypris, C.; Löhe, W.; Meudt, M.; Sander, R.; von Sothen, F.; Isermeyer, F.; Schefski, A.; Schleef, K.-H.; Neander, E.; et al. *Entwicklung Eines Gesamtdeutschen Agrarsektormodells RAUMIS96. Endbericht zum Kooperationsprojekt*; Forschungsbericht für das BML (94 HS 021); Vervielfältigtes Manuskript: Bonn/Braunschweig, Germany, 1996.
40. Kreins, P.; Gömann, H.; Herrmann, S.; Kunkel, R.; Wendland, F. Integrated agricultural and hydrological modeling within an intensive livestock region. *Adv. Econ. Environ. Res.* **2007**, *7*, 113–142.

41. Heidecke, C.; Hirt, U.; Kreins, P.; Kuhr, P.; Kunkel, R.; Mahnkopf, J.; Schott, M.; Tetzlaff, B.; Venohr, M.; Wagner, A.; et al. *Endbericht zum Forschungsprojekt "Entwicklung Eines Instrumentes für ein Flussgebietsweites Nährstoffmanagement in der Flussgebietseinheit Weser" AGRUM+-Weser*; Thünen Report 21; Johann Heinrich von Thünen-Institut: Braunschweig, Germany, 2015; 380p. [CrossRef]
42. Wendland, F. *Die Nitratbelastung in den Grundwasserlandschaften der "Alten" Bundesländer (BRD)*; Berichte aus der Ökologischen Forschung 8; Forschungszentrum Jülich Germany: Juelich, Germany, 1992; p. 150.
43. Wienhaus, S.; Höper, H.; Eisele, M.; Meesenburg, H.; Schäfer, W. *Nutzung Bodenkundlich-Hydrogeologischer Informationen zur Ausweisung von Zielgebieten für den Grundwasser-Schutz—Ergebnisse Eines Modellprojektes (NOLIMP) zur Umsetzung der EG—Wasserrahmenrichtlinie*; GeoBerichte 9; Landesamt für Bergbau, Energie und Geologie: Hannover, Germany, 2008; 56p.
44. Herrmann, F.; Keller, L.; Kunkel, R.; Vereecken, H.; Wendland, F. Determination of spatially differentiated water balance components including groundwater recharge on the federal state level—A case study using the mGROWA model in North Rhine-Westphalia (Germany). *J. Hydrol. Reg. Stud.* **2015**, *4*, 294–312. [CrossRef]
45. Herrmann, F.; Kunkel, R.; Ostermann, U.; Vereecken, H.; Wendland, F. Projected impact of climate change on irrigation needs and groundwater resources in the metropolitan area of Hamburg (Germany). *Environ. Earth Sci.* **2016**, *75*, 1104. [CrossRef]
46. Engel, N.; Müller, U.; Schäfer, W. BOWAB—Ein Mehrschicht-Bodenwasserhaushaltsmodell. *GeoBerichte* **2012**, *20*, 85–98.
47. Wendland, F.; Kunkel, R.; Grimvall, A.; Kronvang, B.; Muller-Wohlfeil, D.I. Model system for the management of nitrogen leaching at the scale of river basins and regions. *Water Sci. Technol.* **2001**, *43*, 215–222. [CrossRef] [PubMed]
48. Kunkel, R.; Wendland, F. WEKU—A GIS-supported stochastic model of groundwater residence times in upper aquifers for the supraregional groundwater management. *J. Environ. Geol.* **1997**, *1–2*, 1–9. [CrossRef]
49. Böttcher, J.; Strebel, O.; Duynisveld, W.H.M. Kinetik und Modellierung gekoppelter Stoffumsetzungen im Grundwasser eines Lockergesteinsaquifers. *Geol. Jahrb.* **1989**, *51*, 3–40.
50. van Beek, C.G.E.M. *Landdbouw en Drinkwatervoorziening, orientierend Onderzoek Naar de Beinvloeding can de Grondwaterkwaliteit Door Bemesting en Het Gebruik van Bestrijdings-Middelen*; Keuringsinstituut Voor Waterleidingsartikelen Kiwa NV: Nieuwegein, The Netherlands, 1987.
51. Walther, W.; Reinstorf, F.; Pätsch, M.; Weller, D. Management tools to minimize nitrogen emissions into groundwater in agricultural used catchment areas, northern low plain of Germany. In Proceedings of the XXX IAHR Congress—Water Engineering and Research in a Learning Society, Thessaloniki, Greece, 24–29 August 2003; pp. 747–754.
52. Venohr, M.; Hirt, U.; Hofmann, J.; Opitz, D.; Gericke, A.; Wetzig, A.; Natho, S.; Neumann, F.; Hürdler, J.; Matranga, M.; et al. Modelling of nutrient emissions in river systems—MONERIS—Methods and Background. *Int. J. Hydrobiol.* **2011**, *96*, 435–483. [CrossRef]
53. Behrendt, H.; Opitz, D. Retention of nutrients in river systems: Dependence of specific runoff and hydraulic load. *Hydrobiologia* **2000**, *410*, 111–122. [CrossRef]
54. OSPAR Commission. *Principles of the Comprehensive Study on Riverine Inputs and Direct Discharges (RID)*; OSPAR Commission: Southhampton, UK, 1998; 16p.
55. Schaap, M.; Hendriks, C.; Kranenburg, R.; Kuenen, J.; Segers, A.; Schlutow, A.; Nagel, H.D.; Ritter, A.; Banzhaf, S. *PINETI-3: Modellierung Atmosphärischer Stoffeinträge von 2000 bis 2015 zur Bewertung der ökosystem-Spezifischen Gefährdung von Biodiversität Durch Luftschadstoffe in Deutschland*; UBA-Texte 79/2018; Umweltbundesamt: Dessau, Germany, 2018; 149p.
56. Kunkel, R.; Wendland, F. The GROWA98 model for Water balance analysis in large river basins—The river Elbe case study. *J. Hydrol.* **2002**, *259*, 152–162. [CrossRef]
57. Schäfer, W.; Höper, H.; Müller, U. Diffuse nitrat- und phosphatbelastung—Ergebnisse der bestandsaufnahme der EUWRR in Niedersachsen. *Geoberichte* **2007**, *2*, 3–32.
58. Schaap, M.; Banzhaf, S.; Scheuschner, T.; Geupel, M.; Hendriks, C.; Kranenburg, R.; Nagel, H.-D.; Segers, A.; von Schlutow, A.; Wichink, R.; et al. Atmospheric nitrogen deposition to terrestrial ecosystems across Germany. *Biogeosciences* **2017**. [CrossRef]

59. MKULNV. *Entwicklung und Stand der Abwasserbeseitigung in Nordrhein-Westfalen*; Ministerium für Klimaschutz, Umwelt, Landwirtschaft, Natur- und Verbraucherschutz Nordrhein-Westfalen: Düsseldorf, Germany, 2014; Available online: https://www.umwelt.nrw.de/fileadmin/redaktion/Broschueren/abwasserbeseitigung_entwicklung_kurzfassung.pdf (accessed on 28 November 2019).
60. DEA-Datendrehscheibe. 2016. Available online: https://www.elwasweb.nrw.de/elwas-web/index.jsf# (accessed on 17 July 2018).

Review

How Can Decision Support Tools Help Reduce Nitrate and Pesticide Pollution from Agriculture? A Literature Review and Practical Insights from the EU FAIRWAY Project

Fiona Nicholson [1,*], Rikke Krogshave Laursen [2], Rachel Cassidy [3], Luke Farrow [3], Linda Tendler [4], John Williams [5], Nicolas Surdyk [6] and Gerard Velthof [7]

1. ADAS Gleadthorpe, Meden Vale, Mansfield, Nottinghamshire NG20 9PD, UK
2. Danish Agriculture & Food Council F.m.b.A., SEGES, Agro Food Park 15, 8200 Aarhus N, Denmark; RILA@seges.dk
3. Agri-Environment Branch, Agri-Food and Bioscience Institute (AFBI), Newforge Lane, Belfast BT9 5PX, Northern Ireland, UK; rachel.cassidy@afbini.gov.uk (R.C.); luke.farrow@afbini.gov.uk (L.F.)
4. Chamber of Agriculture of the Federal State of Lower Saxony, Helene-Künne-Allee 5, 38122 Braunschweig, Germany; linda.tendler@lwk-niedersachsen.de
5. ADAS Boxworth, Battlegate Road, Boxworth, Cambridge CB23 4NN, UK; john.williams@adas.co.uk
6. Bureau de Recherches Geologiques et Minieres (BRGM), 3 avenue Claude Guillemin, BP 36009, CEDEX 2, 45060 Orleans, France; n.surdyk@brgm.fr
7. Wageningen Environmental Research, Droevendaalsesteeg 3, 67808 PB Wageningen, The Netherlands; gerard.velthof@wur.nl

* Correspondence: fiona.nicholson@adas.co.uk

Received: 30 January 2020; Accepted: 4 March 2020; Published: 11 March 2020

Abstract: The FAIRWAY project reviewed approaches for protecting drinking water from nitrate and pesticide pollution. A comprehensive assessment of decision support tools (DSTs) used by farmers, advisors, water managers and policy makers across the European Union as an aid to meeting CAP objectives and targets was undertaken, encompassing paper-based guidelines, farm-level and catchment level software, and complex research models. More than 150 DSTs were identified, with 36 ranked for further investigation based on how widely they were used and/or their potential relevance to the FAIRWAY case studies. Of those, most were farm management tools promoting smart nutrient/pesticide use, with only three explicitly considering the impact of mitigation methods on water quality. Following demonstration and evaluation, 12 DSTs were selected for practical testing at nine diverse case study sites, based on their pertinence to local challenges and scales of interest. Barriers to DST exchange between member states were identified and information was collected about user requirements and attitudes. Key obstacles to exchange include differences in legislation, advisory frameworks, country-specific data and calibration requirements, geo-climate and issues around language. Notably, DSTs from different countries using the same input data sometimes delivered very different results. Whilst many countries have developed DSTs to address similar problems, all case study participants were able to draw inspiration from elsewhere. The support and advice provided by skilled advisors was highly valued, empowering end users to most effectively use DST outputs.

Keywords: DST; software; model; drinking water; diffuse pollution; catchment management; farm management; farm advisors

1. Introduction

Safe drinking water is vital for public welfare and is an important driver of a healthy economy. The productivity of agriculture in the EU has greatly increased over recent decades, in part through the increased availability of fertilisers and pesticides, which have boosted crop and animal production. However, the increased inputs of fertilisers (both mineral and organic) and pesticides to agricultural soils have also led to increased losses to the environment, thereby contributing to the pollution of ground and surface drinking water sources.

To deal with this pollution, the EU has developed an extensive set of directives, guidelines and policies. The requirements of the Drinking Water Directive set an overall minimum quality for drinking water within the EU. In addition, the Water Framework Directive, the Groundwater Directive, the Nitrates Directive and the Directive on the Sustainable Use of Pesticides require member states (MS) to protect water resources against pollution in order to ensure the safety of drinking water. While many improvements have been achieved [1], the directives are not achieving a consistent level of implementation and effectiveness across all MS. As a consequence, limits for nitrate (50 mg/L) and pesticides (0.1 μg/L) in groundwater and surface water are still exceeded in many MS [2]. Whilst easy, low cost measures for tackling pollution sources and pathways have been successful, as they become exhausted, further improvements in water quality will require innovative solutions and more targeted mitigation measures if agricultural productivity is not to be affected.

The need for technology in developing bespoke management approaches and decision support for users has been highlighted as a core element of the Common Agricultural Policy (CAP) proposals for 2021–27, and as key to meeting CAP objectives and targets in the future [3]. In their CAP Strategic Plan, MS will have to outline how they will stimulate knowledge exchange and innovation covering aspects, such as farm advisory services and training [4] and encourage the use of big data and new technologies for control and monitoring [5].

There have been many decision support tools (DSTs) developed in the EU and elsewhere to assist end users to make more effective decisions on how to minimise contaminant losses to water, and hence to protect and improve drinking water quality. The target audience of these DSTs ranges widely. At one end are water managers and policy makers who work at the catchment or national scale to set water quality targets and monitor the effectiveness of mitigation measures. At the other end are individual farmers and advisors who may use DSTs to provide guidance on farm management practices (e.g., product choice; timing of pesticide, manure and fertiliser applications) which could have a direct or indirect impact on water quality. DSTs may either lead end users through clear decision stages and present them with the likelihood of various outcomes (social or/and economic) occurring, or they can encourage users to optimise (minimise) their use of e.g., manufactured fertiliser nitrogen (N) and pesticides with respect to a legal framework. Policy makers may also simulate the impacts of excluding harmful substances, specifying usage limits or changing the taxation regime. DSTs may be paper based systems (e.g., fertiliser recommendations) or dynamic software tools, whose recommendations vary according to the user's inputs, and they may suggest an optimal decision path [6]. However, the extent to which DST are transferrable across MS has not been assessed to date and is an impediment to achieving the technological and DST requirements of the CAP going forward.

The EU Horizon 2020 funded "Farm systems that produce good water quality for drinking water supplies (FAIRWAY)" project was established in 2017 to review current approaches and measures for the protection of drinking water resources against pollution caused by pesticides and nitrate from agriculture in the EU, and to identify and develop innovative approaches to more effective drinking water protection [7]. One of the specific project objectives was to review, demonstrate and evaluate DSTs offering advice, training and communication related to the protection of drinking water quality (either directly or indirectly) for a range of end users. As part of the project, 13 case study areas were identified in 10 MS covering different types of drinking water supplies (groundwater and surface water), pedo-climatic zones, type of farming, land use, legal framework and governance (Figure 1). The case studies provided the project with a means to directly contact farmers and other stakeholders.

They also offered a way to establish which DSTs and mitigation measures were being used in practice to safeguard drinking water resources, and acted as a forum for evaluating DSTs, to establish whether there was potential for DST exchange between countries.

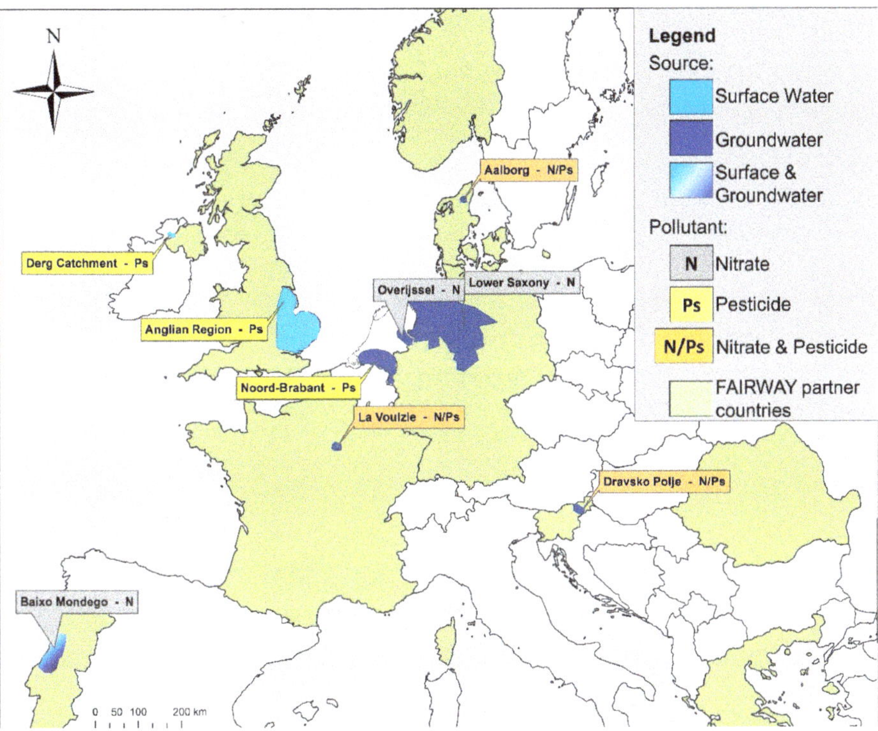

Figure 1. Location and overview of the 9 FAIRWAY case study sites involved in the DST evaluation, including the drinking water source, case study size and the main pollutant(s) of concern.

The objective of this study was therefore to systematically review the literature to identify DSTs that could help to establish awareness of diffuse nitrate and pesticide pollution of vulnerable drinking water resources and/or improve on-farm management practices to protect water quality. The review considered DSTs operating at different scales (e.g., software tools used by individual farmers or advisors; catchment scale policy tools; paper based recommendations), with a different focus (i.e., those dealing with either nitrogen or pesticide management, or both) and which were relevant to protecting the quality of different drinking water sources (i.e., ground water or surface water systems). DSTs dealing solely with phosphorus (P) management were not considered as P concentrations in drinking water are not currently regulated. A subset of the DSTs were selected for demonstration and/or implementation at 9 of the 13 FAIRWAY case study sites to evaluate the potential benefits/opportunities of DST transfer between countries, identify any barriers to exchange and assess stakeholder perception.

2. Methodology

A summary of the different stages of the review, selection and testing of the DSTs in the FAIRWAY case studies is given in Figure 2, and described in more detail below.

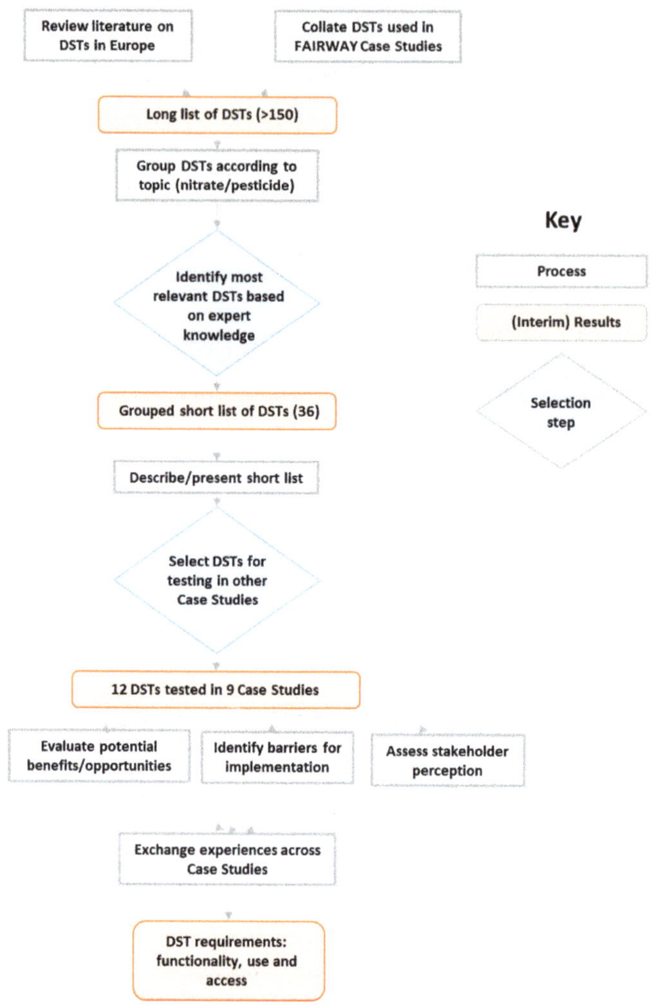

Figure 2. Flowchart showing the stages of the review, selection and testing of DSTs in the FAIRWAY case studies.

2.1. Definitions and Scope

For the purposes of this review, a DST was defined as any bespoke (i.e., custom-made) or generic (i.e., 'off the shelf') software, email/text alerts, online calculator or guidance, phone app, and paper-based guidance that could contribute to an end user decision affecting surface or ground water quality. The definition does not include 'human-based' DSTs, such as advisors or peers. In addition, the DST must have been in practical operation (i.e., in active use) or released by the end of 2017. The DSTs considered were those used by the FAIRWAY project partners and elsewhere in Europe (including Norway, Switzerland and other non-EU countries with similar agroclimatic conditions) on farms and within single catchments, groundwater abstraction areas, regions, countries or larger areas. Demonstration versions of DST's were included if they were functional, had been tested on end-users and were assessed to have a potential for practical use. End users were defined as:

- Farmers

- Agronomists and other farm advisors
- Water quality managers
- Policy makers
- Fertiliser or pesticide manufacturers or suppliers
- Researchers

Water quality was defined in terms of:

- N concentrations in the form of total N and/or nitrate and/or ammonium and/or nitrite.
- Pesticide concentrations, where pesticides are defined as any insecticide, herbicide, fungicide, nematicide, acaricide, slimicide, molluscicide and any product related to any of these including any growth regulator, and their relevant metabolites, degradation and reaction products. Relevant was taken to mean any metabolites, degradation and reaction products that have similar properties to their parent pesticides [8]. The pesticides included were those in current professional use in agriculture in the different countries.

The focus of the review was on DSTs operating at farm, regional or national scale that could be of practical use in reducing N or pesticide pollution in the FAIRWAY project case study areas.

2.2. Identification of Relevant DSTs

The compilation of an initial list of relevant DSTs currently in use was approached in two ways. Firstly, DSTs meeting the definitions given above were identified by undertaking a search of the published scientific literature using the Web of Science Core Collection (1994–2017). The keywords used for the search were discussed and agreed with all task participants, and are listed in Table A1. In addition, each FAIRWAY project participant supplied a list of relevant DSTs used in their respective case studies (Denmark, Germany, France, The Netherlands, Norway, Portugal, Slovenia, England and Northern Ireland), informed by the appropriate national experts. Information for the Republic of Ireland, which is not a FAIRWAY participant country, was supplied by the project partner from Northern Ireland as the case study area is a transboundary river catchment on the Irish border covering both jurisdictions.

The information supplied for each DST is detailed in Table A2, and was collated as a series of 'information capture proformas' in a spreadsheet-based database. Note that information about a DST did not need to have been published in the scientific literature to be included in the database. If documentation was available only in a national language (i.e., not English) then the participants supplied a written summary of the DST in the spreadsheet database.

The DSTs identified in the literature search and by the FAIRWAY project partners were combined into a 'long list'. An assessment was made of the search comprehensiveness by circulating this list to the FAIRWAY project partners, who were then able to identify whether any key DSTs had been omitted and add them to the list as appropriate. The FAIRWAY project partners also confirmed whether the DSTs on the 'long list' were in active use (see Definitions and Scope).

Once the database was complete, the DSTs were grouped according to their broad topic area (i.e., whether they dealt with N or pesticides or both), and the primary users and scale of operation were identified. Recognising that it would not be possible within the framework of the FAIRWAY project to study all the DSTs in depth, the project partners identified (based on their expert knowledge and experience) a 'shortlist' of 3–5 DSTs from their country which they assessed to be the most widely used and/or of most potential relevance in the FAIRWAY case studies. Key information used in the selection process included the:

- number and type of users;
- suitability for use across multiple MS;
- level of complexity;
- ability to meet the needs of actors in the FAIRWAY case study areas.

A series of 3-page 'information sheets' was produced, summarising relevant technical and practical aspects of the shortlisted DSTs, which the FAIRWAY participants had previously agreed should be captured (Table A2). These were made available to the researchers who were leading the FAIRWAY work in the case studies and other project participants. In addition, a classification scheme was devised to better understand the target users of the DSTs and the types of support they were intended to provide.

2.3. Selection, Implementation and Evaluation of DSTs in the Case Studies

Nine of the FAIRWAY case study sites located across the EU (Figure 1) were involved in the evaluation of the DSTs. The case study sites were asked to re-evaluate the shortlisted DSTs (Table A3) and to identify a subset of 1–3 DSTs for testing and/or demonstration. Consideration was given to any barriers for transferring the DST into a new context (e.g., language issues, lack of support/documentation, data availability, specialist skills required, cost) and whether similar DSTs were already available in the country.

Once the selections had been made, bilateral contact with the owners of the DSTs was established and access to the software obtained. An evaluation scheme was designed to help the case study leaders assess the selected DSTs further with regard to scale, data requirements, level of experience/training required, stakeholders etc. Following the pre-testing phase (i.e., the case study site could obtain access to the DST, get support/documentation from the DST owner and provide the required input data), the testing and evaluation of DST outputs commenced. In many case study sites, this also included demonstration of the DST to relevant stakeholders and application of the DST using local data. In the final phase, the participating case study sites evaluated whether it would be possible to implement the DSTs (or parts of the DST) in their country or context, based on the test results and findings. Key objectives of the evaluation of the DSTs in the case studies related to i) evaluating the potential benefits/opportunities presented ii) identifying any barriers to implementation and iii) assessing stakeholder perception.

3. Results and Discussion

3.1. General Remarks

The term 'decision support tool' and its synonyms (Table A1), when entered into a search engine, returns a large number of 'hits'. This is because it can be applied to a wide range of tools encompassing paper-based guidelines, bespoke software and phone apps used by farmers, as well as complex sets of mathematical models intended for modelling and research purposes. All can justifiably claim to aid decision making, albeit for different sets of end users.

We found that the scientific literature searches returned significantly different numbers of 'hits' depending on the intended primary users: papers on DSTs developed for modelling and research purposes have been actively published, whilst only a limited number of papers on tools used by farmers and advisors were found in peer-reviewed journals. By their very nature these tend to be more practical tools intended for routine farm use. They may be based on sound scientific principles, but scientific publications may not necessarily be their main focus. Information on this type of DST is more likely to be made available by the developers or funders (e.g., national government, extension service; fertiliser/pesticide manufacturers) in the form of user guides or other web-based information, and is often only available in the local language. Hence, it was extremely valuable to access the information supplied by the FAIRWAY project partners about the DSTs most widely used in their countries, as these included farm-based tools not captured by the literature searches.

The complexity and competitiveness of the pesticide market can mean that chemical companies will develop product-specific DSTs and will only make these available to users of their product(s); these DSTs are unlikely to appear in the scientific literature and there is limited publicly available information about them. Nevertheless, because of the commercial power of these companies these DSTs can be very popular and widely used. More general pesticide management tools are fewer in number and have usually been developed by academics (e.g., Environmental Yardstick for Pesticides [9]; FarmHedge [10]) to cover a wider range of plant protection products. For example, the Dutch DST Environmental Yardstick for Pesticides [9] offers comparison of 3 crop protection products for free and comparison of an 'unlimited' number on purchase of a subscription.

Many of the DSTs identified in the literature review as tools for N management (i.e., providing N recommendations for optimal crop yield) also deal with the management of other nutrients such as P, therefore these are henceforward referred to under the general term 'nutrient management' DSTs. A number of the nutrient management DSTs identified in this report were also commercial software tools only available at a charge to the end user (e.g. Mark Online [11], Plant Protection Online [12]). In some cases, these DSTs were developed by or in conjunction with academic institutions (e.g., NDICEA [13]); in others, the details of DST development, validation and testing are commercially sensitive and are not publicly available. In the UK, the computer code for nutrient management DSTs such as PLANET [14] and MANNER [15], which were developed using public funding from Defra (the government Department of Environment, Food and Rural Affairs), has now been made freely available and is incorporated with widely-used commercial software tools for farmers; these DSTs are also compatible with advisory information published online as the Nutrient Management Guide (RB209) [16].

Some of the DSTs were either meteorological information services (Agro-Meteorological Service [17]) providing information and advice on when weather conditions are likely to be suitable for pesticide application (and other agricultural operations), or the DST included access to meteorological information (e.g., Plant Protection Online [12]), often via a phone app interface (e.g., FarmHedge [10]), making them suitable for farmers to use in the field.

Table A3 also shows that most of the shortlisted tools were developed in NW Europe, probably reflecting the relative wealth of this region, the availability of funding and investment, and the level of relevant knowledge and expertise.

3.2. Types of DST

Table A3 shows the shortlist of 36 DSTs selected by the FAIRWAY project partners for further consideration and potential practical evaluation in the case studies. Only a few of the selected DSTs addressed both nutrient and pesticide management (Mark Online [11] and Dyrkningsvejledninger [18]; Bodemconditiescore [19]). Mark Online is the most widely used farm information management system in Denmark and covers all aspects of crop management including soil tillage and crop protection [11], whilst another Danish tool, Dyrkningsvejledninger [18], consists of manuals for growing different crops which provide information on Good Agricultural Practice and crop protection. Bodemconditiescore, widely used by dairy farmers in the Netherlands, is a visual evaluation method for grassland, which indicates soil quality problems and provides advice on pest population dynamics [19]. In the UK, widely used commercial farm advice tools, although not on the shortlist in Table A3, often include modules for both nutrient and pesticide planning and management, so that farmers only need to purchase a single software package to cover all their requirements.

The DST classification schemes (Tables 1 and 2) allowed the nutrient and pesticide management DSTs to be separated into those developed to support water quality/agri-environment policy makers operating at a regional or national level, and those intended to support sustainable N management at the farm level. The DSTs were further divided into groups depending on whether they provided support for:

- evaluation of current practices;
- strategic advice for farm management and implementation of nitrate/pesticide mitigation measures;
- on-farm operational management.

Table 1. Classification scheme for nutrient management DSTs (numbers in brackets refer to the DST number in Table A1).

Purpose	Provides Support For: Evaluation Current Practices	Strategic Advice, Farm Management and Implementation of Measures	Operational Management
To support regional (water quality, agri-environment) policy makers	(6) CTtools (7) BEST Kemi (20) STONE (21) CLMN (26) OENBAL (27) GROWA-SI (28) SNGMP (31) FARMSCOPER (35) SCIMAP	(8) TargetEconN (20) STONE (21) CLMN (26) OENBAL (31) FARMSCOPER	
To support sustainable farm nutrient management	(1) Düngeplanung (2) ISIP (3) Mark Online (13) ANCA (16) BWW (17) Bodemconditiescore (25) SSG	(1) Düngeplanung (2) ISIP (3) Mark Online (4) Dyrkningsvejledninger (11) Teagasc NMP Online (13) ANCA (16) BWW (24) Načrtovanje Gnojenja (25) SSG (30) PLANET/MANNER	(12) Farmhedge (14) CBGV (15) BeregeningsWijzer (18) NDICEA (22) Skifteplan (30) PLANET/MANNER

Table 2. Classification scheme for pesticide management DSTs (numbers in brackets refer to the DST number in Table A1).

Purpose	Provides Support for: Evaluation of Current Practices	Strategic Advice on Farm Management and Implementation of Measures	Operational Management
To support regional (water quality, agri-environment) policy makers	(7) BEST Kemi (9) Phytopixal (10) SIRIS (19) Yardstick (28) SNGMP	(9) Phytopixal (10) SIRIS (29) FITO-INFO	
To support sustainable farm pesticide management	(3) Mark Online (9) Phytopixal (10) SIRIS (17) Bodemconditiescore (19) Yardstick (29) FITO-INFO	(3) Mark Online (4) Dyrkningsvejledninger (5) Plant Protection Online (19) Yardstick	(5) Plant Protection Online (12) FarmHedge (32) Check it Out (33) Sentinel Online (34) Procheck. (36) WaterAware

Very few of the selected DST were aimed explicitly at improving water quality or represented water quality directly (e.g., by the calculation of N or pesticide concentrations); Table A3. Many are agronomic tools for farmers and advisors which aim to optimise the use of N and/or pesticides to obtain maximum crop yields. They are effectively farm management tools and their inclusion in this review was based on the assumption that the efficient use of N and pesticides will improve water quality.

Using a fertiliser recommendation system or a manure management tool will facilitate the application of the correct amount of fertiliser/manure to meet crop needs at the appropriate time, thus minimising nutrient losses to water bodies. Most project participants reported using this type of DST. For example, Düngeplanung, which is used in Lower Saxony (Germany), was specifically developed to help farmers in water sensitive areas (e.g., for drinking water abstraction) with fertiliser planning and regulatory compliance [20]. Supported by water suppliers, it indirectly affects water quality by:

- combining all the available information for a farm (soil analyses, crop rotation, fertiliser history, specific restrictions in water protected areas);
- optimising yields and thus the amount of N exported from the field;
- improving N-efficiency;
- providing practical information on amounts and timing of fertiliser applications.

Farmers using Düngeplanung have reported reductions in fertiliser use of roughly 5–10% (L. Tendler, personal communication) Whilst again not specifically designed to represent water quality, the French SIRIS tool allows pesticides to be classified according to their potential to reach surface and ground water, and helps to organize the monitoring of pesticides in waters at the regional or local scale [21].

3.3. Representation of Mitigation Methods

Only three of the shortlisted DSTs were explicitly developed to consider the impact of mitigation methods on water quality: FARMSCOPER, Environmental Yardstick for Pesticides and Catchment Lake Modelling Network (Table A3).

FARMSCOPER, first developed in 2010, is a DST that can be used to assess diffuse agricultural pollutant loads (including nitrate, P, pesticides and sediment) on a farm and quantify the impacts of farm mitigation methods on these pollutants [22]. Inputs are at the farm scale, however the outputs can be scaled up to catchment, regional and national levels. It currently contains over 100 mitigation methods, adapted from the User Guide for England and Wales [23], and they can be tested either individually or in combination for 3 broad soil types, defined according to the probability of having artificial under-drainage for conventional agriculture: (i) not requiring under-drainage; (ii) requiring under-drainage for arable use; and (iii) requiring under-drainage for both arable and grassland. The testable mitigation methods relating to nitrate and pesticide management include:

- Establish cover crops in the autumn;
- Establish riparian buffer strips;
- Extend/reduce grazing season;
- Cultivate land for crops in spring not autumn;
- Cultivate and drill across the slope;
- Early harvesting and establishment of crops in the autumn;
- Do not apply manufactured fertiliser to high-risk areas;
- Calibration of sprayer;
- Fill/mix/clean sprayer in field;
- Avoid plant protection product (PPP) application at high risk timings;
- Drift reduction methods;
- PPP substitution;
- Construct bunded impermeable PPP filling/mixing/cleaning area;
- Treatment of PPP washings through disposal, activated carbon or biobeds.

FARMSCOPER [22] is used primarily by policy makers, scientists and catchment managers, with the potential to be used by advisors on farms. To date, it has been used to study the impacts of various mitigation methods in the Wensum and Avon Demonstration Test Catchments (DTCs) in England [24] and more recently to evaluate the potential economic and environmental impacts of new and revised mitigation measures for nutrient and sediment mitigation across England [25].

The Environmental Yardstick for Pesticides [9] is a DST designed to quantify the environmental impact of the use of pesticides in outdoor and greenhouse crops. The mitigation methods represented are:

- choice of pesticide;
- dose rate;
- application technique (drift);

- width of untreated buffer zone.

For each pesticide, the yardstick assigns environmental impact points for the risk to aquatic organisms, the risk of leaching to groundwater and the risk to soil organisms (depending on the user-specified soil organic matter content and season of application). The yardstick also shows the associated risk to beneficial organisms, including pollinators and natural predators such as ladybirds, mites and parasitoids. It is used in the Netherlands (and Belgium) as a management tool for farmers and technical consultants, a tool for monitoring the environmental performance of farmers, for setting standards for ecolabels, as guidance for the supply chain to preferentially purchase sustainable agricultural products, and as a policy evaluation tool.

The Catchment-Lake Modelling Network, designed specifically for the Lake Vansjø catchment in Southern Norway, consists of a network of process-based, mass-balance models linking climate, hydrology, catchment-scale P dynamics and lake processes [26]. The model network allows the effects of climate change to be disentangled from those of land-use change on lake water quality and phytoplankton growth, and includes the following mitigation methods:

- land use change;
- cultivation change;
- crop rotation;
- erosion risk reduction measures;
- change in fertiliser application.

The model network can thus support decision-making to achieve good water quality and ecological status within the Lake Vansjø catchment. It was developed to model P and suspended sediment loadings, although it is also possible to include nitrate. The model network is transferable to other catchments; however, it is quite time-consuming to set up and calibrate for a new catchment.

Whilst not directly evaluating the effects of mitigation methods, the UK SCIMAP model [27] provides a framework for generating catchment risk maps for contaminant loss via overland flow pathways, so that the areas within a catchment where mitigation methods are most urgently required can be identified. SCIMAP has been used in the River Eden Demonstration Test Catchment project which is investigating the dynamics of water quality (nutrients and sediments) from agricultural land, and by Durham Wildlife Trust to identify areas with high fine sediment pollution risk within the River Wear catchment. In the Netherlands, BedrijfsWaterWijzer is a web application that is currently being developed to provide a starting point for evaluating measures to reduce emissions to water on dairy farms [28], whilst STONE [29] is a modelling tool wherein various policy measures to reduce nutrient emissions to groundwater and surface waters may be specified.

3.4. Representation of Economic and Financial Aspects

The economic and financial implications of implementing mitigation methods were not commonly represented in the shortlisted DSTs. However, FARMSCOPER [22] is able to estimate the cost effectiveness of mitigation methods, singly and in combination, as a cost-efficiency ratio in terms of money (£) saved per % reduction in nitrate, P or sediment loss. In addition, the Danish TargetEconN model [30] is an integrated economic and biophysical social planner model, which minimises the costs of meeting a nutrient load reduction target in a specific water body. Some other DSTs do also have the capability to represent economic aspects, e.g., Düngeplanung [20] allows cost comparison of different fertiliser use scenarios.

Future outputs from the FAIRWAY project will provide a more in-depth assessment of the costs and benefits for farmers and society as a whole of using DSTs that directly or indirectly affect water quality.

3.5. Commentary on DST Uptake and Usage

There was a wide variation in the number and sophistication of the DSTs available in the different participant countries, reflecting the degree of investment and funding provided. In some countries, such as Denmark, a number of different computer-based and online DSTs have been developed aimed at both farmers/advisors (e.g., Mark Online [11], Plant Protection Online [12], Dyrkningsvejledninger [18]) and water quality managers (e.g., CTzoom/Cttools [31], TargetEconN [30]). In contrast, the only DSTs available in Portugal are paper-based manuals and guidelines [32,33], although some of these are also available online. In the absence of other tools capable of modelling agri-environmental measures, Slovenia employs the OECD/Eurostat methodology [34] to calculate N (and P) balances as the basis for reporting Nitrate Directive implementation to the EU, and for the preparation of national policy/legislation and recommendations for farmers on measures for drinking water protection. Slovenia also uses the regional water balance model GROWA-SI developed in Germany for reporting Nitrate Directive implementation at a country wide level [35,36]. Some of the DSTs developed as part of academic research projects are also used internationally, with the results published in the scientific literature. For example, the SCIMAP model developed in the UK has been used in Indonesia to target reforestation to reduce diffuse pollution risks [37].

Figure 3 shows the numbers of users of the shortlisted DSTs (Table A3) for which data was available. In some countries, farmers are obliged under regulations or commercial pressures to use DSTs, and this will clearly affect take-up and user numbers. For example, dairy farmers in the Netherlands who provide milk to Friesland Campina have to use ANCA (Annual Nutrient Cycling Assessment) [38] to analyse nutrient flows and emissions, hence indirectly improving water quality. If they meet certain performance indicators, calculated by ANCA, they can get a bonus for their milk; there are currently c.16,000 users. In the UK, farmers in nitrate vulnerable zones (NVZs) can use a DST such as PLANET [14] to demonstrate compliance with NVZ rules, and it is widely used for this purpose (Figure 3). Düngeplanung, developed in Lower Saxony, Germany [20], is becoming more widely used (currently around 6000 users), following recent changes to regulations which require farmers to produce a fertiliser plan and nutrient balances. In contrast, the number of users is often small for specialised DSTs such as the Norwegian 'Catchment Lake Modelling Network', which comprises a series of process-based, mass-balance models for P and is designed primarily as a catchment management tool, rather than for general use [26].

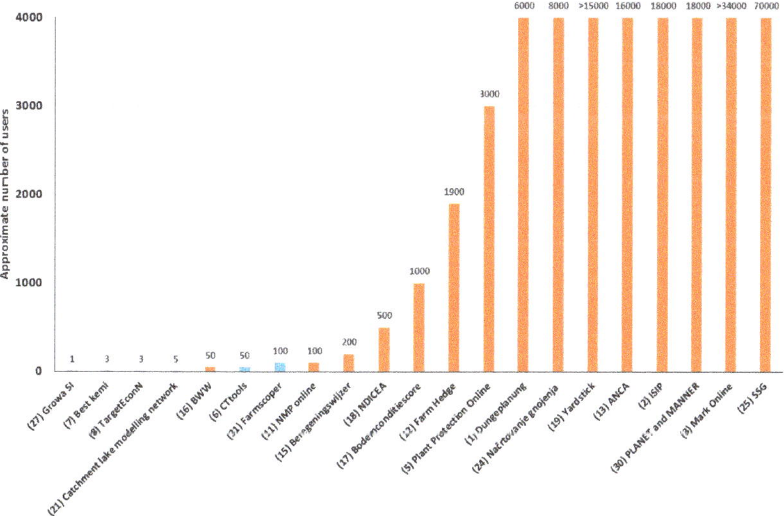

Figure 3. Approximate number of users of the shortlisted DSTs (where data was available). Numbers in brackets on the x-axis refer to the DST number in Table A3.

3.6. Barriers to Uptake

Although some DSTs that cover both nutrients and pesticides (and other aspects of farm management) are available for farmers, some may opt to use more than one DST (or none), depending on their particular needs and requirements, and the legislative and economic environment in which they are operating. DSTs often deal with complex issues, so it is not always easy for farmers to understand and use them—using multiple tools of different types does not always lead to a better decision, as it can be difficult to decide which tools to use under which circumstances.

A recent project undertaken in the UK [39] looked in detail at farmers' usage of the fertiliser recommendations for grassland published in one of the key UK paper-based decision support tools, the Fertiliser Manual (RB209) [40]. Whilst the majority of respondents did not use The Fertiliser Manual (RB209), they described it as 'adequate' as a reference guide. Drawing on information from in-depth interviews and focus groups, the study found that users:

- needed to supplement the information provided with their own information and experience;
- wanted the tool to be more user friendly and flexible; it should be written in 'farmers language';
- thought that potential economic gain should be explicitly demonstrated.

Similarly in Denmark [41], users and non-users of the pesticide DST Plant Protection Online identified several barriers to uptake, including:

- time consuming
- too complicated
- lack of user knowledge (on how to identify weeds and diseases)
- competition from human consultants
- lack of confidence
- only chemical solutions recommended

Another UK study reviewed tools for decision making in agriculture and found that despite their availability in a wide range of formats, uptake in the UK and many other countries has been low [6]. Using a combination of qualitative interviews and quantitative surveys, the authors identified fifteen factors that are influential in convincing farmers and advisers to use DSTs, including:

- usability
- cost-effectiveness
- performance
- relevance to user
- compatibility with compliance demands.

The authors concluded that a better understanding of these factors should lead to more effective DST design and delivery in the future. These authors followed up this work with a study on how stakeholders could be more effectively involved to improve DST design [42]. DST use was explored in a series of 78 interviews and 5 focus groups. Their main suggestion was to assess the 'decision support context' before building a product. Other requirements were better knowledge of user-centered design practices, a clear understanding of advice systems and greater collaboration with human-computer interaction researchers.

DSTs aimed at policy makers, water quality managers or catchment managers tend to be more complex and require more data. However, the drivers for using such tools are often legislative or policy focused; thus, potential users of a particular DST should be provided with an appropriate level of training and have access to the relevant datasets in order to do so.

3.7. Evaluation of the DSTs in the Case Studies

Twelve DSTs were selected for testing and/or demonstration by the participating FAIRWAY case study sites (Table 3). The selected DSTs were a mixture of farm level tools designed to improve nutrient or pesticide management on individual farms, and DSTs designed to be used by water quality managers or policy makers at the catchment/regional level, to identify high-risk areas for losses and prioritise mitigation measures or to identify cost-effective management options to decrease nitrate or pesticide pollution.

Table 3. DSTs selected for testing and/or demonstration by the nine participating FAIRWAY case study sites.

Case Study Site (Country)	DSTs Selected (Owning Country)	Level	Target Pollutant
Aalborg (DK)	Environmental Yardstick for Pesticides (NL)	Farm	Pesticides
	SIRIS (FR)	Catchment/ region	
	TargetEconN (DK)	Catchment/ region	
Anglian Region (UK)	Environmental Yardstick for Pesticides (NL)	Farm	Pesticides
La Voulzie (FR)	SIRIS (FR)	Catchment	Pesticides
Lower Saxony (DE)	Mark Online (DK)	Farm	Nutrients
	NDICEA (NL)	Farm	
Derg catchment (IE)	SCIMAP (UK)	Catchment/ region	Pesticides
	Phytopixal (FR);	Catchment/ region	
	FARMSCOPER (UK)	Farm	
Overijssel (NL)	Düngeplanung (DE)	Farm	Nutrients
Noord Brabant (NL)	Plant Protection Online (DK)	Farm	Pesticides
Baixo Mondego (PT)	MANNER-NPK (UK)	Farm	Nutrients
Dravsko Polje (SI)	ANCA (NL)	Farm	Nutrients

During the selection of DSTs for evaluation in the FAIRWAY case study sites, it became clear that many countries have developed similar DSTs to address similar problems. Thus, an important part of the exchange process was to better understand the range of DSTs available in the EU, to compare DSTs from other countries with those already being used in the case study sites and to gain inspiration for enhancing existing DSTs. The information collected using the evaluation schemes for all the DSTs tested in each case study was collated and summarised in terms of: (1) barriers to exchange (2) requirements of a DST in terms of functionality, use and access and (3) stakeholder attitudes to DSTs.

3.7.1. Barriers to DST Exchange between Countries

During the selection and testing of the DSTs in the case studies, it became clear that there were a number of barriers which may prevent or limit the use of a DST in a country other than the one for which it was developed.

All the case studies identified language as a key barrier to transferring DSTs between countries, because often the DST and supporting information are only available in the local language. This is especially important when dealing with DSTs which are designed to be used by farmers or advisors who are unlikely to have the time or ability to translate the technical or specialist language which is often used. Nevertheless, it is possible to successfully use a DST developed elsewhere. For example, the Dutch Environmental Yardstick for Pesticides [9] is used in both the Netherlands and Belgium, and is currently being tested with data from US farms; the tool and supporting documentation are available in English. In addition, Plant Protection Online [12], which was developed in Denmark, is being used in the Baltics and Poland, with user information available in Danish, English and German.

Some DSTs have been developed based on country-specific legislation, which prevents direct exchange with another country, although there may be potential for exchange of part of the DST or to draw on the principles. For example, Mark Online [11] was tested in the Lower Saxony case study in order to assess how Danish legislation might affect practice in Germany. Danish agricultural legislation is quite restrictive in terms of fertiliser practices, but has been shown to produce positive environmental effects (e.g., nitrate concentrations in the upper oxic groundwater have been decreasing significantly in the past 30 years in different Danish reference catchments [43]). The concept of introducing similar strict limitations on N fertiliser use has been discussed in Germany, but the case study aimed to assess the farm-level implications of such restrictions (i.e., introducing an N quota) by comparing results from Mark Online with those from Düngeplanung [20], which was already being used. Results indicated that most farmers in the German case study would comply quite well with the Danish legislation, although some would face problems with their current management practices if they had to meet certain requirements (e.g., the obligation to establish cover crops, restricted fertiliser use in autumn, strict soil P-levels). Participants liked several elements of Mark Online, including the modular design, the ability to compile farm management information, and the inclusion of cross-compliance checks; they also thought that linking soil type to yield level would have benefits in the future.

There was great variation in the data requirements for the different DSTs depending on their degree of sophistication, and most case study sites reported this as a barrier to DST exchange. For example, in the Derg case study (Northern Ireland) the farm level data required to populate the some of the spreadsheets in the FARMSCOPER model [22] are not freely available and cannot be used without consent. In addition, data conversion to the required format is often required and this might be difficult for a farmer. For example, when using Mark Online [11] in the Lower Saxony case study, German soil types had to be converted into their Danish equivalents.

Regional differences can also present a barrier for exchange. In the Derg case study, for example, the climate, topography and soil type were markedly different from those in England and Wales, where FARMSCOPER [22] was developed. Ireland has, in general, higher runoff rates and monthly rainfall, and evapotranspiration figures differ from those in England and Wales. Most of these coefficients are pre-calculated for FARMSCOPER and recalibration would be required before the DST could be used effectively. Nevertheless, stakeholders found it to be a potentially powerful tool to support water managers in prioritising the most effective mitigation options in drinking water catchments. Similarly, ANCA, developed in the Netherlands for the assessment of nutrient flows on dairy farms [38], was evaluated at Dravsko Polje (Slovenia). One of the problems encountered was that farms in Slovenia are much smaller than farms in the Netherlands and tend not to specialise, making it difficult to adapt and use the model, which was developed specifically for dairy farms.

Other issues identified as being barriers to exchange were a lack of support and supporting documentation, the requirement for specialised personnel to run complex DSTs and interpret the results (e.g., the DST requires expertise in GIS), and financial cost.

Researchers and policy makers in every country often prefer to develop their 'own' DST rather than use an existing DST from another country, and this often leads to 'reinventing the wheel'. Many DSTs are developed using government funding to address a specific need in a particular country or region, or for a specific crop (e.g. vines, potatoes). Funding is not provided for the benefit of other potential users elsewhere in the EU; the additional cost that this would entail cannot be justified. Commercial DSTs face similar limitations, but tend to be less geographically constrained. These findings are very much in line with the previous research [44,45], which concluded that the involvement of stakeholders in DST development is a prerequisite to successful implementation. In the context of the multi-national aspects of DST use and development considered in the FAIRWAY project, a more logical pathway might be for researchers to organize regular knowledge exchange activities, providing an opportunity to transfer ideas to their own scientific and stakeholder communities.

3.7.2. Key DST Requirements in Terms of Functionality, Use, and Access

The case study evaluation identified a number of requirements for DSTs, in terms of their functionality (cost, accessibility, data input/output and compatibility with other DSTs), use and access.

First and foremost, a DST must be simple (user friendly and self-explanatory), not too time-consuming and practical for farmers/advisors to use. However, the level of complexity depends on the target users and the objective of the DST. Sometimes more complexity is required, particularly in the case of DSTs, which operate at the catchment scale and which aim to simulate complex environmental interactions. The environmental impact of a farm on diffuse contaminant loads (e.g., organic fertiliser runoff) to surface waters, for example, is strongly dependent on hydrological connectivity within the landscape. Representing this spatial variability may not be feasible within many of the export-coefficient based models such as FARMSCOPER [22], so interpretation of outputs requires additional expertise. DSTs which can complete complex calculations for the user (e.g. nutrient load calculations, pesticide dosage needs, etc.) without requiring excessive data inputs (or requiring data to be input more than once) are useful. However, the DST should still provide some flexibility to react to specific situations (e.g., extreme weather events, specific regulation in some areas, etc.) and respect user judgement. For example, users in the Lower Saxony case study liked the way that Mark Online [11] allowed farmers some flexibility on how to allocate nutrients in an agronomically sensible way on their farm, whilst still complying with the farm N quota.

It is clearly important that a DST should provide correct information and advice e.g., for cross-compliance checking. The introduction of new regulations (which are usually more complex) must be supported by providing some assistance for those affected; to this end, DSTs developed to ensure that farmers and other end users comply with legislation are helpful. Clear information about the derivation of the outputs produced by the DST should be provided (e.g., data source, assumptions applied, etc.), although it is still important to provide a simple and easy tool for establishing whether the legislation/rules are being followed. Examples include Mark Online [11], which includes cross-compliance checking and MANNER-NPK [15] (evaluated in the Portuguese case study), where users thought that being able to flag manure applications which did not comply with NVZ legislation was useful. A well implemented, simple-to use DST can help to ensure that farmers and other end users comply with legislation. However, when new regulations are introduced or the science improves, DSTs must be updated immediately if they are to retain their relevance and the trust of the end user.

Having consistency in the outputs and advice provided by different DSTs is clearly important, and this was not always found to be case. For example, during testing at the Aalborg case study site in Denmark, the Environmental Yardstick for Pesticides [9], SIRIS [21] and Plant Protection Online [12] all gave different results for the risk from pesticides applied to certain crops. Table 4 shows the results of leaching risk assessments for maize and potatoes. Roundup Bio (glyphosate 360 g/l) was assessed to be 82% using SIRIS (i.e., a high leaching risk), while the Environmental Yardstick for Pesticides indicated a low leaching risk for both soil types and application timings; the Danish assessment was low-medium risk based on the size of the taxes (12 Euros) and the load index to groundwater. One problem was that SIRIS calculates the leaching risk either for groundwater or surface water, making comparison of the outputs difficult. Nevertheless, such differences in results do not inspire stakeholder confidence.

It was found that advisory assistance is often needed to encourage farmers to use DSTs, to assist in their application and to interpret their results. Thus, the success of a DST also crucially depends on the skills and experience of the advisor, who should be able to understand both the science behind and the practical application of the DST and have the skills to communicate complex issues to farmers. Government involvement may, in some cases, increase DST uptake and use (e.g., by making it a legal requirement), although currently adoption is most often driven by market forces. Public recognition of success can also be beneficial, especially for DSTs operating at the catchment level, where a farm or group of farms using a DST could be used to demonstrate best practice. For example, SCIMAP has been used in the co-production of management options to address slurry pollution by the UK Rivers Trust [46], and FARMSCOPER to assess the environmental and economic impacts of revised nutrient and sediment policy measures for catchments across England [25].

Table 4. Leaching risk profiles for groundwater calculated using SIRIS, Environmental Yardstick for Pesticides and Plant Protection Online for pesticides used on maize and potatoes, in comparison with the Danish pesticide tax.

			DST Used for Assessment of Leaching Risk to Groundwater						
Danish Pesticide Name	Active Ingredients	Approved Dosage	Danish Pesticide Tax [1]	Plant Protection Online [2]	SIRIS [3]	Environmental Yardstick for Pesticides [4]			
			(Euro per kg or l)	(Load index)	(Rank-ing %)	(EIP at recommended product dose)			
						1.5% SOM		1.5–3% SOM	
						Spring	Autumn	Spring	Autumn
Callisto	Mesotrion	1.5 l	19	0.10	41	227	2087	58	530
MaisTer	Foramsulfuron/ iodosulfuron-methyl-Na/isoxadifen-ethyl	0.15 kg	2	0.01	55	7	94	6	89
Harmony SX	Thifensulfuron-methyl	15 g/ha	2		63	0	0	0	0
Starane 333 HL	Fluroxypyr	0.811 l	9	0.01	71	27	2701	3	270
Fighter 480	Bentazon	1.04 l	7	0.19	82	58	519	52	324
Roundup Bio	Glyphosate	3.5 l	12	0.17	82	0	0	0	0
Reglone	Diquat dibromid	4 l	79		82	0	0	0	0
Proman	Metobromuron	4 l			91	2740	9340	212	456
Boxer	Prosulfocarb	3.5 l	58	1.20	60	0	172	0	3
Agil 100 EC	Propaquizafop	1.25 l	13	0.07	47	1	2	0	0
Focus Ultra	Cycloxydim	5 l	29	0.01		1300	8000	650	1050
Ranman	Cyazofamid	0.2 l	1		23	87	62	6	6
Shirlan	Fluazinam	0.4 l	7		31	381	442	54	66
Acrobat New	Dimethomorph/ mancozeb	1.68 kg	23		91	118	715	8	475
Cyperb 100	Cypermethrin	0.4 l	85	0.03	18	0	0	0	0
Karate 2,5 WG	Lambda-cyhalothrin	0.3 kg	13	0.01		0	0	0	0

[1] The Danish Pesticide Tax scheme is based on the pesticide load index (see footnote 2) for three sub-indicators for human health, ecotoxicology and environmental fate, respectively. The tax increases as the risk of the pesticide increases; [2] Plant Protection Online classifies pesticides according to the pesticide load index, which is an indicator of pesticide risk [47]. In Denmark, the pesticide load is based on three sub-indicators for human health, ecotoxicology and environmental fate. In the table, only the pesticide load index for the environmental fate sub-indicator is presented, so that comparisons between DSTs assessment of pesticide leaching to groundwater could be made. Red indicates high risk with regard to environmental fate (i.e., a measure of the degradation time of a pesticide in soil and its potential for accumulation in food chains and for transport to groundwater), orange indicates medium risk and green low risk; [3] SIRIS classifies pesticides according to their potential to reach surface or groundwater. Results are expressed as rankings representing risk. In this example, the risk of the various pesticides leaching to groundwater is shown; [4] The Environmental Yardstick for Pesticides assigns environment impact points (EIP) to a pesticide: EIP >1000 indicates high risk (red); EIP (100–1000) indicates medium risk (orange); EIP < 100 indicates low risk (green). In this example, risk was calculated for soils with 1.5% and 1.5–3.0% soil organic matter (SOM), for applications in spring and autumn.

Conversely, where advisory services are limited or non-existent, a well-designed DST can provide useful guidance for protecting drinking water quality. This is particularly relevant for pesticides, where farmers rarely have detailed knowledge of the environmental behaviour of specific products.

3.7.3. Attitudes towards Decision Support Tools

In the case studies, DST users and other stakeholders were asked about their attitudes towards and main requirements for DSTs. 'Ease of use' and 'assistance with compliance' were frequently mentioned as being important in determining whether or not a DST was used. 'Ease of use' encompasses a range of reported user requirements such as:

Intuitive design;

- A centralized and holistic approach, where data only needs to be entered once;
- Checks to avoid data input errors and avoid wasting time;
- Data input and presentation of results (web-interface, excel-sheet, pdf, etc.) designed to suit user preferences;
- Clear results and outputs (e.g., graphical representations).

Similarly, in terms of 'assistance with compliance' users identified requirements such as:

- Trustworthy and reliable results (farmers cannot verify the data themselves, so must have confidence not only in the DST, but also in the developers of the DST);
- Availability of supplementary supporting information;
- Frequent updates to ensure compliance with legal requirements (if a farmer is to invest in a DST, then a 'future proof' tool is needed).

The feedback from the FAIRWAY case studies adds to findings from previous studies. For example, ease of use, defined as the ability of DSTs to present information to the user in a clear and familiar way with rapid comprehension, was highlighted as a concern by Inman et al. [48] Similarly, an EIP-AGRI workshop on "Tools for Environmental Farm Performance" [49] found that the reasons for poor uptake of DSTs among farmers include:

- the tool is not found to be useful by the farmer;
- the tool might be difficult to understand;
- the DST may require the farmer to spend a lot of time setting it up or learning how to use it;
- the costs outweigh the perceived benefits.

In summary, a DST that is acceptable to the majority of end users should fulfil most if not all of the criteria shown in Figure 4. A DST that fulfils these criteria and can deliver a range of functions is more likely to be successful, as end users prefer to limit the number of DSTs that they are required to use. Additionally, the provision of good advisory assistance was highlighted as being of great importance. The results produced by a DST are only as good as the input data, therefore support and advice from skilled advisors is highly valued by end users to enable them to make the right decisions.

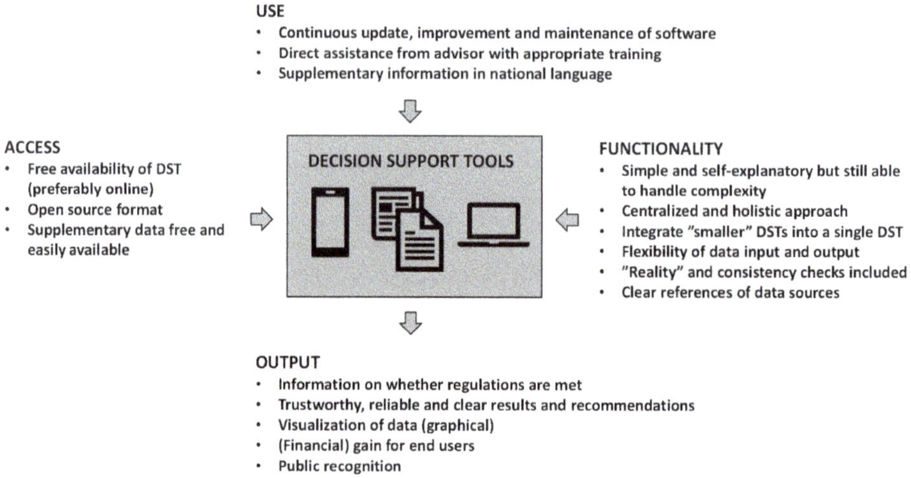

Figure 4. Criteria to be fulfilled by DSTs as identified by end users in nine FAIRWAY case studies.

3.8. Wider Perspectives and Future Developments

Despite the difficulties with the international exchange of DSTs, there are initiatives underway in the EU where partners from different MS are collaborating to develop tools which will have a direct or indirect impact on water quality.

An EU-funded project with similar aims to FAIRWAY is WaterProtect [50]. In this project, the aim is to "create or develop collaborative webtools and mobile apps that facilitate and more effectively implement farm management practices and measures for the protection of the drinking water sources". The project partners are building collaborative webtools for their Actionlabs (equivalent to the case studies implemented this study) that will help local actors (farmers, environment agencies, drinking water producers, and others) to monitor water quality, and to decide where to effectively implement management practices and measures to protect water sources. The tools will include information from the participatory monitoring, landscape information, best management practices and their socio-economic benefits and costs, although at present, the only tool available is for Belgium [50]. This is an online webtool presenting results from pesticide water sampling catchments in the form of online maps, so that local communities are able to access information pertaining to their area. As such, it appears to be a 'citizen science' application rather than a farm decision making or policy tool, such as those considered in our study.

Additionally, there is a tool which is under development as part of the future CAP preparations, and this is the Farm Sustainability Tool for Nutrients (FaST), an EU-wide tool which aims to help all farmers in the EU, without access to equivalent national DSTs, to manage the use of nutrients on their farms [51]. The FaST tool was not available for assessment as part of the FAIRWAY project and is still under development, however it will be interesting to follow its uptake, performance and implementation, as it is the first farm nutrient management DST that will have full EU coverage, and to see how the developers address the challenges identified in this study.

Over the next decade, the development of the DSTs and technologies to support the objectives within the CAP will be critical considerations. Extending and tightening further blanket measures (e.g., closed spreading periods) has limited the potential within some MS to deliver improvements without negatively impacting production. Measures to improve water quality must therefore transform toward more targeted farm- and catchment-specific approaches, using best available data and knowledge. Tools which can simplify these often complex decisions and matrices of scenarios will be critical if uptake targets are to be reached and maximum benefits delivered for users. The problem is whether there is the political will to collect the required data, and whether it is possible to collect it in such a way that any trans-national DSTs can be used consistently across the EU.

4. Summary and Conclusions

The EU FAIRWAY project was established to review approaches for protecting drinking water from nitrate and pesticide pollution. As part of the project, a comprehensive overview of DSTs used by farmers, farm advisors, water managers and policy makers for water, nutrient and pesticide management was undertaken, encompassing paper-based guidelines, farm-level software and phone apps, and complex research models. The unique combination of expertise and practical experience of the FAIRWAY project participants was used to identify farm-scale tools and other locally developed DSTs of importance in a national context. Of the >150 DSTs identified, 36 were selected for further investigation based on their national importance and relevance to the project aim. The majority of the selected DSTs were farm management tools and were included under the assumption that smart use of nutrients/pesticides indirectly improves water quality by reducing losses to the water environment. Only three of the selected DSTs were explicitly developed to consider the impact of mitigation methods on water quality.

Following demonstration and evaluation, 12 DSTs were selected for practical testing at nine of the FAIRWAY case study sites, to best match the challenges faced at each location. During the trial period, the DSTs were populated using local data, and meetings and demonstrations to stakeholders were organised. Barriers to DST exchange between countries were identified, and information was collected about farmer/stakeholder requirements and attitudes.

The FAIRWAY project provided a unique opportunity to exchange and test DSTs in very different environments across several MS. The evaluations indicated that the exchange of DSTs between MS is challenging because of the different legislative environments, advisory frameworks, country-specific data/calibration requirements, regional climate/soil differences, and issues around language. Notably, DSTs from different countries using the same input data sometimes delivered very different results. Whilst many countries have developed comparable DSTs designed to address similar problems, all the case study participants were able to draw inspiration and ideas from elsewhere. The support and advice provided by skilled advisors was highly valued, empowering end users to use the outputs from DSTs to make decisions for protecting water quality.

The case study evaluations will aid the development of a framework to highlight how DSTs can be used to improve awareness of diffuse pollution of vulnerable drinking water resources among farmers and other stakeholders, and will contribute to the wider Fairway project objective of identifying and developing innovative approaches for more effective drinking water protection.

Author Contributions: Methodology, F.N., R.K.L., R.C., L.T., N.S. and J.W.; Data Curation, F.N. and R.K.L.; Writing —Original Draft Preparation, F.N.; Writing—Review and Editing, R.C., L.F., L.T., R.K.L., N.S., and G.V.; Visualization, L.T., R.C. and R.K.L.; Supervision, R.K.L. and J.W.; Funding Acquisition, J.W. and G.V. All authors have read and agreed to the published version of the manuscript.

Funding: This research was funded by the European Union's Horizon 2020 research and innovation programme under grant agreement No 727984.

Acknowledgments: We are grateful to the following people who participated and contributed to the FAIRWAY project, and to the work presented in this paper: D. Doody (AFBI), A. Ferriera (IPC/ESAC), A. Jamsek (KGZ Maribor), Ø. Kaste (NIVA), S. Langas (NIVA), P. Schipper (WUR), J. Van Vliet (CLM), K. Verloop (WUR), F. Bondgaard (SEGES), M. Jørgensen (SEGES), I. Wright (UoL), J. Rowbottom (UoL), I. A. Leitão (IPC/ESAC), B. Hasler (AU), M. Glavan (UL), P. Leendertse (CLM), M. Hoogendoorn (CLM) and L. Jackson-Blake (NIVA).

Conflicts of Interest: The authors declare no conflict of interest. The funders had no role in the design of the study; in the collection, analyses, or interpretation of data; in the writing of the manuscript, or in the decision to publish the results.

Appendix A

Table A1. Keywords used for the online literature search (Web of Science).

Search Term	Keywords		
DST	Decision support tool OR Software tool OR Guidance tool OR AND	Guidance software OR Decision support software OR Decision support system OR	Decision management system OR Decision assistance tool OR Calculator OR App*
Pollutant/effect	Agricultur* OR Farm* OR Financial cost* OR Social cost* OR Cost-effective* OR Welfare* OR Cost-benefit OR Policy* OR Water quality OR Water* OR Groundwater OR Aquifer OR Soil* OR Fertili* OR Rush* OR Nitrogen OR Nutrient* OR	Nitrate* OR Nitrite* OR Ammonium OR Pesticide OR Herbicide OR Fungicide OR Molluscicide OR Insecticide OR Weed control OR Weed manage* OR Growth regulat* OR Metaldehyde OR Organophosphate OR Carbamate OR Diazine OR Phenoxyacetic acid OR MCPA OR	Glyphosate OR Bentazon OR Organochlor* OR Tryazine OR Dinitroaniline OR Bipiridil OR Dithiocarbamate OR Triazole OR Pyrethroid OR Amide OR Sulfonylurea OR Uracil OR Benzimidazole OR Nematicide OR Acaricide OR Slimicide

Table A2. Details for each DST supplied by FAIRWAY project partners.

Explanation of acronym
Brief description
Platform (e.g., paper-based tool, phone app, bespoke software)
Author name(s)
Author institute(s)
Date developed/released (or planned release date)
Member state(s) where developed
Member state(s) where currently used
Intended end user(s) (e.g., farmer, water quality manager, policy maker)
Temporal resolution (e.g., daily, annual, long-term)
Real-time component (e.g., incorporating live weather data, soil moisture data feeds etc.)
Geographical resolution (e.g., field, catchment, national)
Contaminant(s) covered (e.g., nitrate, metaldehyde etc.)
Number and type of mitigation measures included
Age/provenance of supporting data used to develop the DST
Details of validation and testing
Frequency of updates
Number of users or number of copies distributed/downloaded/purchased
Cost/availability
Full publication reference
Publication URL
Links to any other relevant documentation (e.g., user guides)
Demo material
Additional comments (e.g., shortcomings, obstacles)
The level of expertise or training required to use the DST
Input data required to run the DST
Outputs (including links to water quality and economic or financial aspects)
Country-specific calibration or data requirements (including restrictions on use)
The language of the DST and any supporting documentation
Other useful information (e.g., screenshots of inputs/outputs; how the DST is used in practice)

Table A3. Shortlist of DSTs taken forward for further consideration, including summary description and publication reference (where applicable).

No.	Country	DST Name	Type of DST[1]	WQI[1]	WQ[2]	Mitigation[3]	Brief Description	Reference
1	DE	Düngeplanung	Nutrient	Y			Farm-holistic DST to guide fertiliser purchasing and field-specific distribution. Combines on-farm data (soil nutrient contents, farm manure analysis, etc.), information on crop cultivation (crop rotation, yield level, etc.) with economic factors.	[20]
2	DE	ISIP	Nutrient		Y		Process-oriented model which simulates N-mineralisation in the soil and adjusts real-time recommendation for N-fertilisers in winter wheat accordingly.	[52]
3	DK	Mark Online	Nutrient/ Pesticide	Y			Used by farmers and advisors for fertiliser planning, optimization and documentation. Covers all aspects of crop management including soil tillage and crop protection. Ensures that pesticides and nutrients are used according to legislation based on data obtained via field trials.	[11]
4	DK	Dyrknings-vejledninger	Nutrient/ Pesticide				Manuals for growing a broad range of agricultural crops based on results from field trials. Updated at least annually to inform farmers/advisors on all aspects of Good Agricultural Practice.	[18]
5	DK	Plant Protection Online	Pesticide				Used by farmers and advisors to reduce pesticide use and ensure that only legal pesticides are used. The tool gives recommendations on whether or not to spray, dosage and spraying time.	[12]
6	DK	CTzoom/CTtools	Nutrient		Y		Estimates nitrate leaching based on N surplus calculations for individual fields. The results are used to define current practices.	[31]

Table A3. *Cont.*

No.	Country	DST Name	Type of DST[1]	WQI[1]	WQ[2]	Mitigation[3]	Brief Description	Reference
7	DK	BEST Kemi	Nutrient		Y		Groundwater management and forecasting DST used by the state and utilities to assess aquifer nitrate and pesticide concentrations. Can also monitor trends in groundwater quality.	NA
8	DK	TargetEconN	Nutrient	Y			Integrated economic and biophysical social planning model which minimises the costs of meeting a nutrient load reduction target. Calibrated for the the Danish Fjord Limfjorden watershed and currently being set up for Denmark as a whole. Used for WFD policy advice.	[30]
9	FR	PHYTOPIXAL	Pesticide	Y			GIS model based on a combination of indicators relating to the environmental vulnerability of the surface water environment and the agricultural pressure to estimate contamination risk.	[53]
10	FR	SIRIS	Pesticide				Classifies pesticides according to their potential to reach surface or groundwater. Results expressed as rankings representing risk. Helps farmers select the best product according environment parameters. Aids pesticide monitoring in waters at regional or local scale.	[21]
11	IE	Teagasc NMP online	Nutrient				System for developing farm-scale nutrient management plans for environmental and regulatory purposes. Likely to be used by agricultural consultants on behalf of most farmers.	[54]
12	IE	FarmHedge	Pesticide				Commercial phone app to manage feed/fertiliser purchases with a secondary component using farm location to create a set of weather alerts and advice on minimising environmental impact.	[10]

Table A3. *Cont.*

No.	Country	DST Name	Type of DST[1]	WQI[1]	WQ[2]	Mitigation[3]	Brief Description	Reference
13	NL	ANCA	Nutrient	Y			Farm specific assessment of nutrient inputs/outputs and emissions to the environment used by advisors to improve on-farm nutrient efficiency. Used as a monitoring tool to evaluate the effects of mitigation measures and by policy makers to estimate catchment N & P surplus reductions..	[38]
14	NL	Adviesbasis CBGV	Nutrient				Recommendations for fertilisation of grassland and maize. N rates specified for different growing conditions e.g., soil type, N release in soil by mineralisation and hydrology (water availability).	[55]
15	NL	Beregeningswijzer	Nutrient				Online meteorological data and field data are used to provide optimum irrigation requirements for individual fields. Prevents excess irrigation which could enhance leaching. Preserves optimal soil water content, resulting in higher N uptake and better fertiliser N utilization.	[56]
16	NL	BedrijfsWaterWijzer (BWW)	Nutrient		Y		Identifies risks to water quality specific to dairy farms and suggests measures for improvement. The risks are scored qualitatively (Good, Moderate, Insufficient, Bad). Enables farmers to indirectly evaluate the effect of mitigation measures.	[28]
17	NL	Bodemconditiescore	Nutrient/ Pesticide				Evaluation method for visual observations of sod density, botanical composition of grass sod, soil density, biological activity, abundance of macro fauna, rooting depth. Optionally also chemical quality of the grass and maize silage. Supports farmers by indicating soil problems.	[19]

Table A3. *Cont.*

No.	Country	DST Name	Type of DST[1]	WQI[1]	WQ[2]	Mitigation[3]	Brief Description	Reference
18	NL	NDICEA	Nutrient				Planner for integrated assessment of N availability for crops. Accounts for crop N demand, expected N availability from artificial fertilisers and manures, crop residues, green manures and soil, as well as leaching and denitrification losses.	[13]
19	NL	Environmental Yardstick for Pesticides (Yardstick)	Pesticide		Y	Y	Online version and crop information sheets to support on-farm integrated pesticide management. An offline version evaluates current practices and the effect of mitigation measures. Spraying schemes are evaluated in terms of environmental impact. Provides water utilities with a proxy for the value of programs designed to reduce impacts on groundwater.	[9]
20	NL	STONE	Nutrient				Calculates nutrient emissions to water and evaluates the effects of fertiliser policy measures on runoff and leaching of N and P to waters at national and regional levels. Can distinguish processes and sources of runoff and leaching to water. Used by policy makers to introduce effective mitigation measures and allocate source reduction targets	[29]
21	NO	Catchment-Lake Modelling Network (CLMN)	Nutrient		Y	Y	Network of process-based, mass-balance models linking climate, hydrology, catchment-scale nutrient dynamics and lake processes. Allows disentangling of the effects of climate change from those of land-use change on lake water quality and phytoplankton growth. Supports decision-making to achieve good water quality and ecological status.	[26]
22	NO	Skifteplan	Nutrient	Y			Most commonly used farm level DST for fertiliser application on agricultural fields. Calculates optimal fertilisation rates, to avoid excess N and P in soils and runoff. Also used to record crops grown in each field and year, and other treatments/measures implemented.	[57]

Table A3. *Cont.*

No.	Country	DST Name	Type of DST[1]	WQI[1]	WQ[2]	Mitigation[3]	Brief Description	Reference
23	NO	Agro-Meteo-rological Service	-				Run by NIBIO in collaboration with the Norwegian Met. Office. Provides meteorological data for better management of risks from farm operations in important agricultural districts.	[17]
24	SI	Načrtovanje Gnojenja	Nutrient				Assists agricultural advisers/farmers to optimise fertiliser use in all agricultural sectors as annual or multi-year fertiliser plans. Advises on crop rotations.	NA
25	SI	Smernice za Strokovno Gnojenje (SSG)	Nutrient				Fertiliser use guidelines which comply with the regulations and requirements for the quality of crops and the preservation of a clean environment. Intended to set a broader framework based on rational expert findings.	[58]
26	SI	OECD/EUROSTAT N Balance (OENBAL)	Nutrient	Y			Paper based handbook of methodology for calculating N and P balances in OECD/EU MS. Provides a consistent indicator based on harmonised methodology and definitions. Used to report on Nitrate Directive implantation and prepare legislation/measures for drinking water protection.	[32]
27	SI	GROWA-SI	Nutrient		Y		This regional water balance model is the official state model for reporting of Nitrate Directive implementation at a country level. Can calculate groundwater recharge rates and has the capability to account for N balances.	[35,36]
28	SI	State Network of Groundwater Monitoring Points (SNGMP)	Nutrient/ Pesticide		Y		State approved water quality monitoring network used by key decision makers. Measured values and their trends over the years serve as one of the base indicators for introducing new measures and indicate the success of previously introduced measures.	[59]

Table A3. *Cont.*

No.	Country	DST Name	Type of DST[1]	WQI[1]	WQ[2]	Mitigation[3]	Brief Description	Reference
29	SI	FITO-INFO	Pesticide				Slovenian public information system for plant protection which includes: plant protection products; related legislation; organism names, descriptions, pictures, weather forecasts etc.	[60]
30	UK	PLANET/MANNER	Nutrient	Y			Nutrient management DSTs for use by farmers/advisers for field level nutrient planning and demonstrating compliance with NVZ rules.	[14,15]
31	UK	FARMSCOPER	Nutrient/pesticide		Y	Y	Assesses diffuse agricultural pollutant loads on a farm and quantifies the impacts of mitigation methods. Can be customised to reflect management and environmental conditions representative of farming across England & Wales. Contains over 100 mitigation methods.	[22]
32	UK	Check it Out	Pesticide				Helps farmers/sprayer operators review and improve spraying practices and reduce the risk of pesticides reaching water. Each aspect of the spraying operation is scored and a total provided.	[61]
33	UK	Sentinel Online	Pesticide				Online information system to make key decisions in crop management. Includes: The Pesticide Database; Library; Decision support for crop nutrition, NVZ rules and recommendations; Technical updates; Weed/pests/disease identification etc..	NA
34	UK	Procheck	Pesticide				Interactive DST consisting of a pesticide database containing details of product information updated daily. Allows in-field use; includes a multi-criteria search engine for product selection.	NA

Table A3. *Cont.*

No.	Country	DST Name	Type of DST[1]	WQI[1]	WQ[2]	Mitigation[3]	Brief Description	Reference
35	UK	SCIMAP	Nutrient		Y		Helps decision-makers to prioritise activities that protect the water environment. Generates probabilistic risk maps for diffuse pollution from surface pathways within catchments.	[27]
36	UK	WaterAware	Pesticide				Phone app estimating risk of selected pesticide movement based on soil type, soil moisture deficit and weather conditions. Uses a traffic light system to advise farmers/sprayer operators when it is safe to apply chemicals or slug pellets.	[62]

[1] Represents indicators of water quality such as inputs (use of fertiliser/pesticides), nutrient balance/surplus/efficiency; [2] Water quality is explicitly represented (e.g., amount or risk of nitrate/pesticide leaching); [3] Mitigation methods are specifically represented; NA: no published information or website.

References

1. Steinebach, Y. Water Quality and the Effectiveness of European Union Policies. *Water* **2019**, *11*, 2244. [CrossRef]
2. European Environment Agency. *European Waters. Assessment of Status and Pressures 2018. EEA Report 7/2018*; Publications Office of the European Union: Luxembourg, 2018.
3. Directorate-General for Agriculture and Rural Development (European Commission). *EU Budget: The CAP after 2020*; Directorate-General for Agriculture and Rural Development: Brussels, Belgium, 2018. [CrossRef]
4. Erjavec, E.; Jongeneel, R.A.; Garcia Azcárate, T. *Research for AGRI Committee—The CAP Strategic Plans beyond 2020: Appraisal of the EC Legislative Proposals*; European Parliament, Policy Department for Structural and Cohesion Policies: Brussels, Belgium, 2018.
5. Hogan, P. Common Agricultural Policy Post-2020: Simplification and Modernisation. Speech of Commissioner Phil Hogan at the Agriculture and Fisheries Council on 16 July 2018. Available online: https://ec.europa.eu/info/sites/info/files/food-farming-fisheries/key_policies/presentations/simplification_modernisation_cap.pdf (accessed on 23 January 2020).
6. Rose, D.C.; Sutherland, W.J.; Parker, C.; Lobley, M.; Winter, M.; Morris, C.; Twining, S.; Ffoulkes, C.; Amono, T.; Dicks, L.V. Decision support tools for agriculture: Towards effective design and delivery. *Agric. Syst.* **2016**, *149*, 165–174. [CrossRef]
7. FAIRWAY: Farm Systems Management and Governance for Producing Good Water Quality for Drinking Water Supplies. Available online: https://www.fairway-project.eu/index.php (accessed on 21 January 2010).
8. Drinking Water Inspectorate. Guidance on the Implementation of the Water Supply (Water Quality) Regulations 2000 (as Amended) in England. Version 1.1—March 2012. Available online: http://dwi.defra.gov.uk/stakeholders/guidance-and-codes-of-practice/WS(WQ)-regs-england2010.pdf#page=18 (accessed on 22 January 2020).
9. Reus, J.A.W.A.; Leendertse, P.C. The environmental yardstick for pesticides: A practical indicator used in the Netherlands. *Crop Prot.* **2000**, *19*, 637–641. [CrossRef]
10. FarmHedge. Available online: http://farmhedge.io/ (accessed on 21 January 2020).
11. Bligaard, J. Mark Online, a Full Scale GIS-based Danish Farm Management Information System. *Int. J. Food Syst. Dyn.* **2014**, *5*, 190–195.
12. Planteværn Online. Available online: https://plantevaernonline.dlbr.dk//cp/menu/Menu.asp?SubjectID=1&ID=djf&MenuID=10009999&Language=en (accessed on 27 January 2020).
13. Burgt, G.J.H.M.; van der Oomen, G.J.M.; Habets, A.S.J.; Rossing, W.A.H. The NDICEA model, a tool to improve nitrogen use efficiency in cropping systems. *Nutr. Cycl. Agroecosyst.* **2006**, *74*, 275–294. [CrossRef]
14. Gibbons, M.M.; Fawcett, C.P.; Waring, R.J.; Dearne, K.; Dampney, P.M.R.; Richardson, S.J. PLANET Nutrient Management Decision Support System—A Standard Approach to Fertiliser Recommendations. In Proceedings of the EFITA/WCCA Conference, Vila Real, Portugal, 25–28 July 2005; pp. 89–94.
15. Nicholson, F.A.; Bhogal, A.; Chadwick, D.; Gill, E.; Gooday, R.D.; Lord, E.; Misselbrook, T.; Rollett, A.J.; Sagoo, E.; Smith, K.A.; et al. An enhanced software tool to support better use of manure nutrients: MANNER-NPK. *Soil Use Manag.* **2013**, *29*, 473–484. [CrossRef]
16. AHDB. Nutrient Management Guide (RB209). Available online: https://ahdb.org.uk/RB209 (accessed on 22 January 2020).
17. Nordskog, B.; Rafoss, T.; Nærstad, R.; Hofgaard, I.S.; Hole, H.; Elen, O.; Brodal, G. Farm scale weather data in plant pest forecasting. In Proceedings of the 10th Conference of the European Foundation for Plant Pathology (EFPP), Wageningen, The Netherlands, 1–5 October 2012.
18. SEGES. Dyrkningsvejledninger. Available online: https://dyrk-plant.dlbr.dk/Web/(S(rtwv44tjpqjr40sgewuih5k3))/forms/Afgroeder.aspx (accessed on 22 January 2020).
19. Sonneveld, M.P.W.; Heuvelink, G.B.M.; Moolenaar, S.W. Application of a visual soil examination and evaluation technique at site and farm level. *Soil Use Manag.* **2014**, *30*, 263–271. [CrossRef]
20. Landwirtschaftskammer Niedersachsen. Düngeplanungsprogramm. Available online: https://www.lwk-niedersachsen.de/index.cfm/portal/96/nav/2208/article/31583.html (accessed on 22 January 2020).
21. Le Gall, A.-G.; Jouglet, P.; Morot, A.; Guerbet, M. SIRIS-Pesticides: Update and validation of a decision support system for pesticides monitoring in freshwater. In Proceedings of the 17th SETAC Europe Annual Meeting, Porto, Portugal, 20–24 May 2007.

22. Gooday, R.; Anthony, S.; Chadwick, D.; Newell-Price, P.; Harris, D.; Duethmann, D.; Fish, R.; Collins, A.; Winter, M. Modelling the cost-effectiveness of mitigation methods for multiple pollutants at farm scale. *Sci. Total Environ.* **2014**, *468–469*, 1198–1209. [CrossRef]
23. Newell-Price, J.P.; Harris, D.; Taylor, M.; Williams, J.R.; Anthony, S.G.; Duethmann, D.; Gooday, R.D.; Lord, E.I.; Chambers, B.J.; Chadwick, D.R.; et al. An Inventory of Methods and Their Effects on Diffuse Water Pollution, Greenhouse Gas Emissions and Ammonia Emissions from Agriculture. User Guide. Final Report for Defra Project WQ0106 (Module 5). Available online: http://randd.defra.gov.uk/Default.aspx?Menu=Menu&Module=More&Location=None&ProjectID=14421&FromSearch=Y&Publisher=1&SearchText=wq0106&SortString=ProjectCode&SortOrder=Asc&Paging=10#Description (accessed on 24 January 2020).
24. Zhang, Y.; Collins, A.L.; Gooday, R.D. Application of the FARMSCOPER tool for assessing agricultural diffuse pollution mitigation methods across the Hampshire Avon Demonstration Test Catchment, UK. *Environ. Sci. Policy* **2012**, *24*, 120–131. [CrossRef]
25. Collins, A.L.; Newell-Price, J.P.; Zhang, Y.; Gooday, R.; Naden, P.S.; Skirvin, D. Assessing the potential impacts of a revised set of on-farm nutrient and sediment 'basic' control measures for reducing agricultural diffuse pollution across England. *Sci. Total Environ.* **2018**, *621*, 1499–1511. [CrossRef]
26. Couture, R.M.; Tominaga, K.; Starrfelt, J.; Moe, J.; Kaste, O.; Wright, R.F. Modelling phosphorus loading and algal blooms in a Nordic agricultural catchment-lake system under changing land-use and climate. *Environ. Sci. Process. Impacts* **2014**, *16*, 1588–1599. [CrossRef] [PubMed]
27. Perks, M.T.; Warburton, J.; Bracken, L.J.; Reaney, S.M.; Emery, S.B.; Hirst, S. Use of spatially distributed time-integrated sediment sampling networks and distributed fine sediment modelling to inform catchment management. *J. Environ. Manag.* **2017**, *202*, 469–478. [CrossRef] [PubMed]
28. Agriwijzer. BedrijfsWaterWijzer. Available online: https://www.agriwijzer.nl/product/17 (accessed on 22 January 2020).
29. Wolf, J.; Beusen, A.H.W.; Groenendijk, P.; Kroon, T.; Rötter, R.; van Zeijts, H. The integrated modeling system STONE for calculating nutrient emissions from agriculture in the Netherlands. *Environ. Model. Softw.* **2003**, *18*, 597–617. [CrossRef]
30. dNmark Research Alliance. Fact Sheet: The TargetEconN Framework—A Cost-Minimization Model for Nitrogen Management. Available online: http://dnmark.org/wp-content/uploads/2017/03/Fact-sheet-TargetEconN-modelling-framework_Final.pdf (accessed on 22 January 2020).
31. ConTerra. CT-Tools. Available online: http://www.conterra.dk/index.php?action=text_pages_show&id=158&menu=36 (accessed on 25 January 2020).
32. Instituto Nacional de Investigação Agrarian e Veterinaria (INIAV). *Manual de Fertilização das Culturas*; Laboratório Químico Agrícola Rebelo da Silva: Lisbon, Portugal, 2006.
33. Simões, J.S. *Utilização de Produtos Fitofarmaceuticos na Agricultura*; Principia: Porto, Portugal, 2005.
34. Kremer, A.M. Methodology and Handbook Eurostat/OECD. Nutrient Budgets. EU-27, Norway, Switzerland. Available online: http://ec.europa.eu/eurostat/cache/metadata/Annexes/aei_pr_gnb_esms_an1.pdf (accessed on 22 January 2020).
35. Andelov, M.; Kunkel, R.; Uhan, J.; Wendland, F. Determination of nitrogen reduction levels necessary to reach groundwater quality targets in Slovenia. *J. Environ. Sci.* **2014**, *26*, 1806–1817. [CrossRef]
36. Tetzlaff, B.; Andjelov, M.; Kuhr, P.; Uhan, J.; Wendland, F. Model-based assessment of groundwater recharge in Slovenia. *Environ. Earth Sci.* **2015**, *74*, 6177–6192. [CrossRef]
37. Curry, C. A New Era of Targeted Reforestation for Diffuse Pollution Risk Reduction Using SCIMAP and UAV Technology to Determine the Spatial Distribution of Diffuse Pollution Risk in Lake Rawa Pening, Central Java, Indonesia—SCIMAP-UGM16. Available online: http://www.scimap.org.uk/2016/10/a-new-era-of-targeted-reforestation-for-diffuse-pollution-risk-reduction-using-scimap-and-uav-technology-to-determine-the-spatial-distribution-of-diffuse-pollution-risk-in-lake-rawa-pening-central-ja/ (accessed on 2 January 2020).
38. Aarts, H.F.M.; de Haan, M.H.A.; Schroder, J.J.; Holster, H.C.; de Boer, J.A.; Reijs, J.; Oenema, J.; Hilhorst, G.J.; Sebek, L.B.; Verhoeven, F.P.M.; et al. Quantifying the environmental performance of individual dairy farms—The Annual Nutrient Cycling Assessment (ANCA). In Proceedings of the 18th Symposium of the European Grassland and Forages in High Output Dairy Farming Systems, Wageningen, The Netherlands, 15–17 June 2015; pp. 377–380.

39. Department of Environment Food and Rural Affairs (Defra). Use of "Fertiliser Manual (RB209)" Recommendations for Grassland Draft Case Study and Focus Groups Report. Final Report for Defra Project IF01121. Available online: http://sciencesearch.defra.gov.uk/Default.aspx?Menu=Menu&Module=More&Location=None&ProjectID=17803&FromSearch=Y&Publisher=1&SearchText=rb209&SortString=ProjectCode&SortOrder=Asc&Paging=10#Description (accessed on 24 January 2020).
40. Department of Environment Food and Rural Affairs (Defra). *The Fertiliser Manual (RB209)*, 8th ed.; TSO: Norwich, UK, 2015.
41. Axelsen, J.; Munk, L.; Sigsgaard, L.; Ørum, J.E.; Streibig, J.C.; Navntoft, S.; Christensen, T.; Branth, A.; Elkjær, K.; Korsgaard, M.; et al. *Udredning om Moniterings, Varslings- og Beslutningsstøttesystemer for Skadevoldere i Planteproduktionen i Landbrug, Gartneri og Frugtavl. Miljøprojekt nr. 1407*; Miljøstyrelsen: Odense, Denmark, 2012; Available online: https://www2.mst.dk/Udgiv/publikationer/2012/06/978-87-92779-73-1.pdf (accessed on 24 January 2020).
42. Rose, D.; Parker, C.; Fodey, J.; Park, C.; Sutherland, W.; Dicks, L. Involving Stakeholders in Agricultural Decision Support Systems: Improving User-Centred Design. *Int. J. Agric. Manag.* **2018**, *6*, 80–89.
43. Thorling, L.; Ditlefsen, C.; Ernstsen, V.; Hansen, B.; Johnsen, A.R.; Troldborg, L. Grundvandsovervågning. Status og Udvikling 1989–2018. GEUS, 2019. Available online: https://www.geus.dk/media/22654/grundvand1989-2018-endelig.pdf (accessed on 27 January 2020).
44. Rose, D.C.; Bruce, T.J.A. Finding the right connection: What makes a successful decision support system? *Food Energy Secur.* **2018**, *7*, e00123. [CrossRef]
45. Lundström, C.; Lindblom, J. Considering farmers' situated knowledge of using agricultural decision support systems (AgriDSS) to Foster farming practices: The case of CropSAT. *Agric. Syst.* **2018**, *159*, 9–20. [CrossRef]
46. Whitman, G.P.; Pain, R.; Milledge, D.G. Going with the flow? Using participatory action research in physical geography. *Prog. Phys. Geogr. Earth Environ.* **2015**, *39*, 622–639. [CrossRef]
47. Kudsk, P.; Jørgensen, L.N.; Ørum, J.E. Pesticide Load—A new Danish pesticide risk indicator with multiple applications. *Land Use Policy* **2018**, *70*, 384–393. [CrossRef]
48. Inman, D.; Blind, M.; Ribarova, I.; Krause, A.; Roosenschoon, O.; Kassahun, A.; Scholten, H.; Arampatzis, G.; Abrami, G.; McIntosh, B.; et al. Perceived effectiveness of environmental decision support systems in participatory planning: Evidence from small groups of end-users. *Environ. Model. Softw.* **2011**, *26*, 302–309. [CrossRef]
49. EIP-AGRI Workshop. Tools for Environmental Farm Performance. Available online: https://ec.europa.eu/eip/agriculture/sites/agri-eip/files/eip-agri_ws_tools_for_environmental_farm_performance_final_report_2017_en.pdf (accessed on 24 January 2020).
50. WaterProtect. Available online: https://water-protect.eu/ (accessed on 24 January 2020).
51. A New Tool to Increase the Sustainable Use of Nutrients across the EU. Available online: https://ec.europa.eu/info/news/new-tool-increase-sustainable-use-nutrients-across-eu-2019-feb-19_en (accessed on 24 January 2020).
52. ISIP—Das Informationssystem für die Integrierte Pflanzenproduktion. Available online: https://www.isip.de/isip/servlet/isip-de (accessed on 24 January 2020).
53. Macary, F.; Morin, S.; Probst, J.-L.; Saudubray, F. A multi-scale method to assess pesticide contamination risks in agricultural watersheds. *Ecol. Indic.* **2014**, *36*, 624–639. [CrossRef]
54. Murphy, P. NMP Online. In Proceedings of the Teagasc Soil Fertility Conference, Clonmel, Ireland, 16 October 2015; pp. 8–9.
55. Commissie Bemesting Grasland en Voedergewassen. Bemestingsadvies. Available online: https://edepot.wur.nl/413891 (accessed on 25 January 2020).
56. Everts, H.; Hoving, I.E.; van der Schans, D.A. Beregeningswijzer. Achtergrondinformatie Gebruiksaanwijzing Toelichting vochtboekhouding. Available online: http://edepot.wur.nl/24356 (accessed on 25 January 2020).
57. Skifteplan. Available online: http://www.agromatic.no/skifteplan.html (accessed on 25 January 2020).
58. Mihelič, R.; Čop, J.; Jakše, M.; Štampar, F.; Majer, D.; Tojnko, S.; Vršič, S. *SMERNICE Za Strokovno Utemeljeno Gnojenje. Ministrstvo za Kmetijstvo*; Gozdarstvo in Prehrano: Ljubljana, Slovenia, 2010; Available online: https://www.program-podezelja.si/sl/knjiznica/26-smernice-za-strokovno-utemeljeno-gnojenje/file (accessed on 25 January 2020).
59. Tehonnik, M.D. *Water Quality in Slovenia*; Bograf: Ljubljana, Slovenia, 2008. Available online: http://www.arso.gov.si/en/water/reports%20and%20publications (accessed on 25 January 2020).

60. Fito-Info: Slovenian Information System for Plant Protection. Available online: http://www.fito-info.si/E_index.asp (accessed on 25 January 2020).
61. Check it Out Tool. Available online: http://checkitout.voluntaryinitiative.org.uk/tool/ (accessed on 25 January 2020).
62. WaterAware App. Available online: http://www.adama.com/uk/en/wateraware/ (accessed on 25 January 2020).

Article

Influence of Farming Intensity and Climate on Lowland Stream Nitrogen

Guillermo Goyenola [1,*], Daniel Graeber [2], Mariana Meerhoff [1,3], Erik Jeppesen [3,4,5], Franco Teixeira-de Mello [1], Nicolás Vidal [1], Claudia Fosalba [1], Niels Bering Ovesen [3], Joerg Gelbrecht [6], Néstor Mazzeo [1,7] and Brian Kronvang [3]

1 Departamento de Ecología y Gestión Ambiental, Centro Universitario Regional Este (CURE), Universidad de la República, Punta del Este 20100, Uruguay
2 Department Aquatic Ecosystem Analysis, Helmholtz Centre for Environmental Research—UFZ, 39114 Magdeburg, Germany
3 Department of Bioscience and Arctic Research Centre, Aarhus University, 8600 Silkeborg, Denmark
4 Sino-Danish Centre for Education and Research, Beijing 100000, China
5 Limnology Laboratory, Department of Biological Sciences and EKOSAM, Middle East Technical University, 06800 Ankara, Turkey
6 Leibniz-Institute of Freshwater Ecology and Inland Fisheries, 12587 Berlin, Germany
7 South American Institute for Resilience and Sustainability Studies (SARAS), Bella Vista 20302, Uruguay
* Correspondence: ggoyenola@cure.edu.uy or goyenola@gmail.com; Tel.: +598-9565-5636

Received: 17 February 2020; Accepted: 30 March 2020; Published: 2 April 2020

Abstract: Nitrogen lost from agriculture has altered the geochemistry of the biosphere, with pronounced impacts on aquatic ecosystems. We aim to elucidate the patterns and driving factors behind the N fluxes in lowland stream ecosystems differing about land-use and climatic-hydrological conditions. The climate-hydrology areas represented humid cold temperate/stable discharge conditions, and humid subtropical climate/flashy conditions. Three complementary monitoring sampling characteristics were selected, including a total of 43 streams under contrasting farming intensities. Farming intensity determined total dissolved N (TDN), nitrate concentrations, and total N concentration and loss to streams, despite differences in soil and climatic-hydrological conditions between and within regions. However, ammonium (NH_4^+) and dissolved organic N concentrations did not show significant responses to the farming intensity or climatic/hydrological conditions. A high dissolved inorganic N to TDN ratio was associated with the temperate climate and high base flow conditions, but not with farming intensity. In the absence of a significant increase in farming N use efficiency (or the introduction of other palliative measures), the expected farming intensification would result in a stronger increase in NO_3^-, TDN, and TN concentrations as well as in rising flow-weighted concentrations and loss in temperate and subtropical streams, which will further exacerbate eutrophication.

Keywords: agricultural impact; stream; nitrogen concentration; nitrogen losses; eutrophication

---

1. Introduction

The changes promoted by agriculture in terrestrial and freshwater ecosystems can be quite dramatic [1,2]. During the 20th and 21st centuries, the global nitrogen (N) cycle has accelerated due to the artificial fixation of atmospheric N_2 (g) and extensive use of N fertilizers to boost agricultural production [3,4]. A high amount of reactive N (~100 Tg N·year^{-1}) is used in global agriculture, but it is estimated that less than 1 out of every 5 N-atoms used as fertilizer is finally consumed by humans [3]. The N lost from agricultural production ultimately alters the geochemistry of the biosphere and particularly impacts aquatic ecosystems, thus contributing to eutrophication, degradation of water quality, and biodiversity loss [5–9].

At the catchment scale, the biogeochemical processes determining the natural fluxes of N from land to aquatic ecosystems mainly depend on climatic and hydrological regimes, and their interaction with local soil and geological conditions [10]. Consequently, stream N concentrations are dependent on variations in water temperature and discharge [11–13]. Moreover, farming intensity alters N mass balance and fluxes, leading to changes in the hydrochemistry of the streams [14]. One of the main changes induced by agriculture is the enhanced N level [4] due to the higher input of dissolved inorganic N (DIN), resulting in increased nitrate exports to aquatic ecosystems [15,16].

Intensive farming is a worldwide phenomenon [1,17]. Europe has a much longer history of intensive farming than tropics and subtropics, where the expansion of cropland areas became particularly rapid after 1850 [18]. Furthermore, 50% of the N fertilizer (inorganic and manure) used is confined to approximately 10% of the fertilized land, in tropical and subtropical areas of south and southeast Asia and southeast South America [19,20]. So far, most studies of the environmental consequences of N use in farming on streams have been conducted in developed countries, particularly in areas characterized by a temperate climate within Europe and North America [21], whereas only a few investigations have focused on developing countries and/or warm climate conditions e.g., [22–24]. Better knowledge of the use, transformation, and transport of N in subtropical and tropical climate regions—and thus of the N cycle—is urgently needed to generate more nutrient-efficient ways of producing food, while simultaneously reducing the negative side effects on the environment [18,25–27].

We aim to elucidate the patterns and driving factors behind the N fluxes in lowland stream ecosystems with contrasting land-uses and climate. Specifically, we aimed to analyze to what extent natural variations in soil characteristics and climate/hydrology would influence the effect of catchment farming on concentrations and N losses in lowland streams. We expected that streams draining microcatchments with high-intensity farming would have the highest concentrations of total N (TN), total dissolved N (TDN), nitrate (NO_3^-), ammonium (NH_4^+), and dissolved organic N (DON) and a higher DIN/TDN ratio than streams draining low-intensity farming microcatchments, independently of variations in edaphic variability or climatic-hydrological conditions.

2. Materials and Methods

To obtain a suitable balance between the different scales of analysis [28], the sampling strategy included three complementary monitoring schemes using different combinations of sampling frequencies and the associated number of streams (considered as replicates of farming intensity within each region).

A total of 43 lowland streams draining microcatchments under two contrasting conditions of agricultural intensity (hereafter farming intensity) were selected in two distinctive climate areas: humid cold temperate (Dfb sensu: [29]; n = 21 in summer; n = 20 in winter) and humid subtropical (Cfa sensu: [29]; n = 22; Table 1). The topography of both selected landscapes was characterized by gently rolling plains (mean slope < 5%) and the hydrographic catchment size varied around 10 km^2 (Denmark 9 ± 11 km^2, Uruguay 12 ± 7 km^2, average and standard deviation, respectively).

One stream per combination of farming intensity/climate-hydrology condition ($n_{streams}$ = 4, Table 1) was described in detail relative to hydrology, hydrochemistry, meteorological conditions, and catchment land-use. These streams acted as "benchmark streams" for each condition, and water samples with two different temporal resolutions were taken for a 2-year period (see Section 2.2). Both Danish benchmark streams are part of the Gudenå River basin, while the Uruguayan benchmark streams are part of the Santa Lucía Chico River basin. Besides the benchmark streams, a larger set of streams grouped according to farming intensity/climate-hydrology conditions were sampled with a "snapshot" approach ($n_{streams}$ = 39 winter, 38 summer; 1 sample per season; Table 1). More detailed information can be found in Goyenola et al. [30] and Graeber et al. [31]. Automatized gauging stations were established in the four benchmark microcatchments. Hydrometric data were recorded every 10 minutes using CR10X data loggers (Campbell Scientific Ltd., Shepshed, UK). In the subtropical streams, we used CS450 Submersible Pressure Transducers (Campbell Scientific Ltd., Shepshed, UK) for water stage monitoring and Rain-O-Matic Professional rainfall automatized gauges (Pronamic). In temperate catchments, the water level was registered with PDCR 1830 pressure sensors (Druck), while meteorological information was obtained from the Danish Meteorological Institute monitoring the network based on a 10 × 10 km grid. Periodic instantaneous flow measurements were taken using a C2-OTT Kleinflügel, transferring data to software for the calculation of instantaneous discharge (VB-Vinge 3.0, Mølgaard Hydrometri).

Table 1. Characteristics of the soil and land-use of all the studied stream catchments according to the sampling method used. Benchmark streams were sampled fortnightly using grab and automatized pooled sampling.

Climate & Farming Intensity	Benchmark Streams	Snapshot Grab-Sampling

TEMP Low	Granslev stream
	Haplic Luvisols # O.M. = 5%
	F.A.= 29%; mean L.U. = 0.25 ha^{-1}
	N fertilizer = 45 kg $N{\cdot}ha^{-1}{\cdot}year^{-1}$
	(45 % fertilizers; 55% manure)
		Mainly Luvisols and Podsols; Arenosols #
		O.M. < 5%. Range F.A. = 0%–26%

TEMP High	Gelbæk stream
	Gleyic Luvisols # O.M. < 5%
	F.A.= 92%; mean L.U. = 0.79 ha^{-1}
	N fertilizer = 143 kg $N{\cdot}ha^{-1}{\cdot}year^{-1}$
	(45 % fertilizers; 55% manure)
		Mainly Luvisols and Podsols, some
		Albeluvisols Arenosols and Cambisols #
		O.M. < 5%. Range F.A. = 74%–93%

SUBT Low	Chal-Chal stream Luvic Phaeozem and Eutric Vertisols * O.M. = 5.2% F.A.= 30%; mean L.U. = 0.62 ha^{-1} N fertilizer = 76 kg $N{\cdot}ha^{-1}{\cdot}year^{-1}$ (18% fertilizers; 82% manure)	Phaeozem and Vertisols * O.M. = 5% ± 1.5 Range F.A. = 0%–25%
SUBT High	Pintado Stream Eutric Regosols * O.M. = 4% to 5% F.A.= 90%; mean L.U. = 2.00 ha^{-1} N fertilizer = 242 kg $N{\cdot}ha^{-1}{\cdot}year^{-1}$ (17% fertilizers; 83% manure)	Mainly Phaeozem * O.M. = 5% ± 1.5 Range F.A. = 75%–100%
Total n	$n_{streams} = 4$	$n_{streams}$ = 39 (w), 38 (s)
Sampling	grab and pooled sampling 2 years	1 sample in (w) and 1 in (s)

Abbreviations: TEMP: temperate streams; SUBT: subtropical streams; Low and High: low and high-farming intensity; O.M.: organic content of soils (%); F.A.: percentage of the Farming area; mean L.U.: mean livestock units by ha; N fertilizer: total N inputs by year, discriminated in contributions of fertilizers and manure. (w): winter; (s): summer. Source: (#) World Reference Soil Database classification, European Commission and European Soil Bureau Network (2004); (*) SOTERLAC database, ISRIC Foundation (www.isric.org).

Nitrogen catchment input by hectare and year in Danish catchments was estimated considering the surface of agriculture land in the catchment multiplied by the national average of chemical fertilizer use for the years 2011–2012 (total input 69 kg $N{\cdot}ha^{-1}{\cdot}year^{-1}$), and the livestock density multiplied by the average of N production in manure by livestock (86 kg $N{\cdot}ha^{-1}{\cdot}year^{-1}$ [32]). Nitrogen catchment input by hectare and year in Uruguayan catchments was estimated through interviews with the technical managers of the establishments. Based on the best available knowledge considering empirical data [33], the average of N production in manure for the Uruguayan was assumed to equal than for Danish catchments. The total N input was divided by the total catchment area to be able to make quantitative comparisons with the TN losses.

2.1. Farming Intensity

Land-use intensity is a complex multidimensional concept and is difficult to measure [34,35], and therefore we applied explicit operational definitions (Table 1). Low-intensity farming catchments represent the condition with minimal anthropogenic pressures for each region. The subtropical low-intensity farming catchments (n = 9, both summer and winter) were dominated by the natural grasslands of the Pampa Biome [36] and sustained low-density cattle production (below one head per hectare), while a mixture of deciduous and coniferous forests dominated the temperate low-intensity farming catchments (n_{summer} = 8; n_{winter} = 9; Figure 1; Table 1). Less than 30% of the selected low-intensity farming catchments in both countries were influenced by arable cropping systems.

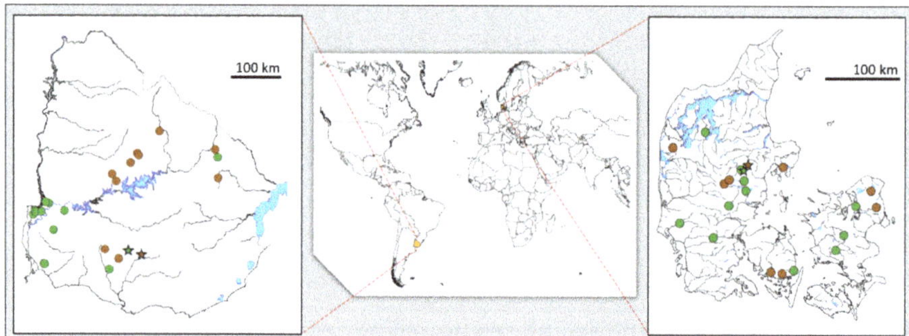

Figure 1. The geographic location of sampled streams. Left: subtropical/Uruguayan streams. Right: temperate/Danish streams. Circles: Snapshot grab-sampling. Stars: Benchmark streams. Green: high-farming intensity; brown: low-farming intensity. LAT/LONG of benchmark streams: TEMP Low: 56.2837/9.8975; TEMP High: 56.2254/9.8117; SUBT Low: −33.8256/−56.2821; SUBT Low: −33.9036/−56.0064. All the catchments fall within polygons limited by the following coordinates: Denmark 55.09 to 56.64 N; 8.38 to 12.44 E, Uruguay −31.79 to −34.18; −54.41 to −58.31 (decimal degrees; WGS84).

The criteria for the selection of high-intensity farming catchments were: (1) that arable cropping systems with intensive use of fertilizers affected more than 70% of the total area and (2) that they represented real and typical high-intensity farming catchments in the climatic-hydrological regions of the two countries (n_{TEMP} = 12; n_{SUBT} = 13; Table 1).

2.2. Hydrochemistry Monitoring

The complementary strategy allowed us to evaluate whether the patterns observed were generalizable for each climate/hydrological region or even common to both. The monitoring approach included:

- Fortnightly grab-sampling in benchmark streams: Sub-surface grab samples were taken from a well-mixed section with no macrophytes in the center of the stream channel during the daytime. This instantaneous sampling was used for the analysis of conservative and non-conservative N fractions (i.e., TN, TDN, NO_3^-, DON, and NH_4^+).
- Automatic pooled sampling in benchmark streams: High-frequency monitoring using automated equipment was conducted during the same two-year period. Glacier refrigerated automatic samplers (ISCO-Teledyne) collected an equal water volume every four hours from the same sampling point, and the pooled samples were collected fortnightly. The final nutrient concentration in the only sampler carboy thus represented a time-proportional average for the fortnightly sampling period. As this sampling involved refrigerated storage of pooled samples for up to two weeks, the emphasis was placed on the analysis of TN.
- Snapshot grab-sampling in the series of streams was made once in winter and once in summer. Sub-surface grab samples were taken in a well-mixed section with no macrophytes from the center of the stream channels during the daytime. This instantaneous sampling was used for the analysis of different N fractions, with emphasis on dissolved compounds (i.e., TDN, NO_3^-, DON, and NH_4^+).

2.3. Laboratory Measurements

All pooled water samples from the high frequency monitored streams were analyzed for total N (TN), total dissolved N (TDN), and nitrate (NO_3^-). Analysis of fortnightly and snapshot samples included TN, TDN, NO_3^-, NH_4^+, and DON (dissolved organic N). Different techniques were applied to guarantee accuracy, address the different concentration ranges, and assure the inter-comparability of results between countries.

Water samples for the determination of dissolved N fractions were filtered through 0.45-μm membrane filters pre-rinsed with ultrapure water (Milli-Q water). TN was converted to nitrate following the protocol of Valderrama [37] and analyzed as NO_3^-. In Uruguay, the standard sodium salicylate method was used for NO_3^- determination [38,39]. For the Danish samples, the sum of both nitrate and nitrite was determined by flow analysis (CFA and FIA) and spectrometric detection ([40], Danish Standard 223). Additionally, the samples were analyzed by segmented flow analysis including an additional channel to measure NO_2^-. As the NO_2^- concentration was always below the quantification limit of the technique (<0.01 mg $N \cdot L^{-1}$), it was not considered in the analysis; instead we assumed that (NO_3^-) + (NO_2^-) = (NO_3^-).

The analysis of the Uruguayan samples used for total TDN determination followed the same approach as for the TN samples. The TDN concentrations of the Danish samples were measured using high-temperature catalytic oxidation (HTCO, multi N/C 3100, Jena Analytik, Jena, Germany) after acidifying the samples to pH 2–3 with HCl and sparging with synthetic air for 5 min. The samples were oxidized with a platinum catalyst at 700 °C in a synthetic air stream, and TDN was measured as NO gas with a chemiluminescence detector [41]. For the Danish water samples with high NO_3^- levels (from high-intensity farming), the HTCO method led to significant underestimation of TDN, likely because the HTCO method did not permit oxidization of all the N [42,43]. When oxidation was not possible, TDN was estimated as the addition of DIN + DON, DIN being in turn estimated as the addition of NO_3^- and NH_4^+ and DON being measured by size-exclusion chromatography [43,44]. DON samples taken in subtropical streams were acidified with hydrochloric acid and frozen, following Hudson, et al. [45], and sent for analysis at the Leibniz-Institute of Freshwater Ecology and Inland Fisheries laboratory in Berlin. Before measurements, the samples were brought to the same target pH level of 7.5 ± 0.5 by neutralization with sodium hydroxide. NH_4^+ was measured following the indophenol-blue method [46].

2.4. Data Analysis

Non-linear regressions between stage and discharge at each monitoring station (rating curves) were fitted. Rating curves were used to generate a discharge data series with a 10-min resolution using the software HYMER (www.orbicon.com). Base flow index (BFI) was estimated for the complete data set from daily hydrographs using the automatic routine proposed by Arnold et al. [47] to set the magnitude of the groundwater contribution to the streamflow.

Total N concentrations were determined from the fortnightly samples, while the high frequency automatized pooled data were used to estimate the annual TN transport, loss, and flow-weighted TN concentrations in the subset of the four benchmark streams (2-year period). The TN transport was calculated by multiplying the TN concentration obtained from the pooled samples by the accumulated discharge for the same fortnightly period and summing yearly [48]. The TN loss was calculated dividing the annual transport by the catchment area in hectares [49]. Missing data from the relatively short periods when the automatic samplers were not in operation (e.g., due to freezing in Denmark) were re-generated through linear interpolation of concentrations [50]. The flow-weighted concentration (FWC) was calculated as the annual TN transport divided by the annual runoff. Dissolved N fractions were analyzed from both the fortnightly grab samples in the four benchmark streams (2-year period) and the snapshot samples ($n_{streams}$ = 39 winter, 38 summer; 1 sample per season).

The factorial design relative to climate/hydrology conditions and farming intensity was evaluated using two-way nested ANOVA with farming intensity nested within climate/hydrology, followed by a post hoc pairwise multiple comparison when appropriate [51]. Variability in the high frequency-automatized pooled and fortnightly samples represents the temporal variation within each intensively sampled stream, while variability in the snapshot sampling expresses spatial variation among comparable systems. The relationship between TN concentrations from the fortnightly instantaneous grab samples, discharge, and water temperature were analyzed by Spearman rank-order correlations.

3. Results

3.1. Climate and Hydrology

The climatic characteristics in the study period (2010 to early 2012) can be considered typical for both Denmark and Uruguay (Table 2; [30]). The annual average air temperature did not exhibit any anomaly [52,53] and corresponded to the mean for the corresponding region recorded by national meteorological services based on recent historical information [54,55]. Mean air temperature was 8.8 °C and ranged between −7.0 and 20.4 °C in the temperate sites and was ca. 17.5 °C, ranging between 3.7 to 32.2 °C, in the subtropical sites. No dry or wet seasons occurred in either country, but marked differences in frequency and intensity of rainfall were detected. Total annual precipitation was lower in the Danish catchments than in the Uruguayan catchments (Table 2), while rain events were less intense but more frequent in the Danish than in the Uruguayan catchments [30,31]. Thus, hydrologically, the Danish streams are more stable than the Uruguayan catchments, the latter being described as "flashy" in previous publications (Richards-Baker Flashinnes Index < 0.3 for Danish streams and > 0.9 for comparable Uruguayan streams; [30]). The Danish stable streams have much higher contribution of groundwater to water flow (higher base flow index, Table 2).

Table 2. Main climatic and hydrological characteristics of the four benchmark catchments monitored at high frequency (nstreams = 4, 2-year period), showing annual accumulated rainfall in mm for each study year (sources: [a] [56], [b] [55]). Abbreviations: TEMP: temperate streams; SUBT: subtropical streams; Low and High: low and high-farming intensity.

Characteristic	TEMP Low	TEMP High	SUBT Low	SUBT High
Accumulated rainfall of each study year ($mm{\cdot}y^{-1}$)	756–770	766–778	1010–1030	1196–1405
Mean regional accumulated rainfall ($mm{\cdot}y^{-1}$)	765 [a]		1100–1200 [b]	
Base Flow Index (BFI)	0.88	0.64	0.39	0.29

3.2. Total Nitrogen Concentrations and Losses in Benchmark Streams

The farming intensity was, as expected, a strong determinant factor of stream TN concentrations, while climate/hydrology had no significant effects on the TN concentrations in the benchmark streams (Figure 2). This was expressed by low and not statistically different TN concentrations in the low-farming intensity streams (varying around 1.0 mg $N{\cdot}L^{-1}$) and significantly higher average annual TN concentrations in high-intensity farming streams (2.2 ± 1.4 mg $N{\cdot}L^{-1}$ and 4.3 ± 2.5 mg $N{\cdot}L^{-1}$ for the subtropical and temperate high-intensity farming streams, respectively; mean ± SD; Figure 2). Total N loss and flow-weighted concentrations of TN (TN-FWC) obtained through high frequency-automatized pooled sampling were also higher in the highest intensity farming streams for both climatic/hydrological conditions (annual estimations for two years, no statistical testing possible; Table 3).

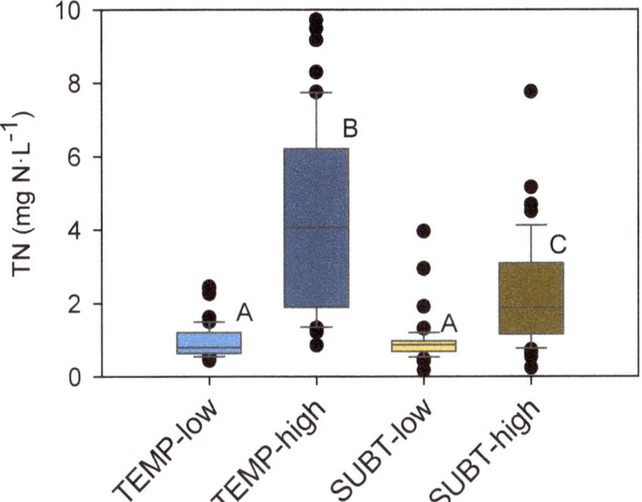

Figure 2. Variability in total nitrogen (TN) concentrations for each of the four fortnightly grabs sampled benchmark streams. Data correspond to a 2-year period. ANOVA results: F TEMP vs. SUBT (1,190) = 23.73, Farming intensity conditions nested in climate/hydrology (2,190) = 70.44, F Interaction (1,190) = 378.9. $P < 0.001$ for all cases. A, B, and C describe statistically similar groups according to Bonferroni post hoc tests. The upper and lower boundaries of the box mark the 25th and 75th percentile, whiskers above and below the box indicate the 90th and 10th percentiles and the line within the box marks the median. Black dots display outliers. Abbreviations: TEMP: temperate streams; SUBT: subtropical streams; low and high: low and high-farming intensity in the catchments.

Table 3. Total nitrogen losses by hectare (kg N·ha^{-1}·year^{-1}) and flow-weighted concentrations (FWC; mg N·L^{-1}) estimated annually using high frequency automatized pooled sampling of the four benchmark streams (water samples taken every 4 h and accumulated fortnightly). Abbreviations: TEMP: temperate streams; SUBT: subtropical streams; low and high: low and high-farming intensity in the catchments.

Region	Year	Low-Farming Intensity		High-Farming Intensity	
		TN Loss	FWC TN	TN Loss	FWC TN
SUBT	1	1.39	0.82	4.67	1.99
SUBT	2	2.12	0.72	9.17	2.13
TEMP	1	6.11	1.2	13.16	6.28
TEMP	2	4.68	0.98	12.65	6.23

Without trends associated with farming intensity, the total N lost from subtropical catchments was between 2% and 4% of the total annual inputs as fertilizers and manure (Tables 1 and 3). For the case of temperate catchments, the total N lost by the stream was 9% for high-intensity farming streams and between 10% and 14% (year 1 and 2 of monitoring) in low-intensity farming streams (Tables 1 and 3).

3.3. *Influence of Temperature and Discharge on Total Nitrogen Concentrations*

Stream TN concentrations tended to decrease with increasing water temperature and decreasing discharge, as reflected in the set of four fortnightly sampled benchmark streams (Figure 3). In both climates, the relationship between TN concentrations and discharge showed a higher explained variance for high-intensity farming than for low-intensity farming (Figure 3). The TN concentration of subtropical low-intensity farming stream, however, did not exhibit statistical relationships with temperature and discharge (Figure 3).

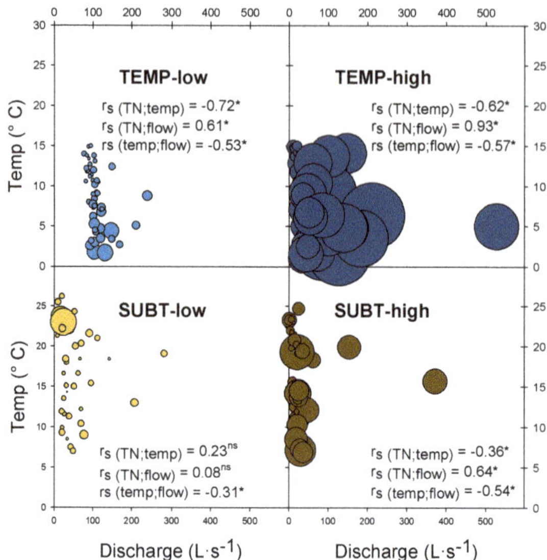

Figure 3. Total nitrogen concentrations vs. water temperature and discharge in the subset of the four benchmark streams under high-frequency monitoring. Bubble size represents the concentration of TN from fortnightly grab sampling (two-year data). The graphs show the main environmental gradients for each stream. Spearman rank-order correlations (rs) are marked with * when significant ($0.01 \leq p \leq 0.05$). ns: non-significant. Abbreviations: TEMP: temperate streams; SUBT: subtropical streams; low and high: low and high-farming intensity in the catchments.

3.4. Influence of Climate/Hydrology and Farming Intensity on Nitrogen Species

Total dissolved N (TDN) constituted the main fraction of TN in all the studied streams (Figures 2 and 4). It was also affected by farming intensity regardless of climate/hydrological conditions and monitoring method/sampling time (Table 4; Figure 4). These assertions are based on the comparable results obtained in all the sampled streams (n = 43), including both monitoring schemes: fortnightly grab sampling in benchmark streams and snapshot sampling (the analyses of this section include all these data sets; see Figure 4).

Relatively low (range of averages: 0.4 to 0.9 mg N·L^{-1}), and not statistically different average TDN concentrations were found in all low-intensity farming streams, intermediate concentrations were found in the subtropical-high intensity farming streams (1.2 to 1.8 mg N·L^{-1}), and the highest concentrations occurred in the temperate high-intensity farming streams (3.5 to 5.2 mg N·L^{-1}; Table 4; Figure 4 upper panels). The pattern of intermediate TDN concentrations in the subtropical high-intensity farming streams was similar for the two sampling strategies, the only significant difference with low-intensity farming streams occurring for the fortnightly samples (Table 4; Figure 4).

Nitrate (NO_3^-) concentrations resembled the above TN and TDN: low and not statistically different NO_3^- concentrations in the low-intensity farming streams, intermediate in the subtropical high-intensity farming streams, and the highest in the temperate high-intensity farming streams (complete data set; Table 4; Figure 4). Average annual NO_3^- concentrations ranged between values as low as 0.05 and 0.3 mg N·L^{-1} in the low-intensity farming streams in both climates (Figure 4). Average nitrate concentrations in the subtropical high-intensity farming streams ranged between 0.4 and 0.8 mg N·L^{-1}. In contrast, in the temperate high-intensity farming streams, nitrate varied between 3.2 and 4.9 mg N·L^{-1} on average, depending on the sampling method and season (Figure 4).

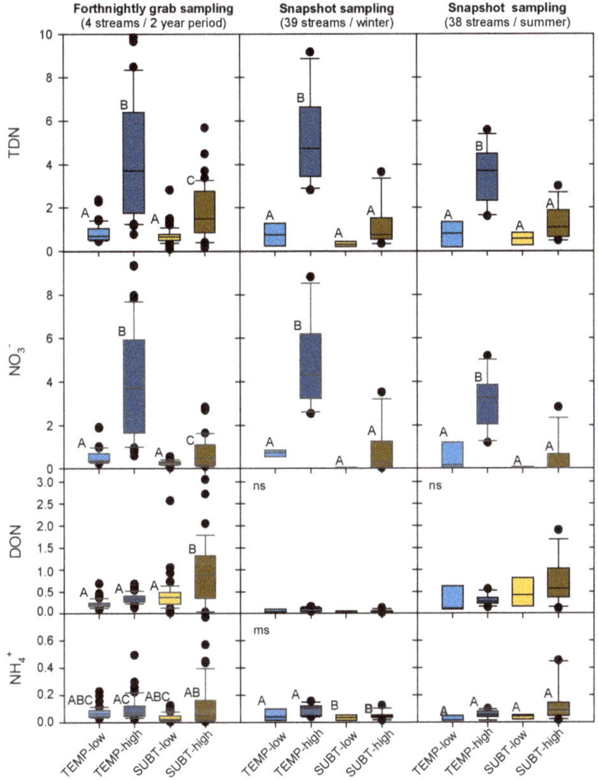

Figure 4. The concentration of dissolved nitrogen fractions in fortnightly grab sampling in benchmark streams and snapshot samples. Significance level P < 0.05. A, B, and C describe statistical groups according to post hoc Bonferroni tests. We indicate non-significant results as ns (p > 0.1), and marginally significant as ms (0.05 < P < 0.1). Abbreviations: TEMP: temperate streams; SUBT: subtropical streams; low- and high-FI: low and high-farming intensity. Vertical axes are concentrations expressed in mg $N{\cdot}L^{-1}$. The upper and lower boundaries of the box mark the 25th and 75th percentile, whiskers above and below the box indicate the 90th and 10th percentiles and the line within the box marks the median. Black dots display outliers. Note that the scale varies among fractions.

We found no significant association between DON concentrations and climate/hydrological conditions or farming intensity, except for significantly higher levels of DON in the subtropical high-intensity farming streams (benchmark streams only, average = 0.9 mg $N{\cdot}L^{-1}$; Figure 4). All the other streams representing both climatic conditions had average levels below 0.7 mg $N{\cdot}L^{-1}$ (Table 4; Figure 4).

No significant relationships were found between ammonium (NH_4^+) concentrations and climate/hydrological conditions and farming intensity (fortnightly grab sampling in benchmark streams and snapshot sampling in summer; Table 4; Figure 4). Average NH_4^+ concentrations were always < 0.1 mg $N{\cdot}L^{-1}$, regardless of climate and farming intensity. In the wintry snapshot sampling, average NH_4^+ concentrations were significantly higher in the temperate (average 0.08 mg $N{\cdot}L^{-1}$) than in the subtropical streams (0.04 mg $N{\cdot}L^{-1}$; Table 4; Figure 4).

The DIN/TDN ratio was higher in the temperate (average ranging from 0.5 to almost 1) than in the subtropical streams (average ranging from 0.2 to 0.5) for the fortnightly and snapshot sampled streams (Table 4; Figure 5). The temperate high-intensity farming streams exhibited the highest DIN/TDN ratios (average between 0.89 and 0.95), which was linked to the strong predominance of

NO_3^- (Figures 4 and 5). In general, higher variability in the DIN/TDN ratio was observed in the subtropical streams (SD ranging from 0.2 to 0.3; Figure 5) than in the temperate streams.

Table 4. Summary of 2-way nested ANOVA tests for the concentration of N forms, indicating the origin (sampling method) of the data. Above: Main effects of climate/hydrology conditions (two levels). Below: Main effects of farming intensity (two levels: low- and high-intensity farming) nested within climate/hydrology conditions and interaction between factors. F values and the respective degrees of freedom are indicated. Significance level: P > 0.1 ns, 0.05 < P < 0.1 ms, P < 0.05 *, P < 0.01 **, P < 0.001 ***. Results of post hoc pairwise multiple comparisons are shown in Figures 3–5. Abbreviations: TEMP: temperate streams; SUBT: subtropical streams; low and high: low- and high-farming intensity in the catchments.

N Form	Benchmark Streams	Snapshot Sampling	
	Fortnightly Sampling	**Winter**	**Summer**
	Comparison between Climate/Hydrology Conditions (TEMP vs. SUBT)		
TDN	F(1, 181) = 38.16 ***	F(1, 35) = 29.97 ***	F(1, 34) = 19.60 ***
NO_3^-	F(1, 186) = 78.25 ***	F(1, 35) = 29.99 ***	F(1, 34) = 30.58 ***
DON	F(1, 181) = 27.27 ***	F(1, 35) = 1.44, p = 0.24 ns	F(1, 34) = 4.90 *
NH_4^+	F(1, 186) = 0.19, p = 0.66 ns	F(1 35) = 7.72 **	F(1, 34) = 2.31, p = 0.14 ns
DIN/TDN	F(1, 181) = 179.67 ***	F(1, 35) = 30.06 ***	F(1, 34) = 23.67 ***
	Comparison between Farming Intensity Conditions (Low & High) Nested in Climate/Hydrology		
TDN	F(2, 181) = 71.85 ***	F(2, 35) = 30.30 ***	F(2, 34) = 20.78 ***
NO_3^-	F(2, 186) = 80.12 ***	F(2, 35) = 28.48 ***	F(2, 34) = 20.49 ***
DON	F(2, 181) = 24.42 ***	F(2, 35) = 0.232, p = 0.79 ns	F(2, 34) = 1.19, p = 0.31 ns
NH_4^+	F(2, 186) = 10.05 ***	F(2, 35) = 2.75, p = 0.08 ms	F(2, 34) = 3.09, p = 0.08 ms
DIN/TDN	F(2, 181) = 14.67 ***	F(2, 35) = 6.26 **	F(2, 34) = 5.50**
	Interaction between Farming Intensity and Climate/Hydrology		
TDN	F(1, 181) = 307.1 ***	F(1, 35) = 85.0 ***	F(1, 34) = 116.4 ***
NO_3^-	F(1, 186) = 193.6 ***	F(1, 35) = 57.5 ***	F(1, 34) = 52.33 ***
DON	F(1, 181) = 255.1 ***	F(1 35) = 103.0 ***	F(1, 34) = 59.9 ***
NH_4^+	F(1, 186) = 140.3 ***	F(1, 35) = 106.0 ***	F(1, 34) = 21.6 ***
DIN/TDN	F(1, 181) = 2744.1 ***	F(1, 35) = 299.3 ***	F(1, 34) = 132.9***

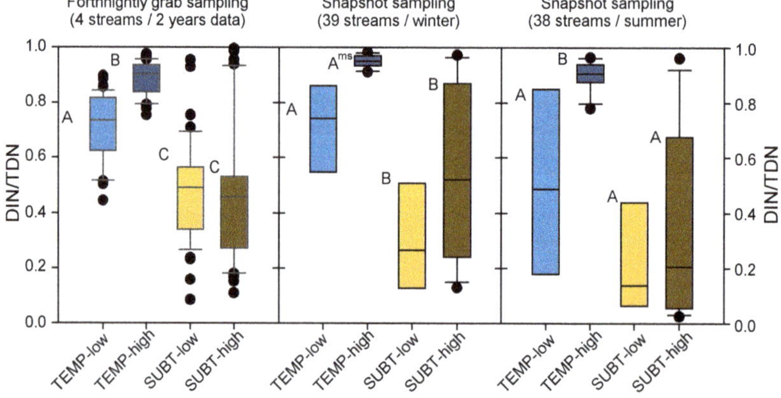

Figure 5. DIN/TDN ratio for the fortnightly grab sampling in benchmark streams and snapshot samples. Significance level P < 0.05. A, B, and C describe statistical groups according to post hoc Bonferroni tests. We indicate marginally significant as ms (0.05 < P < 0.1). The upper and lower boundaries of the box mark the 25th and 75th percentile, whiskers above and below the box indicate the 90th and 10th percentiles and the line within the box marks the median. Black dots display outliers. Note that the scale varies among fractions. Abbreviations: TEMP: temperate streams; SUBT: subtropical streams; low and high: low and high-farming intensity.

4. Discussion

4.1. Influence of Farming Intensity

Our analysis of streams draining microcatchments under low-intensity farming conditions in contrasting climatic-hydrological settings revealed low and quite comparable TN, TDN, NO_3^-, NH_4^+, and DON concentrations. In addition, NO_3^- concentrations in low-intensity farming streams at both climate/hydrological conditions exhibited levels that were considered as background concentrations in a recent and independent study conducted for streams draining relatively undisturbed catchments in Denmark and elsewhere [57]. No reference data for background concentrations in Uruguayan or subtropical streams have previously been reported in the scientific literature.

In contrast, the highest concentrations of NO_3^- were found in all sampled streams draining microcatchments impacted by high-intensity farming, irrespective of the monitoring method. Annual flow-weighted concentrations of TN never exceeded 1.2 mg $N{\cdot}L^{-1}$ in the two benchmark streams draining low intensity farmed catchments, but they were always higher than 2.0 mg $N{\cdot}L^{-1}$ in the two benchmark streams draining high intensity farmed catchments. The higher NO_3^- concentrations in the water (leading to higher concentrations of TDN and TN) can, therefore, be attributed to the impact of intensive farming in the catchments.

The TN, TDN, NO_3^-, and TN-FWC concentrations in the streams draining high-intensity farming catchments were significantly higher in the temperate climate, with stable discharge conditions than in the subtropical climate with flashy discharge conditions.

The N input to the catchments as fertilizer and manure, was higher in subtropical than in temperate catchments, particularly by the higher contribution of manure derived from higher livestock loads. Contrarily, the N loss/N input fraction was higher in temperate catchments (9% to 14%), respect to subtropical ones (2% to 4%). Further, more detailed studies must be done to establish if assumptions made about N content of manure for Uruguay are correct, or if our results could be biased by it. Notwithstanding, these results are consistent with the much longer history of intensive farming in central and northern Europe than in South America, creating a potentially high N legacy in the groundwater feeding the streams [58–60]. Moreover, the widespread use of artificial drainage practices via tile drains in Danish productive catchments is a shortcut pathway for nitrate from the soils to surface waters; thus avoiding attenuation processes in groundwater [61–64]. Accordingly, the streams draining the temperate high intensity farmed catchments were characterized by higher NO_3^- concentrations and a larger contribution of groundwater to the total flow measured (higher base flow index, BFI) than in similar subtropical streams.

The concentrations of NH_4^+ and DON in streams did not show any clear relationship with the analyzed environmental factors. The proportional contribution of DON to TDN in the subtropical streams was, however, higher than in the temperate streams, which might be explained by the moderate to low levels of NO_3^- observed in the subtropical streams.

The global use of N fertilizers increases steadily [65], and the trend is forecasted to continue for the next decades despite more efficient management practices [66]. In the absence of a significant increase in N use efficiency (or the introduction of other retention or mitigation measures), the expected farming intensification in the future will result in an increase in N concentrations and losses in streams, which will further exacerbate eutrophication in surface freshwater bodies.

4.2. Influence of Climate

Benchmark streams showed statistically significant relationships between N concentrations and water temperature (negative) and discharge (positive). The former may be linked with high temperature-driven denitrification and higher N assimilation by aquatic macrophytes and periphyton in summer [65–67]. The latter, in contrast, may be explained by the N legacy in groundwaters in the intensively farmed catchments together with the annual N surplus and hence diffuse N contributions from agriculture in the catchments (leading to higher NO_3^- and TN concentrations in streams with increasing discharges). The lack of a significant relationship between N concentrations and temperature and discharge in the subtropical stream draining the low intensity farmed catchment was probably caused by the extremely low or lack of N surplus and N legacy, together with high denitrification and biological N uptake promoted by the higher temperatures.

Our results suggest that in a stationary scenario of farming intensity and management, warming alone might promote a reduction of TN concentrations in lowland low-order streams driven mainly by a reduction in NO_3^- concentrations. In contrast, the predicted increase in annual precipitation and the enhanced intensity of precipitation events in both countries [67–69] will probably increase the risk of diffuse N losses to streams, at least in intensively farmed catchments. Consequently, given these contradictory trends, it is uncertain what will be the resulting impact of climate change on N concentrations and losses in lowland streams in different climates and agricultural systems [70]. Several model scenario studies of climate change effects on N cycling in catchments have, however, suggested increases in exported N in the temperate climate regions [71,72]. The sense of the changes that the water tables suffer (e.g., height, residence time) probably will be one of the most influential factors regarding the N loss towards the streams. If the increase in flow regime variability, flashiness, and enhanced evapotranspiration results in a decrease in the contribution of groundwater [12], a lowering of the NO_3^- concentration and DIN/TDN ratio could be expected. Nevertheless, if the increasing precipitations get more infiltration, higher groundwater tables, and longer periods with tile drain flow, the effects could be the contrary.

5. Conclusions

The results from our three complementary monitoring approaches were broadly comparable and support that farming intensity is of key importance for determining N concentrations and losses in lowland streams, despite differences in soil and climatic-hydrological conditions between and within regions.

Overall, farming intensity determines the concentrations of TN, TDN, and NO_3^-, flow-weighted TN, and TN exported to streams, but not those of ammonium (NH_4^+) and dissolved organic N (DON).

In the absence of a significant increase in farming N use efficiency (or the introduction of other palliative measures), the expected farming intensification will result in a stronger increase in NO_3^-, TDN, and TN concentrations as well as rising flow-weighted N concentrations and N losses in temperate and subtropical streams, further exacerbating eutrophication.

In contrast to our expectations, a high dissolved inorganic N (DIN) to TDN ratio was associated with temperate climate and high base flow conditions but not with farming intensity.

The consequences of changes in climate for the streams in the studied countries are hard to predict as higher temperatures and higher precipitation had contrasting effects on TN concentrations in our study.

Author Contributions: Conceptualization, B.K., E.J., M.M., and G.G.; data curation, G.G. and D.G.; formal analysis, G.G.; funding acquisition, B.K., E.J., and M.M.; investigation, all authors; project administration, M.M.; supervision, B.K., E.J., and M.M.; visualization, G.G.; writing—original draft preparation, G.G.; writing—review and editing, D.G., M.M., E.J., F.T.-d.M., N.V., N.M., and B.K. All authors have read and agreed to the published version of the manuscript.

Funding: The study was funded by the Danish Council for Independent Research, a grant by ANII-FCE (2009-2749) Uruguay, and the National L'Oreal-UNESCO Award for Women in Science-Uruguay with support from DICyT granted to MM. GG, FTM, MM, IGB, and NM received support from the SNI (Agencia Nacional de Investigación e Inovación, ANII, Uruguay). GG was supported by a Ph.D. scholarship from PEDECIBA. E.J. and D.G. were further supported by MARS (Managing Aquatic ecosystems and water Resources under multiple Stress) funded under the 7th EU Framework Programme, Theme 6 (Environment including Climate Change), Contract No.: 603378 (http://www.mars-project.eu). DG was also supported by a grant from the Danish Centre for Environment and Energy.

Acknowledgments: We thank the technicians of AU in Silkeborg for their work in the field and lab. In Uruguay, we are grateful for the field assistance from Iván González-Bergonzoni, Natalie Corrales, Anahí López-Rodríguez, Clementina Calvo, Gonzalo Moreno, Alfonsina López, Carlos Iglesias, Juan Clemente, Mariana Vianna, and Juan Pablo Pacheco, and also to Verónica Ciganda, Diego Michelini and Carlos Perdomo for their generosity sharing agronomic information. We thank especially Daniella Agrati and family and the landowners (Mendiverri & Laturre) in Uruguay.

Conflicts of Interest: The authors declare no conflict of interest. The funders had no role in the design of the study; in the collection, analyses, or interpretation of data; in the writing of the manuscript, or in the decision to publish the results.

References

1. Foley, J.A.; DeFries, R.; Asner, G.P.; Barford, C.; Bonan, G.; Carpenter, S.R.; Chapin, F.S.; Coe, M.T.; Daily, G.C.; Gibbs, H.K.; et al. Global consequences of land use. *Science* **2005**, *309*, 570–574. [CrossRef] [PubMed]
2. Gordon, L.J.; Peterson, G.D.; Bennet, E.M. Agricultural modifications of hydrological flows create ecological surprises. *Trends Ecol. Evol.* **2008**, *23*, 211–219. [CrossRef] [PubMed]
3. Erisman, J.W.; Sutton, M.A.; Galloway, J.; Klimont, Z.; Winiwarter, W. How a century of ammonia synthesis changed the world. *Nat. Geosci.* **2008**, *1*, 636–639. [CrossRef]
4. Gruber, N.; Galloway, J.N. An Earth-system perspective of the global nitrogen cycle. *Nature* **2008**, *451*, 293–296. [CrossRef]
5. Vitousek, P.M. Beyond global warming: Ecology and global change. *Ecology* **1994**, *75*, 1861–1876. [CrossRef]
6. Carpenter, S.R.; Caraco, N.F.; Correll, D.L.; Howarth, R.W.; Sharpley, A.N.; Smith, V.H. Nonpoint pollution of surface waters with phosphorus and nitrogen. *Ecol. Appl.* **1998**, *8*, 559–568. [CrossRef]
7. González-Sagrario, M.A.; Jeppesen, E.; Goma, J.; Søndergaard, M.; Jensen, J.; Lauridsen, T.L.; Lankildehus, F. Does high nitrogen loading prevent clear-water conditions in shallow lakes at moderately high phosphorus concentrations? *Freshw. Biol.* **2005**, *50*, 27–41. [CrossRef]
8. James, C.; Fisher, J.; Russell, V.; Collings, S.; Moss, B. Nitrate availability and plant species richness: Implications for management of freshwater lakes. *Freshw. Biol.* **2005**, *50*, 49–63. [CrossRef]
9. Steffen, W.; Richardson, K.; Rockström, J.; Cornell, S.E.; Fetzer, I.; Bennett, E.M.; Biggs, R.; Carpenter, S.R.; de Vries, W.; de Wit, C.A.; et al. Planetary boundaries: Guiding human development on a changing planet. *Science* **2015**, *347*, 1259855. [CrossRef]
10. Bormann, F.H.; Likens, G.E. Nutrient cycling. *Science* **1967**, *155*, 424–428. [CrossRef]
11. Boulêtreau, S.; Salvo, E.; Lyautey, E.; Mastrorillo, S.; Garabetian, F. Temperature dependence of denitrification in phototrophic river biofilms. *Sci. Total Environ.* **2012**, *416*, 323–328. [CrossRef] [PubMed]
12. Deelstra, J.; Iital, A.; Povilaitis, A.; Kyllmar, K.; Greipsland, I.; Blicher-Mathiesen, G.; Jansons, V.; Koskiaho, J.; Lagzdins, A. Hydrological pathways and nitrogen runoff in agricultural dominated catchments in Nordic and Baltic countries. *Agric. Ecosyst. Environ.* **2014**, *195*, 211–219. [CrossRef]
13. Exner-Kittridge, M.; Strauss, P.; Blöschl, G.; Eder, A.; Saracevic, E.; Zessner, M. The seasonal dynamics of the stream sources and input flow paths of water and nitrogen of an Austrian headwater agricultural catchment. *Sci. Total Environ.* **2016**, *542 Pt A*, 935–945. [CrossRef] [PubMed]
14. Durand, P.; Breuer, L.; Johnes, P.J.; Billen, G.; Butturini, A.; Pinay, G.; van Grinsven, H.; Garnier, J.; Rivett, M.; Reay, D.S.; et al. Nitrogen processes in aquatic ecosystems. In *The European Nitrogen Assessment—Sources, Effects and Policy Perspectives*; Sutton, M.A., Howard, C.M., Erisman, J.W., Billen, G., Bleeker, A., Grennfelt, P., Eds.; Cambridge University Press: Cambridge, UK, 2011; p. 664.
15. Huang, J.C.; Lee, T.Y.; Lin, T.C.; Hein, T.; Lee, L.C.; Shih, Y.T.; Kao, S.J.; Shiah, F.K.; Lin, N.H. Effects of different N sources on riverine DIN export and retention in a subtropical high-standing island, Taiwan. *Biogeosciences* **2016**, *13*, 1787–1800. [CrossRef]
16. Sharpley, A.N.; Smith, S.J.; Naney, J.W. Environmental impact of agricultural nitrogen and phosphorus use. *J. Agric. Food Chem.* **1987**, *35*, 812–817. [CrossRef]
17. Alexandratos, N.; Bruinsma, J. *World Agriculture Towards 2030/2050: The 2012 Revision*; FAO: Rome, Italy, 2012; Volume 12-03, p. 153.
18. Lambin, E.F.; Geist, H.J.; Lepers, E. Dynamics of land-use and land-cover change in tropical regions. *Annu. Rev. Environ. Res.* **2003**, *28*, 205–241. [CrossRef]
19. Potter, P.; Ramankutty, N.; Bennett, E.M.; Donner, S.D. Characterizing the spatial patterns of global fertilizer application and manure production. *Earth Interact.* **2010**, *14*, 1–22. [CrossRef]
20. Potter, P.; Ramankutty, N.; Bennett, E.M.; Donner, S.D. *Global Fertilizer and Manure, Version 1: Nitrogen Fertilizer Application*; NASA Socioeconomic Data and Applications Center (SEDAC): Palisades, NY, USA, 2011.

21. Allan, J.D.; Castillo, M.M. *Stream Ecology: Structure and Function of Running Waters*; Springer: Dordrecht, The Netherlands, 2007; p. 436.
22. Dudgeon, D. *Tropical Stream Ecology*; Academic Press/Elsevier: London, UK, 2008; p. 324.
23. Bustamante, M.M.C.; Martinelli, L.A.; Pérez, T.; Rasse, R.; Ometto, J.P.; Pacheco, F.S.; Lins, S.R.M.; Marquina, S. Nitrogen management challenges in major watersheds of South America. *Environ. Res. Lett.* **2015**, *10*, 065007. [CrossRef]
24. Gücker, B.; Silva, R.C.S.; Graeber, D.; Monteiro, J.A.F.; Brookshire, E.N.J.; Chaves, R.C.; Boëchat, I.G. Dissolved nutrient exports from natural and human-impacted Neotropical catchments. *Glob. Ecol. Biogeogr.* **2016**, *25*, 378–390. [CrossRef]
25. Follett, R. Chapter 2. Transformation and transport processes of nitrogen in agricultural systems. In *Nitrogen in the Environment: Sources, Problems, and Management*, 2nd ed.; Hatfield, J.L., Follett, R.F., Eds.; Academic Press/Elsevier: San Diego, CA, USA, 2008; pp. 19–50.
26. Phalan, B.; Bertzky, M.; Butchart, S.H.M.; Donald, P.F.; Scharlemann, J.P.W.; Stattersfield, A.J.; Balmford, A. Crop expansion and conservation priorities in tropical countries. *PLoS ONE* **2013**, *8*, e51759. [CrossRef]
27. Laurance, W.F.; Sayer, J.; Cassman, K.G. Agricultural expansion and its impacts on tropical nature. *Trends Ecol. Evol.* **2014**, *29*, 107–116. [CrossRef] [PubMed]
28. Schindler, D.W. Replication versus realism: The need for ecosystem-scale experiments. *Ecosystems* **1998**, *1*, 323–334. [CrossRef]
29. Peel, M.C.; Finlayson, B.L.; McMahon, T.A. Updated world map of the Köppen-Geiger climate classification. *Hydrol. Earth Syst. Sci.* **2007**, *11*, 1633–1644. [CrossRef]
30. Goyenola, G.; Meerhoff, M.; Teixeira-de Mello, F.; González-Bergonzoni, I.; Graeber, D.; Fosalba, C.; Vidal, N.; Mazzeo, N.; Ovesen, N.B.; Jeppesen, E.; et al. Monitoring strategies of stream phosphorus under contrasting climate-driven flow regimes. *Hydrol. Earth Syst. Sci.* **2015**, *19*, 4099–4111. [CrossRef]
31. Graeber, D.; Goyenola, G.; Meerhoff, M.; Zwirnmann, E.; Ovesen, N.B.; Glendell, M.; Gelbrecht, J.; Teixeira de Mello, F.; González-Bergonzoni, I.; Jeppesen, E.; et al. Interacting effects of climate and agriculture on fluvial DOM in temperate and subtropical catchments. *Hydrol. Earth Syst. Sci.* **2015**, *19*, 2377–2394. [CrossRef]
32. Blicher-Mathiesen, G.; Rasmussen, A.; Grant, R.; Jensen, P.G.; Hansen, B.; Thorling, L. *Landovervågningsoplande 2012. NOVANA*; Aarhus Universitet, DCE—Nationalt Center for Miljø og Energi, 2013; p. 154, Videnskabelig Rapport fra DCE—Nationalt Center for Miljø og Energi nr. 174; Available online: http://dce152.au.dk/pub/SR74.pdf (accessed on 1 April 2020).
33. Ciganda, V. *Instituto Nacional de Investigación Agropecuaria, La Estanzuela*; Personal Communication: Colonia, Uruguay, 2020.
34. Brown, M.T.; Vivas, M.B. Landscape development intensity index. *Environ. Monit. Assess.* **2005**, *101*, 289–309. [CrossRef]
35. Blüthgen, N.; Dormann, C.F.; Prati, D.; Klaus, V.H.; Kleinebecker, T.; Hölzel, N.; Alt, F.; Boch, S.; Gockel, S.; Hemp, A.; et al. A quantitative index of land-use intensity in grasslands: Integrating mowing, grazing and fertilization. *Basic Appl. Ecol.* **2012**, *13*, 207–220. [CrossRef]
36. Allaby, M. *Grasslands*; Chelsea House: New York, NY, USA, 2006; p. 270.
37. Valderrama, J.C. The simultaneous analysis of total nitrogen and total phosphorus in natural waters. *Mar. Chem.* **1981**, *10*, 109–122. [CrossRef]
38. Müller, R.; Weidemann, O. Die bestimmung des Nitrat-ions in wasser. *Wasser* **1955**, *22*, 247.
39. Monteiro, M.I.C.; Ferreira, F.N.; de Oliveira, N.M.M.; Avila, A.K. Simplified version of the sodium salicylate method for analysis of nitrate in drinking waters. *Anal. Chim. Acta* **2003**, *477*, 125–129. [CrossRef]
40. ISO13395. Water Quality—Determination of Nitrite Nitrogen and Nitrate Nitrogen and the Sum of Both by Flow Analysis (CFA and FIA) and Spectrometric Detection. Available online: http://www.iso.org/iso/catalogue_detail.htm?csnumber=21870 (accessed on 26 April 2016).
41. Zwirnmann, E.; Krüger, A.; Gelbrecht, J. *Analytik im Zentralen Chemielabor des IGB, Berichte des IGB, Heft 9*; Institut Für Gewa¨Ssero¨Kologie und Binnenfischerei: Berlin, Germany, 1999.
42. Lee, W.; Westerhoff, P. Dissolved organic nitrogen measurement using dialysis pretreatment. *Environ. Sci. Technol.* **2005**, *39*, 879–884. [CrossRef] [PubMed]

43. Graeber, D.; Gelbrecht, J.; Kronvang, B.; Gücker, B.; Pusch, M.T.; Zwirnmann, E. Technical Note: Comparison between a direct and the standard, indirect method for dissolved organic nitrogen determination in freshwater environments with high dissolved inorganic nitrogen concentrations. *Biogeosciences* **2012**, *9*, 4873–4884. [CrossRef]
44. Huber, S.; Balz, A.; Abert, M.; Pronk, W. Characterisation of aquatic humic and non-humic matter with size-exclusion chromatography—Organic carbon detection—Organic nitrogen detection (LC-OCD-OND). *Water Res.* **2011**, *45*, 879–885. [CrossRef] [PubMed]
45. Hudson, N.; Baker, A.; Reynolds, D.; Carliell-Marquet, C.; Ward, D. Changes in freshwater organic matter fluorescence intensity with freezing/ thawing and dehydration/rehydration. *J. Geophys. Res. Biogeosci.* **2009**, *114*, G00F08. [CrossRef]
46. Koroleff, F. *Direct Determination of Ammonia in Natural Water as Indophenol-Blue*; ICES: Copenhagen, Denmark, 1970; pp. 19–22.
47. Arnold, J.G.; Allen, P.M.; Muttiah, R.; Bernhardt, G. Automated base flow separation and recession analysis techniques. *Ground Water* **1995**, *33*, 1010–1018. [CrossRef]
48. Kronvang, B.; Bruhn, A.J. Choice of sampling strategy and estimation method for calculating nitrogen and phosphorus transport in small lowland streams. *Hydrol. Process.* **1996**, *10*, 1483–1501. [CrossRef]
49. Smith, S.V.; Swaney, D.P.; Buddemeier, R.W.; Scarsbrook, M.R.; Weatherhead, M.A.; Humborg, C.; Eriksson, H.; Hannerz, F. River nutrient loads and catchment size. *Biogeochemistry* **2005**, *75*, 83–107. [CrossRef]
50. Jones, A.S.; Horsburgh, J.S.; Mesner, N.O.; Ryel, R.J.; Stevens, D.K. Influence of sampling frequency on estimation of annual total phosphorus and total suspended solids loads. *J. Am. Water Resour. Assoc.* **2012**, *48*, 1258–1275. [CrossRef]
51. Bonferroni, C.E. Teoria statistica delle classi e calcolo delle probabilità. *Pubbl. Ist. Super. Sci. Econ. Commer. Firenze* **1936**, *8*, 3–62.
52. Hansen, J.; Ruedy, R.; Sato, M.; Lo, K. Global surface temperature change. *Rev. Geophys.* **2010**, *48*, RG4004. [CrossRef]
53. GISTEMP Team. GISS Surface Temperature Analysis (GISTEMP). Available online: http://data.giss.nasa.gov/gistemp/ (accessed on 16 May 2016).
54. DMI. Decadal Mean Weather. Available online: http://www.dmi.dk/en/vejr/arkiver/decadal-mean-weather/ (accessed on 26 April 2016).
55. INUMET. Cilimatological Statistics. Available online: http://www.meteorologia.com.uy/ServCli/caracteristicasClimaticas (accessed on 26 April 2016).
56. DMI. Denmark's Future Climate. Available online: https://www.dmi.dk/en/klima/fremtidens-klima/danmark-og-groenland/ (accessed on 4 November 2016).
57. Kronvang, B.; Windolf, J.; Larsen, S.E.; Bøgestrand, J. Background concentrations and loadings of nitrogen in Danish surface waters. *Acta Agric. Scand. B Soil Plant* **2015**, *65*, 155–163. [CrossRef]
58. Hansen, B.; Dalgaard, T.; Thorling, L.; Sørensen, B.; Erlandsen, M. Regional analysis of groundwater nitrate concentrations and trends in Denmark in regard to agricultural influence. *Biogeosciences* **2012**, *9*, 3277–3286. [CrossRef]
59. McLauchlan, K. The nature and longevity of agricultural impacts on soil carbon and nutrients: A review. *Ecosystems* **2006**, *9*, 1364–1382. [CrossRef]
60. Tesoriero, A.J.; Duff, J.H.; Saad, D.A.; Spahr, N.E.; Wolock, D.M. Vulnerability of streams to legacy nitrate sources. *Environ. Sci. Technol.* **2013**, *47*, 3623–3629. [CrossRef]
61. Grant, R.; Laubel, A.; Kronvang, B.; Andersen, H.E.; Svendsen, L.M.; Fuglsang, A. Loss of dissolved and particulate phosphorus from arable catchments by subsurface drainage. *Water Res.* **1996**, *30*, 2633–2642. [CrossRef]
62. Andersen, H.E.; Windolf, J.; Kronvang, B. Leaching of dissolved phosphorus from tile-drained agricultural areas. *Water Sci. Technol.* **2016**, *73*, 2953–2958. [CrossRef]
63. Schoumans, O.F.; Chardon, W.J.; Bechmann, M.E.; Gascuel-Odoux, C.; Hofman, G.; Kronvang, B.; Rubæk, G.H.; Ulén, B.; Dorioz, J.M. Mitigation options to reduce phosphorus losses from the agricultural sector and improve surface water quality: A review. *Sci. Total Environ.* **2014**, *468–469*, 1255–1266. [CrossRef]
64. Windolf, J.; Thodsen, H.; Troldborg, L.; Larsen, S.E.; Bøgestrand, J.; Ovesen, N.B.; Kronvang, B. A distributed modelling system for simulation of monthly runoff and nitrogen sources, loads and sinks for ungauged catchments in Denmark. *J. Environ. Monit.* **2011**, *13*, 2645–2658. [CrossRef]

65. FAO. *World Fertilizer Trends and Outlook to 2018*; FAO: Rome, Italy, 2015; p. 53.
66. Tilman, D.; Balzer, C.; Hill, J.; Befort, B.L. Global food demand and the sustainable intensification of agriculture. *Proc. Natl. Acad. Sci. USA* **2011**, *108*, 20260–20264. [CrossRef]
67. CDKN. The IPCC's Fifth Assessment Report What's in it for Latin America? Executive Summary. Available online: https://cdkn.org/wp-content/uploads/2014/11/IPCC-AR5-Whats-in-it-for-Latin-America.pdf (accessed on 1 April 2020).
68. Madsen, H.; Arnbjerg-Nielsen, K.; Steen-Mikkelsen, P. Update of regional intensity–duration–frequency curves in Denmark: Tendency towards increased storm intensities. *Atmos. Res.* **2009**, *92*, 343–349. [CrossRef]
69. Marengo, J.; Rusticucci, M.; Penalba, O.; Renom, M.; Laborbe, R. An intercomparison of observed and simulated extreme rainfall and temperature events during the last half of the twentieth century: Part 2: Historical trends. *Clim. Chang.* **2010**, *98*, 509–529. [CrossRef]
70. Olesen, J.E.; Carter, T.R.; Díaz-Ambrona, C.H.; Fronzek, S.; Heidmann, T.; Hickler, T.; Holt, T.; Minguez, M.I.; Morales, P.; Palutikof, J.P.; et al. Uncertainties in projected impacts of climate change on European agriculture and terrestrial ecosystems based on scenarios from regional climate models. *Clim. Chang.* **2007**, *81*, 123–143. [CrossRef]
71. Andersen, H.E.; Kronvang, B.; Larsen, S.E.; Hoffmann, C.C.; Jensen, T.S.; Rasmussen, E.K. Climate-change impacts on hydrology and nutrients in a Danish lowland river basin. *Sci. Total Environ.* **2006**, *365*, 223–237. [CrossRef] [PubMed]
72. Jeppesen, E.; Kronvang, B.; Olesen, J.; Audet, J.; Søndergaard, M.; Hoffmann, C.; Andersen, H.; Lauridsen, T.; Liboriussen, L.; Larsen, S.; et al. Climate change effects on nitrogen loading from cultivated catchments in Europe: Implications for nitrogen retention, ecological state of lakes and adaptation. *Hydrobiologia* **2011**, *663*, 1–21. [CrossRef]

Article

Spatiotemporal Dynamics of Nitrogen Transport in the Qiandao Lake Basin, a Large Hilly Monsoon Basin of Southeastern China

Dongqiang Chen [1,2], Hengpeng Li [2,*], Wangshou Zhang [2], Steven G. Pueppke [3,4], Jiaping Pang [2] and Yaqin Diao [1,2]

1 University of Chinese Academy of Sciences, Beijing 100049, China; dqchen@niglas.ac.cn (D.C.); yqdiao@niglas.ac.cn (Y.D.)
2 Key Laboratory of Watershed Geographic Sciences, Nanjing Institute of Geography and Limnology, Chinese Academy of Sciences, Nanjing 210008, China; wszhang@niglas.ac.cn (W.Z.); pangjp@niglas.ac.cn (J.P.)
3 Center for Global Change and Earth Observations, Michigan State University, 1405 South Harrison Road, East Lansing, MI 48823, USA; pueppke@msu.edu
4 Asia Hub, Nanjing Agricultural University, Nanjing 210095, China
* Correspondence: hpli@niglas.ac.cn; Tel.: +86-025-87714759

Received: 17 March 2020; Accepted: 4 April 2020; Published: 9 April 2020

Abstract: The Qiandao Lake Basin (QLB), which occupies low hilly terrain in the monsoon region of southeastern China, is facing serious environmental challenges due to human activities and climate change. Here, we investigated source attribution, transport processes, and the spatiotemporal dynamics of nitrogen (N) movement in the QLB using the Soil and Water Assessment Tool (SWAT), a physical-based model. The goal was to generate key localized vegetative parameters and agronomic variables to serve as credible information on N sources and as a reference for basin management. The simulation indicated that the basin's annual average total nitrogen (TN) load between 2007 and 2016 was 11,474 tons. Steep slopes with low vegetation coverage significantly influenced the spatiotemporal distribution of N and its transport process. Monthly average TN loads peaked in June due to intensive fertilization of tea plantations and other agricultural areas and then dropped rapidly in July. Subsurface flow is the key transport pathway, with approximately 70% of N loads originating within Anhui Province, which occupies just 58% of the basin area. The TN yields of sub-basins vary considerably and have strong spatial effects on incremental loads entering the basin' major stream, the Xin'anjiang River. The largest contributor to N loads was domestic sewage (21.8%), followed by livestock production (20.8%), cropland (18.6%), tea land (15.5%), forest land (10.9%), atmospheric deposition (5.6%), orchards (4.6%), industry (1.4%), and other land (0.8%). Our simulation underscores the urgency of increasing the efficiency of the wastewater treatment, conserving slope land, and optimizing agricultural management as components of a comprehensive policy to control N pollution in the basin.

Keywords: N sources; spatiotemporal patterns of pollution; N transport processes; Qiandao Lake Basin; fertilization

1. Introduction

The nitrogen (N) cycle is dynamic and strongly influenced by both human activities and physical conditions, especially in environmentally sensitive areas that are prone to nutrient pollution [1,2]. Agricultural practices and domestic sewage account for a substantial proportion of released N, which subsequently enters waterways via pathways that are diverse and poorly understood. Climatic features and geomorphological factors, including intense rainfall, steep slopes, and the presence of easily erodible soils, exacerbate N emission, transport, and export, adding complexity to the problem of pollution control [1–3]. Against this backdrop, access to comprehensive information about spatiotemporal patterns of N distribution becomes a primary concern for pollution management.

Physical-based models are powerful tools to provide detailed information about the key driving factors of N release and transport. Available models include the Hydrological Simulation Program—FORTRAN [4], the Spatially Referenced Regressions On Basin Attributes model [5], the Regional Nutrient Management model [6], the Annualized Agricultural Nonpoint Source model [7], and the Soil and Water Assessment Tool (SWAT) [8]. As confirmed by many studies, SWAT has performed particularly well for nutrient source analysis and the interpretation of transport processes in spatially heterogeneous areas of China [9,10] and elsewhere where agriculture predominates [11,12]. Moreover, SWAT can comprehensively estimate the attribution patterns of nutrients from various diffuse pollution sources [13,14]. Previous applications of SWAT have mainly focused on basins in semi-humid or semiarid zones [15,16], flat regions [17,18], cropland-dominated areas [19,20], and other environments [21]. Much less is known about the ability of this model to provide information on runoff and nutrient cycles in hilly monsoon areas characterized by diverse land use types and the absence of systematic information needed for science-based policy making [2,22].

We address these issues here by employing SWAT to precisely simulate N loads in one of the largest basins of southeastern China, the Qiandao Lake Basin (QLB). This basin, which lies in a typical low hilly area, experiences a monsoon climate and is a critical source of drinking water for Hangzhou, a prefecture-level city with 10 million residents. Total nitrogen (TN) concentration at the basin's outlet increased from approximately 0.75 to 1.12 mg/L during the past 15 years and now exceeds the target concentration (< 1.0 mg/L) for use as a drinking water source [23,24]. Elevated levels of TN have been attributed to socio-economic development, increasing environmental pressure from rapid expansion of tea plantations, and domestic sewage discharge, all of which have adversely affected water quality in this region over the past decade [22,25–27]. Agriculture is especially important, because fertilizer application rates in the basin, which average 436 kg/ha, are excessive compared to the rest of the world—triple those in the United States and double those in Europe and Japan [28]. Moreover, intensive application of fertilizer occurs from March to May, just before the onset of the rainy season, when erosion is most likely [29,30].

In this study, we applied the SWAT model to the QLB to (i) simulate, estimate, and analyze key localized TN parameters including factors influencing loads, source appointment, and spatiotemporal dynamics, (ii) determine how TN delivery and transport are influenced by complex terrain and monsoon climate, and, (iii) inform a more efficient and targeted strategy for controlling nutrient pollution based on a comprehensive understanding of attribution and spatiotemporal patterns of TN.

2. Materials and Methods

2.1. Study Area

The QLB (29°11′ N–30°02′ N, 118°21′ E–119°20′ E) is centered on Qiandao Lake, which was impounded by a hydroelectric dam constructed in the late 1950s [31]. The basin covers an area of 10,369 km^2, of which 4341 km^2 is in Zhejiang Province and 6028 km^2 is in Anhui Province (Figure 1). The main land use types in the basin are forest land (79.4%), followed by orchards (1.4%), cropland (8.2%), tea land (3.8%), water body (5.4%), and other land (1.8%). The annual average precipitation within the Zhejiang and Anhui portions of the basin is 1498 mm and 1712 mm, respectively. The Xin'anjiang River is the longest and most significant river pathway of the basin. It flows in an eastwardly direction, crossing from Anhui Province to Zhejiang province at Jiekou (location 243.7 km, Figure 1), and proceeding to the basin's outlet at Baqian (location 320.4 km, Figure 1), which lies just above the dam. The middle and lower reaches of the river below Baqian are known locally as the Fuchun River and the Qiantang River; ultimately, the water empties into the East China Sea.

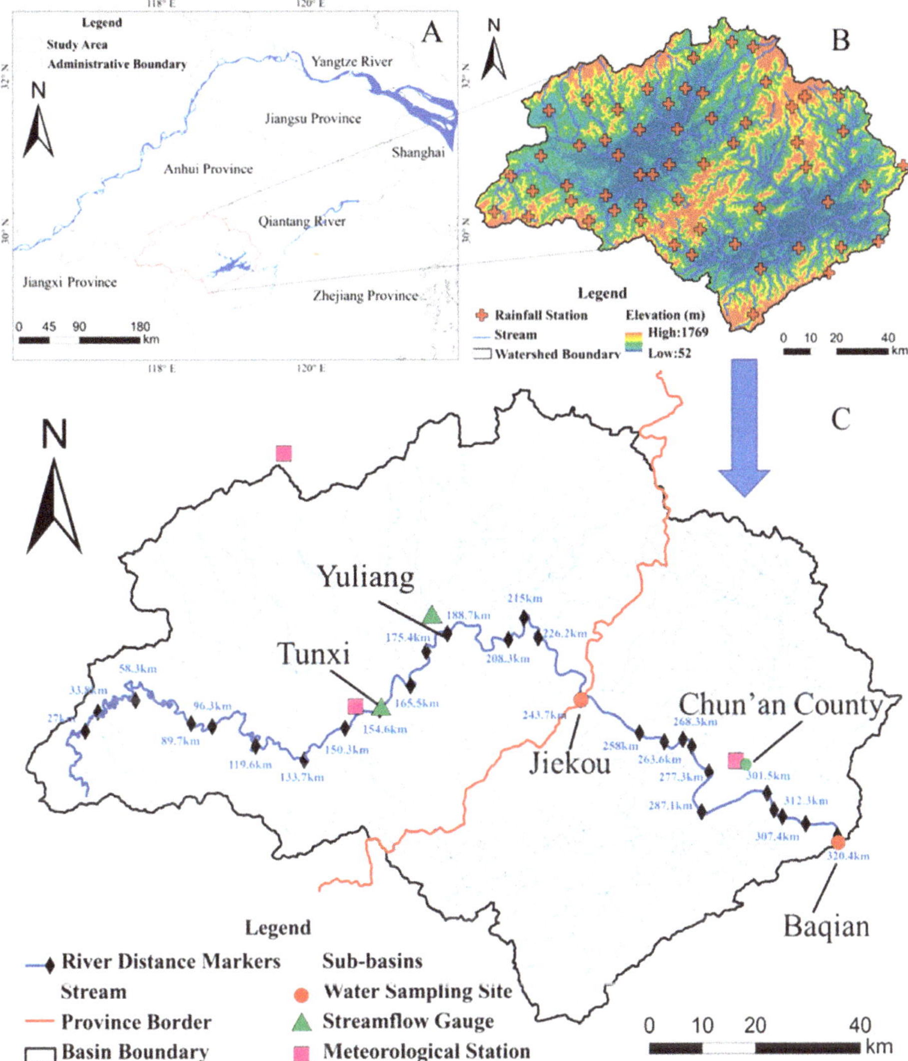

Figure 1. The Qiandao Lake Basin (QLB). (**A**) The Yangtze River delta region showing the orientation of the basin in Anhui and Zhejiang provinces. (**B**) Topography of the basin showing elevations and the location of rainfall stations. (**C**) Detailed map of the basin showing streams, sub-basins, important monitoring sites, and distance markers along the reach of the Xin'anjiang River. The location of the urban core of Chun'an County is also given.

2.2. Data Collection

As shown in Figure 1, the entire QLB was delineated into 96 sub-basins based on the 25-meter digital elevation model. High resolution, 2.1-meter land use data were derived from ZY-3 satellite imagery for 2015, which was purchased from the Chinese Geographical Monitoring Cloud Platform (http://www.dsac.cn/Data/Product/Detail/1009). Whenever possible, e.g., for forest land and built land, default parameters from the land use database were used in SWAT. eCognition Developer, an object-based tool for automatic analysis of remotely sensed data (http://www.ecognition.com/suite/ecognition-developer), was employed to identify land use types that are typical for the area, e.g., tea land and orchard land. Multiresolution segmentation was adopted for this purpose, with scope of classification algorithms from sample-based nearest neighbor analysis. Every imagery object was checked by visual interpretation, with verification by field observations at 100 separate sites in the basin. The final land use classification map is shown in Figure S1 (see Supplementary Materials for Figure S1).

Soil types and their physical properties were extracted from databases of the Nanjing Institute of Soil Science and, when necessary, verified by field sampling to generate the map shown in Figure S2 (see Supplementary Materials for Figure S2). Distributed rainfall data were collected from 54 rainfall gauges within the QLB (Figure 1). Daily streamflow data from two hydrological stations (Tunxi and Yuliang, see Figure 1) were obtained from the Huangshan Hydrology Bureau and published in Chinese Hydrological Yearbooks [32]. Monthly water quality data were from records of the Chun'an Environmental Protection Bureau. TN data, expressed as the sum of different forms of nitrogen including ammonia, nitrate, nitrite and organic N, were measured according to the National Surface Water Quality Standard (GB 3838-2002) and obtained from the Chun'an Environmental Protection Bureau, which also supplied precipitation and dry deposition data used to determine atmospheric deposition of TN.

The Chun'an Environmental Protection Bureau and Chun'an County Statistical Yearbooks [33,34] provided additional information for the model, including the locations of industrial and domestic wastewater outlets and discharge from them. The spatial distribution of livestock production facilities, their scales of production, and discharge from them was based on governmental monitoring of wastewater streams. Supplemental Information about cropping systems, agronomic practices, management decisions, and rural sewage treatment was obtained from detailed governmental interviews with 487 farmers from 44 villages in the study region. Sewage in the QLB is collected into centralized treatment facilities that discharge wastewater into nearby rivers, and thus for the purpose of SWAT modeling, domestic, industrial, and animal agricultural pollution sources could be generalized as point sources that discharge into point outlets.

2.3. SWAT Model Configuration

Soil data (Figure S2) were transformed into a modeling format by means of standardized definition and interpretation of SWAT parameters, and discharge and associated TN loads from domestic, industrial, and animal agriculture from each sub-basin were configured into the model as described above. Source attribution of atmospheric N deposition was generalized as the single direct source of pollution into water bodies as follows: deposition of ammonium (0.97 mg/L), deposition of nitrate (0.73 mg/L), dry deposition of ammonium (0.434 kg/ha/yr), and dry deposition of nitrate (0.61 kg/ha/yr).

Information from field surveys of agricultural practices in the QLB was used to guide the scheduling of cropland, tea land, and orchard management for the model. Cropland and orchards were operated with heat unit scheduling, but because of its unique growth and management characteristics, tea land was operated with date scheduling in Table S1 (see Supplementary Materials for Table S1). Localized parameters for tea land were optimized on the basis of multi-year field measurements during the growth season, as well as information obtained in similar areas of southeastern China, and then manually assembled into the final input suite for the SWAT database (Table S1).

The previously used SUFI-2 algorithm was employed to analyze the sensitivity, calibration, validation, and uncertainty of the SWAT model [35,36]. Three statistical indicators were utilized to evaluate the model's performance during calibration and validation: the coefficient of determination (R^2), which is an indicator of the goodness of fit between the observed and simulated results; the Nash–Sutcliffe simulation efficiency (ENS), which determines the ability of the model to predict the 1:1 line of correspondence between the observed and simulated values; and PBIAS (Percent bias), which measures the average tendency of simulated values to be larger or smaller than corresponding observed values [37,38]. Previously defined metrics of model performance were adopted. These assign satisfactory, good, and very good performance to ENS values of 0.5, 0.65, and 0.75, and R^2 values of 0.6, 0.7, and 0.8, respectively [39,40]. PBIAS values of ±25% and ±25% ≤ ±40% are considered to indicate very good and good model performance for N, respectively. The corresponding values for streamflow are ±10% and ±10% ≤ ±15%, respectively.

The SWAT model was used for simulations during the interval from 2003 to 2016. The first four-year interval, from 2003 to 2006, was employed as a warm-up period to allow the streamflow variables to reach a limited range of initial values (see Table S2 for details in Supplementary Materials). Observed daily data from 2007 to 2012 were subsequently used for calibration and those from 2013 to 2016 for validation. N load simulations at the two sampling sites, e.g., Jiekou and Baqian, were executed simultaneously, and iterations of the entire model were performed until satisfactory results were obtained.

3. Results

3.1. Calibration and Sensitivity Analysis of the SWAT Model

Parameters such as cropping systems, fertilization protocols, and soil conditions can vary significantly in different basins, with substantial impacts on modeling outcomes [15,18,41]. Appropriate calibration with parameters that are optimized for the QLB is consequently important, especially for tea, which is intensively managed, known to have undergone considerable expansion during the study period, and recognized to be an important source of nutrient pollution [42,43]. Twenty-one optimized SWAT parameters for tea land in the QLB are given in Table S2. Three crucial parameters (Table 1) diverge significantly from those provided by the SWAT model for forest land and land used for cultivation of other crops. This indicates that both maximum root depth and harvest index are comparatively low for tea, but demand for N in seed is relatively high.

Table 1. Comparison of key Soil and Water Assessment Tool (SWAT) parameter values derived for tea with those available for other crops grown in the QLB.

Parameter	Definition	Cropland	Orchard	Forest Land	Tea Land
HVSTI	Harvest index for land cover/plant [(kg/ha)/kg/ha)]	0.45	0.10	0.76	0.07
RDMX	Maximum root depth for land cover/plant (m)	2.0	2.0	3.5	0.5
CNYLD	Normal fraction of nitrogen in seed for land cover/plant (kg N/kg seed)	0.0199	0.0019	0.0015	0.0246

The most important and sensitive parameters during the calibration of the SWAT model were groundwater (.gw), soil (.sol), basins (.bsn), and management (.mgt). In agreement with previous studies, the moisture condition II curve of the soil conservation service (SCS) curve number method (CN2) was the most sensitive parameter for streamflow [18,36]. This was also the key parameter for runoff yield, as would be expected in a basin characterized by a hilly terrain, agricultural land, and abundant rainfall. Other important parameters representing land-cover features were OV_N, CANMX, HRU_SLP, GWQMN, SOL_AWC, TRNSRCH, and ALPHA_BNK. The most sensitive parameters during calibration of TN loads included N source, transport process, and transformation. The most sensitive nitrogen source parameters were RCN, CMN, SOL_NO3, and FRT_SURFACE, and the most sensitive N transformation and transport process parameters were NPERCO, LAT_ORGN, ERORGN, and BIOMIX.

Overall attribution patterns of TN pollution sources are quite variable across the QLB (Table 2). CN2, a function of land use and soil permeability and thus runoff yield, was the most sensitive hydrological parameter for streamflow and TN output in the basin. Low calibrated values for the initial nitrate concentration in the soil layer and for the N percolation coefficient provide evidence that lateral flow in the shallow soil layer covering much of the basin's steep topography is the main pathway of TN loads. These observations are consistent with an earlier SWAT modeling effort that was confined to the upper half of the basin in Anhui Province [44]. However, a small Manning's 'n' value and low transmission losses (Table 2) indicate that, in contrast to findings of the earlier study, surface and lateral runoff significantly affect TN transport and loss across the basin. We attribute these differences to the larger, basin-wide scope of the current study area, which alters fundamental topographic (Figure 1) and land use (Figure S1) relationships and includes soil types not prevalent in the upper basin (Figure S2).

Table 2. Key parameter values and sensitivities for streamflow and total nitrogen (TN) load simulation in the QLB.

Parameter	Definition	Range	Calibrated Value
Hydrology			
CN2.mgt (Forest land)	Default SCS curve number for moisture conditions	35–98	97.9261
OV_N.hru	Default Manning's 'n' value for overland flow	0.01–30	0.9790
TRNSRCH.bsn	Fraction of transmission losses partitioned to the deep aquifer	0–1	0.0039
Nitrogen			
RCN.bsn	Concentration of nitrogen in rainfall	0–15	0.0412
NPERCO.bsn	Nitrogen percolation coefficient	0–1	0.0020
SOL_NO3.chm	Initial nitrate conc. in soil layer	0–100	27.5619

3.2. *Validation of Streamflow and TN Loads*

Distributed rainfall data, which are critical for understanding spatiotemporal dimensions of key nutrient transport processes in hilly areas [45,46], exhibited strong annual variation, ranging from a low of 1375 mm to a high of 2373 mm (see Figure S3 in Supplementary Materials). Significant annual fluctuation with a coefficient of variation above 0.1 was recorded for the entire basin, and monthly precipitation was, as expected, characterized by spatial heterogeneity [47]. Rainfall tended to rise rapidly from May to June and decrease sharply from June to July. Average monthly rainfall peaked in June (389 mm), which accounted for over 20% of the total annual accumulation (Figure S3).

Rainfall regulates streamflow and TN transport in the QLB, but other driving forces also occur, and monitoring data are limited. Instead of simply relying on the coupling model [12,48], we applied stepwise calibration and validation to the stimulation to optimize model performance in different regions of the basin. TN data are available from the border between the two provinces at Jiekou and from the basin's outlet at Baqian; streamflow data are also available at Baqian and additionally from two upstream stations at Tunxi and Yuliang (Figure 1). Following the warm-up period, the SWAT model was consequently calibrated and validated for the variables available at each of these sites over a decade-long interval from 2007 to 2016 (Figures 2 and 3).

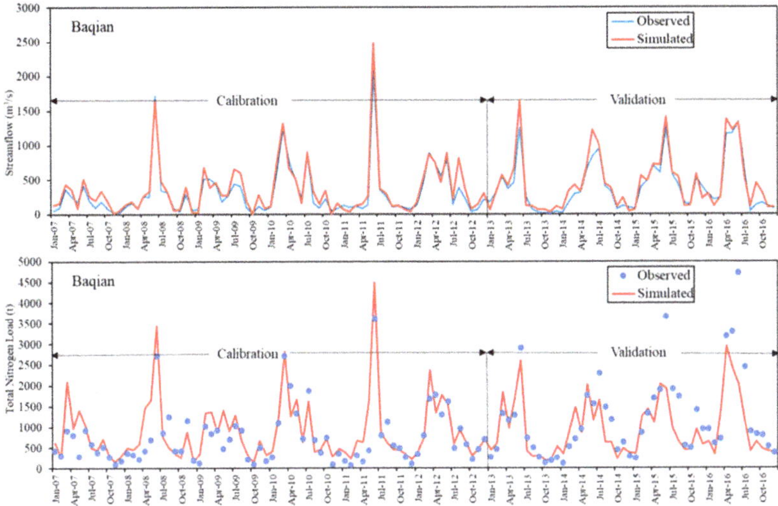

Figure 2. Observed and simulated daily streamflow values and monthly TN loads at the basin's outlet at Baqian.

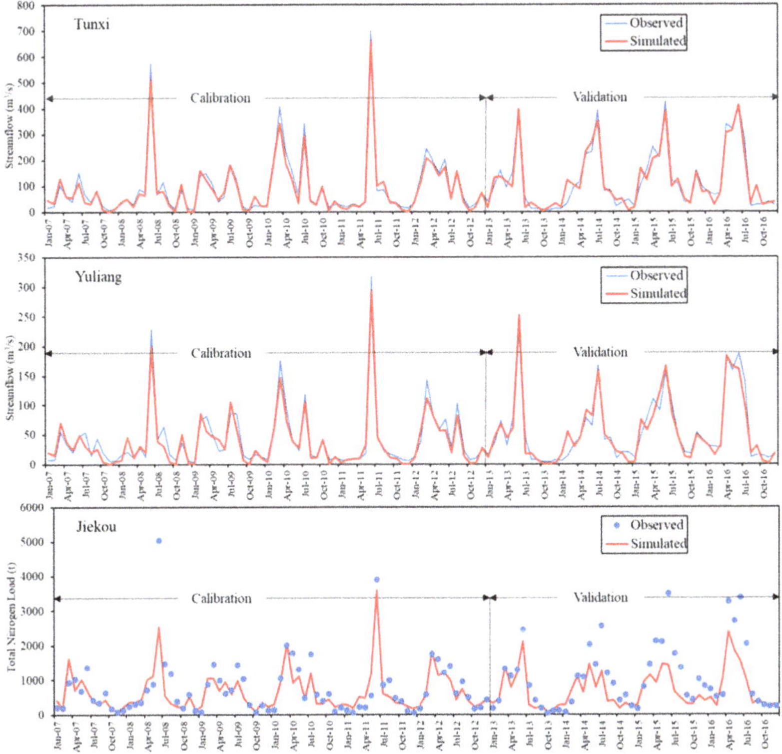

Figure 3. Observed and simulated daily streamflow values at Tunxi and Yuliang, and observed and simulated monthly TN loads at Jiekou.

The correspondence between measured and simulated values for TN and streamflow at Baqian is illustrated in Figure 2. Both variables are characterized by cyclical fluctuations, with annual maxima during the summer rainy season. The TN peaks tend to be displaced to the right compared to corresponding peaks for streamflow, likely because of differences in the frequency of measurement (monthly for TN, daily for streamflow). There is some divergence between observed and simulated TN levels during the peak flow periods of 2015 and 2016, both of which were wet years with extreme rainfall events. R^2 and NSE values (Table 3) are nevertheless indicative of the SWAT model's good and very good performance, respectively, during the calibration and validation periods [35,36].

Table 3. Performance of the SWAT model in simulating streamflow and TN loads in the QLB.

	Station	Streamflow		Station	Total Nitrogen	
		Calibration	Validation		Calibration	Validation
R^2	Baqian	0.97	0.97	Baqian	0.77	0.78
ENS		0.92	0.91		0.65	0.68
PBIAS		−14.4%	−12%		−18.1%	1.3%
R^2	Yuliang	0.95	0.95	Jiekou	0.75	0.71
	Tunxi	0.98	0.97			
ENS	Yuliang	0.94	0.92	Jiekou	0.74	0.62
	Tunxi	0.97	0.93			
PBIAS	Yuliang	9.9%	2.1%	Jiekou	14.6%	22.6%
	Tunxi	8.9%	1.5%			

The SWAT model was also calibrated and validated at sites above the lake and distant from the outlet at Baqian. Observed monthly streamflow at Tunxi and Yuliang was much lower than that at the basin's outlet, and the model accurately simulated this parameter (Figure 3), i.e., the performance of the simulation model at Tunxi and Yuliang was rated as very good according to all three statistical measurements (Table 3). In comparison, performance for streamflow at Baqian was rated as very good by the R^2 and NSE statistics and good by the PBIAS statistic. The model was somewhat less accurate in simulating TN levels at Baqian and Jiekou, especially during the validation period, when TN levels tended to be underestimated (Figure 3). Nevertheless, with just one exception (the NSE statistic at Jiekou during the validation period), model performance at these two sites was uniformly rated as good to very good, indicating that overall model performance was satisfactory.

3.3. *Spatiotemporal Patterns of TN Loads and Yields*

TN loads over the interval from 2007 to 2016 at the outlet of the QLB are strongly correlated with variations in average rainfall in the basin, both when measured annually (Figure 4A) and monthly (Figure 4B). This indicates that rainfall trends are key drivers of N discharge. TN loads spiked early in the annual fertilization cycle, which begins in March, and reached a maximum in June, when precipitation was also greatest and the risk of erosion highest. Precipitation and TN loads declined thereafter.

The mean annual TN load of the basin was 11,474 t, as calculated from simulations spanning the period 2007–2016. This corresponds to an overall mean TN yield of 1.1 t/km^2/yr. The TN point sources that were generalized into three categories for simulation with the SWAT model jointly accounted for 44% of the annual TN load in the basin. The contribution of industrial sources was negligible in comparison to that of domestic sewage and livestock production, which accounted for nearly equivalent amounts of TN loads to the basin (Figure 5A). In addition, nearly 6% of the total could be attributed to atmospheric deposition, with other non-point sources of TN accounting for the remaining half of TN loads in the basin. On a percentage basis, the most important of these sources were cropland and tea land, followed by forest land and orchard land (Figure 5A). These relationships obscure the fact that intensively fertilized tea land contributes 30 times as much TN as forest land does when expressed in terms of annual TN yields per km^2 (Figure 5B). Other agricultural land, including orchards and, to a lesser extent, cropland, also account for disproportionate amounts of non-point source TN pollution when measured this way.

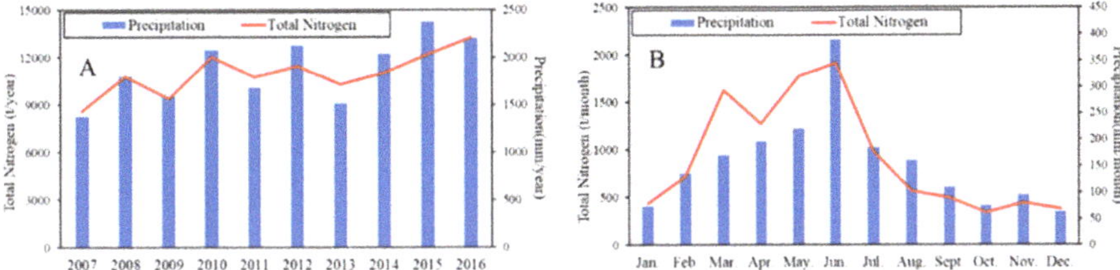

Figure 4. Annual (**A**) and monthly (**B**) variations in precipitation and TN output from the QLB during the interval from 2007 to 2016. Precipitation values were averaged from measurements at 54 sites across the basin, and TN output was measured at the basin's outlet at Baqian.

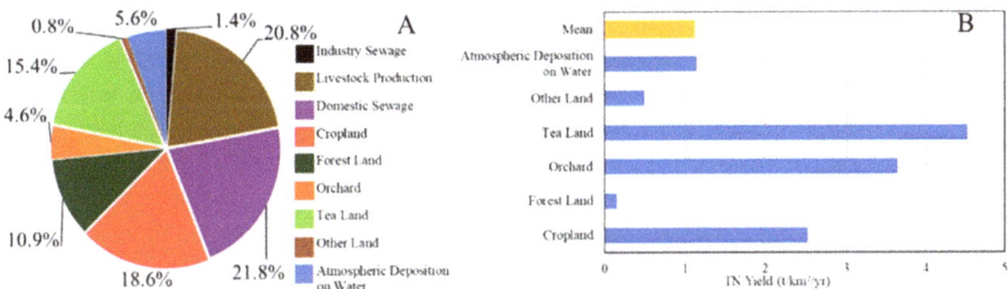

Figure 5. (**A**) Simulated annual contributions of point and non-point sources of pollution to TN loads, as expressed as percentages of the total. (**B**) Simulated annual contributions of non-point sources of pollution, expressed on a per unit area basis, i.e., as TN yields. The yellow bar gives the mean for all land use types.

The spatial patterns of TN loads and TN yields in each of the 96 sub-basins of the QLB are mapped in Figure 6. Upstream sub-basins in Anhui Province jointly account for 70% of the basin's TN loads, but significant sub-basin to sub-basin heterogeneity is obvious, and several hotspots are apparent (Figure 6A). The largest of these lies near the Huangshan City urban core. Downstream riparian areas near the lake also emit significant TN loads, but most sub-basins in Zhejiang Province are relatively minor contributors (Figure 6A). A different pattern emerges when TN yield is considered (Figure 6B). The highest such yields are from a single cluster of sub-basins, and their magnitude tends to progressively diminish with distance from this hotspot. TN yields in other sub-basins, including those near the lake, are generally low. A few additional areas of high TN yield are nevertheless evident, including one near the rapidly urbanizing tourist area in Zhejiang Province's Chun'an County.

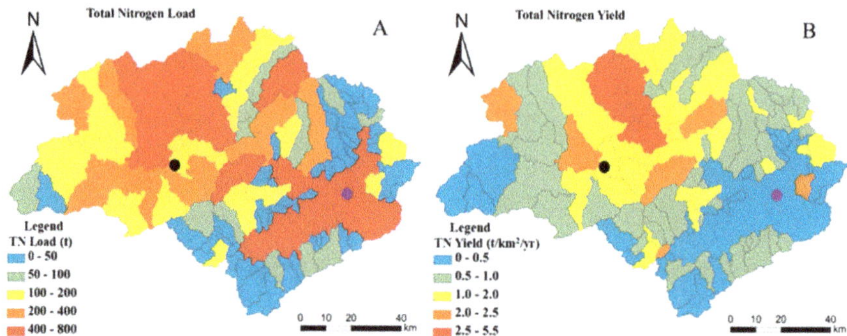

Figure 6. Spatial distribution of annual (**A**) TN loads and (**B**) TN yields among the sub-basins of the QLB. The locations of the urban core of Huangshan City (black dots) and Chun'an County (purple dots) are indicated.

3.4. Sources and Spatial Dynamics of TN Entering the Main Reach of the Xin'anjiang River

Optimization of pollution management requires as much spatiotemporal understanding as possible about sources and transport of pollutants. We consequently used detailed information about TN loads and yields in the sub-basins of the QLB to simulate sources, transport, and entry points of TN pollution into the main stem of the Xin'anjiang River. Figure 7 relates average annual contributions of different pollution sources of TN to their entry along the river's pathway through the QLB. Almost no TN enters the upper 45% of the river's course, but thereafter, three distinct surges of TN become evident. The first, which accounts for 25% of the total TN pollution entering the river (Figure 7), is at Tunxi (location 154.6 km; see Figure 1). Increased discharges of TN from agriculture, especially livestock production, predominate at this site and account for the majority of the surge.

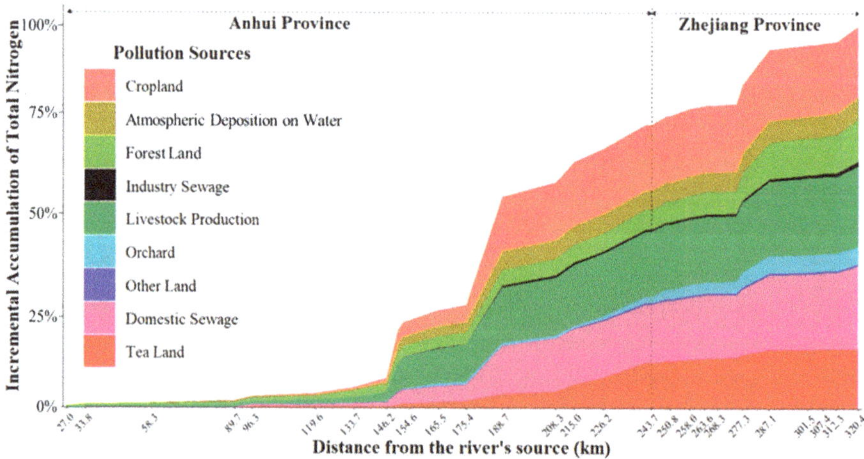

Figure 7. Average annual contributions of point and non-point sources of TN pollution along the main reach of the Xin'anjiang River.

A second, even sharper surge in TN is at location 188.7 km, near a tributary that flows through the municipalities of Shexian and Huizhou (both part of Huangshan City) and nearby highly populated areas before entering the Xin'anjiang River. Significant increases in TN attributable to domestic sewage are evident at this site. Downstream TN pollution from all sources then increases gradually until a third surge appears at location 287.1 km in Zhejiang Province. Elevated TN discharge from highly erodible forested slopes [2], and to a lesser extent from domestic sewage and orchard, are major contributors to the increase at this location. Distinct spatial patterns of TN release from plant-based agriculture are clearly evident along the course of the river. Thus, the accumulation of TN from cropland and tea land reaches its near maximum before the Xin'anjiang River flows out of Anhui Province, but most of the TN accumulation from orchard land occurs later, as the river flows through Zhejiang Province.

Delivery of TN from non-point sources to the Xin'anjiang River can follow three general pathways: lateral flow, groundwater, and surface flow. The contributions of these pathways to TN accumulation in the river are visualized in Figure 8, which reveals four significant surges of TN from non-point sources and identifies spatially explicit contributions of lateral flow, groundwater, and surface water. The first surge, at Tunxi (location 154.6 km), is primarily attributable to increases from surface flow and groundwater, but the second, at location 208.3 km, is primarily attributable to increases in lateral and surface flow. The third and fourth surges peak at Jiekou, on the border between the provinces (location 243.7 km) and within Zhejiang Province (location 287.1 km), respectively. Almost all of the incremental increase at both of these locations is contributed by lateral flow, which ultimately accounted for nearly two-thirds of TN transport in the river.

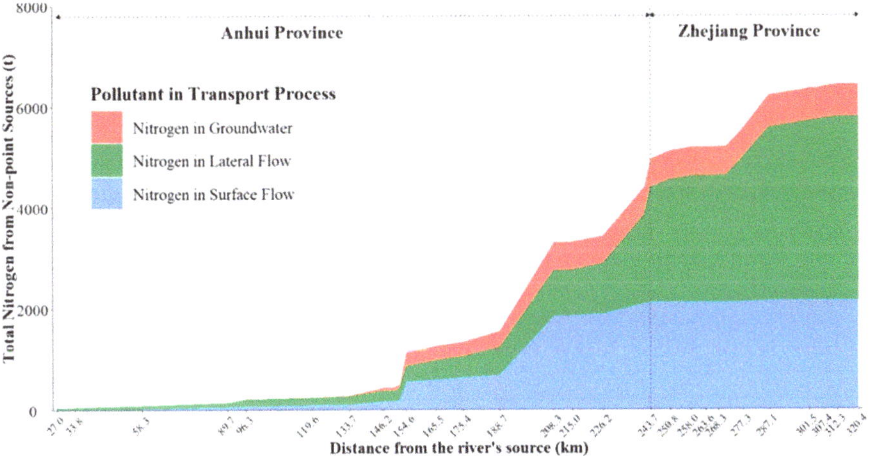

Figure 8. Average annual contributions of non-point sources of TN pollution along the main reach of the Xin'anjiang River.

The monthly dynamics of TN discharge from non-point sources via groundwater, surface flow, and lateral flow (Figure 9) generally correspond to monthly rainfall (Figure 4). Release of TN is most significant from February through August and sharply reduced during the other months, when TN from groundwater and surface flows become insignificant. Lateral flow and surface flow transport roughly equivalent amounts of TN during the period of peak movement, except for February and March, when lateral flow predominates and exceeds the contribution from surface flow by three to four-fold. Groundwater movement of TN, on the other hand, increases progressively during this same period, but is of minor significance compared to the other two pathways.

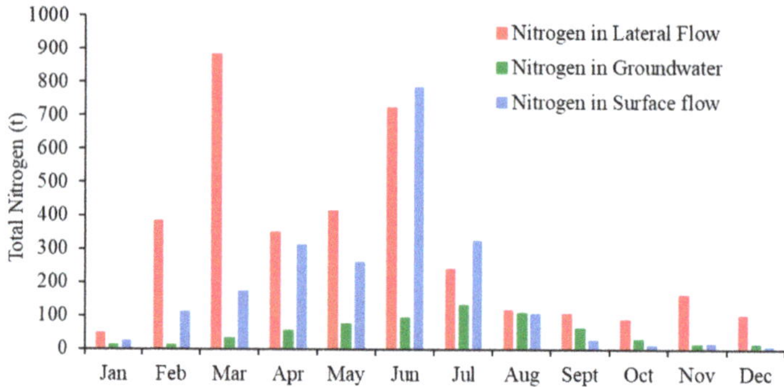

Figure 9. Monthly dynamics of TN movement via three different non-point source routes.

4. Discussion

4.1. Spatio-Temporal Dynamics of TN Transport

The QLB is characterized by hilly topography, diverse, rapidly intensifying agricultural production, and several urban areas where domestic and industrial activities are concentrated [25,31]. It is no wonder, then, that nutrient pollution in the basin is governed by dynamic upstream to downstream transport processes that are spatially and temporally heterogeneous and poorly understood [24,49]. It has been known for several decades that these factors have seriously compromised water quality in the basin [27]. China's first transboundary Payment for Ecosystem Services (PES) plan was consequently established in the QLB in 2012 [50,51]. Although this scheme has generated significant investment to reduce nutrient pollution entering the lower basin from Anhui Province, many fundamental questions persist, and water quality in the Xin'anjiang River and Qiandao Lake remains at risk [31,52].

Here a modeling approach scaled to the entire QLB was employed to understand the spatiotemporal dynamics of TN, a key polluting nutrient in both hilly [44,49,53,54] and lowland basins [55,56]. We establish that TN from point sources, which is almost equally attributable to livestock production and domestic sewage, accounts for nearly half of the aggregate TN loads in the basin. Both of these point sources are longstanding targets for mitigation under the PES plan [51]. On the other hand, seasonal surface and lateral flows from non-point sources account for the majority of TN loads in the QLB, with crop, tea, and orchard lands together comprising nearly 40% of the total. The yield of TN from tea land exceeds that from other crop types, and although crop production is declining in the area [57,58], tea cultivation is rapidly expanding, both in the QLB and in other nearby basins with similar topography [27,31,59,60]. This is important because agricultural fertilizers are commonly applied in excessive amounts to tea and other intensively managed crops in the basin, especially the upper basin in Anhui Province [59–63].

In general, incremental contributions of TN from non-point sources in the lower basin rise modestly and proportionately to one another, but disproportionately high contributions are evident from forest land and orchards in Zhejiang Province [31,59]. The dynamics differ, however, because forest land contributes high TN loads but low TN yields, but the situation is reversed with orchard land, which contributes low TN loads but high TN yields. The significant degree of forest coverage in the QLB contributes to high TN loads [27], but slope and erosion are also major factors. The mean slope of the study area is 24.9°, and the area proportions of slopes < 10°, 10°–20°, 20°–25°, 25°–30°, and >30° are 22%, 13%, 9%, 12%, and 44%, respectively. Sediment loss from such highly erodible forested slopes in the QLB is known to be substantial [44], and the importance of these sediments in the release of TN is documented [57]. In contrast, orchards are currently a minor but growing land use category in the QLB. Our data underscore the need for further study of their role in nutrient pollution in the basin.

Not all sub-basins in a given catchment contribute equally to nutrient loads, even if they are similar [55,56,64]. Nutrient runoff is known to be sensitive to mountainous topography and abundant annual rainfall in eastern China [44], and the differential influence of steeper and flatter lands on seasonal variation of nutrient loads is well documented [55,56]. We found that sub-basin to sub-basin heterogeneity in TN loads and yields was pronounced across the QLB, where four clustered sub-basins located upstream in Anhui Province stand out as a hotspot, because of their extensive and intensive contributions to TN pollution. These adjacent sub-basins deliver high annual TN loads ranging from 423 to 576 t and correspondingly high TN yields that can be as high as 5.3 t/km^2/yr. Similar TN loads characterize a few other areas of the QLB, including the large sub-basin surrounding Qiandao Lake, where the annual contribution is 464 t. TN yields in these areas are nevertheless modest (just 0.4 t/km^2/yr in the case of the sub-basin surrounding Qiandao Lake).

There are two important practical reasons to assign significance to pollution hotspots such as those identified here in the QLB. First, hotspots represent priority areas for detailed analysis of factors that contribute to TN pollution, e.g., soils, land use patterns (including legacy effects), agronomic management procedures, and topographical features such as hillslopes and streams that affect movement of nutrients [65–71]. Second, hotspots are locations where efforts to control pollution are likely to be most efficient and have the greatest impacts [31,59]. Knowledge of their location and characteristics is especially relevant in the QLB, where significant ongoing investments are underway to reduce transport of TN into Qiandao Lake [72,73].

Although pathways of nutrient transport have been examined in the QLB [10,27] and other hilly areas of eastern China [74–76], the proportion of pollution contamination from surface flow, sub-surface flow, and groundwater is difficult to monitor in field experiments [77]. Physical-based modeling with multivariate correlated data as employed here is consequently useful, because it can spatially resolve the relative contributions of TN transport processes. These dynamics are illustrated with the Xin'anjiang River, where, for example, modeling revealed that TN loads from nonpoint sources doubled over a distance of less than 10 km (between locations 146.2 and 154.6 km, see Figure 8). Although groundwater was a relatively significant transport component at this location, surface and lateral flow assumed increasing importance as the river flowed downstream, such that the overall significance of groundwater was eventually marginalized.

The relative contributions of lateral and surface flow to TN also changed as the river flowed downstream, i.e., the incremental contribution of lateral flow rose continuously after the river passed beyond location 154.6 km, but that of surface flow nearly ceased before the river entered Zhejiang Province at location 243.7 km. Thereafter, almost all incremental increase in TN was due to lateral flow, which would be expected to be significant in this area, which is characterized by dense agriculture and disturbed soils capable of enhancing such flows. The above insights from basin-wide analysis underscore the importance of this perspective, which can amplify more detailed information obtained from narrowly focused studies of sub-basins [27].

4.2. The Basin's N Cycle and Implications for Pollution Management

Figure 10 shows the integrated N cycle for the QLB as simulated by the SWAT model used here. Net N intake is assigned to three non-point sources (in decreasing order of magnitude: fertilizer, N stored in the soil, and atmospheric deposition) and three point sources (in decreasing order of magnitude: domestic sewage, livestock production, and industrial sewage). The most significant N outlet in the terrestrial area consists of processes associated with plant growth, e.g., N removal via harvest of agricultural crops, but processes such as ammonia volatilization, denitrification, and the return of N from plant litter to the soil are also important [78]. Thirteen percent of N from non-point sources in the basin is discharged into water bodies (Figure 10). This amount exceeds that discharged from point sources in the QLB and on a percentage basis, it is much greater than the 1.8%–4.5% determined earlier in a nearby low hilly basin with humid climate [79]. Entry of N from non-point sources thus represents a major environmental concern, even though nearly one-quarter of the N is removed from the basin's waterways during transport.

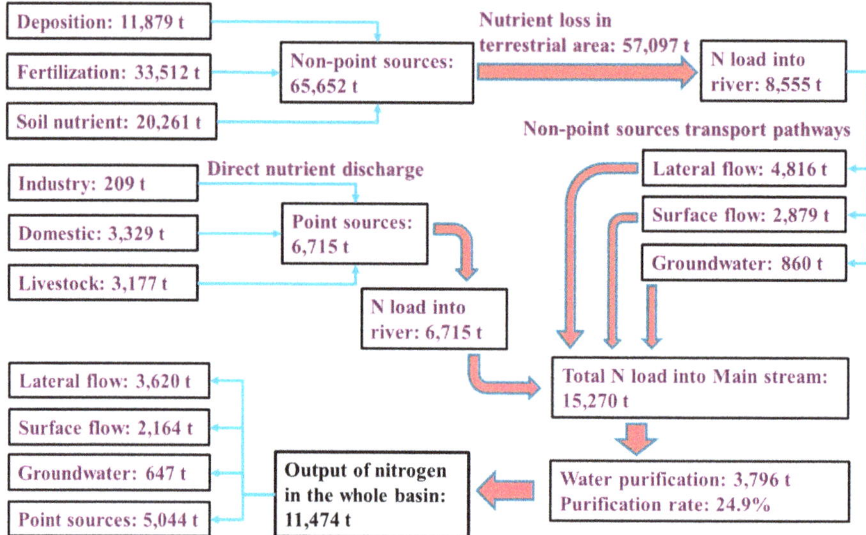

Figure 10. Diagrammatic representation of the area-averaged N budget of the QLB during the period from 2007 to 2016. Blue arrows represent net N inputs and pink arrows represent net N outlets and in-basin consumption. Note that rainfall N deposition here represents all deposition from every land use across the basin, including the water body component.

Key parameters of streamflow and N transport (Table 1), as well as the relatively small Manning 'n' value, the percolation coefficient, and transmission loss to the deep aquifer (Table 2), highlight the significance of subsurface flow in the QLB. Such flow is facilitated by shallow soils and frequent heavy rainfall events in basins with steep topography and convex hillsides [80], and it represents the main N delivery pathway and thus the principal target for reducing N pollution in the QLB. Consistent with the view that agricultural land and associated N fertilizer inputs are key factors for protecting aquatic environments [10,27,81,82], the control of N pollution in the QLB and similar basins currently emphasizes reduced fertilization, application of controlled-release fertilizers, construction of waste water treatment plants, collection of livestock sewage, and restriction of agricultural development [83]. On the basis of SWAT modeling, a strategy which has previously been employed to identify ways to reduce pollutant loads in other regions [84,85], we suggest a shift from controlling N loads to a targeted and comprehensive strategy emphasizing N distribution and transport processes.

Our SWAT analysis draws attention to two key factors for nutrient management in the QLB. The first is temporal and indicates that processes occurring during the period before June, especially those that influence subsurface flows in March, are crucial for annual N output. The second is spatial and emphasizes the importance of hotspots in the basin that contribute disproportionately to N outputs. These factors underscore the importance of restriction of agriculture on steep slopes—especially intensive agriculture that often relies on heavy applications of fertilizers in the spring. They also highlight the need for effective sewage collection—especially strategies that minimize N in tail water discharge, a goal that has proved elusive in the past [2,27,44].

Lakeside buffers [86,87], maintenance of high fractional vegetation cover [88,89], and construction of ponds to slow water movement and intercept nutrients [60] during the rainy season would undoubtedly reduce N loss in the QLB. The health of the basin's riparian and aquatic ecosystem could be further improved by adopting soil and water conservation strategies such as the restoration of vegetation and construction of wetlands. A third option involves the establishment of an ecological interception system to disrupt the N delivery pathway from subsurface flow to the lake and the basin's major waterways. Measures such as vegetative filter strips and ecological interception ditches would exploit chemical and biological mechanisms of self-purification and N retention to promote effective N reduction in the ecosystem [31,59,90].

5. Conclusions

Complex topographical conditions, varying climatic factors, and intensive anthropogenic activities exacerbate N pollution in hilly basins with monsoon climate. Understanding of source apportionment and spatiotemporal dynamics of N distribution and transportation is consequently vital for formulating reasonable strategies to ensure comprehensive water quality management in these basins. The physical-based SWAT model was utilized in this study to simulate the drivers and transport process of N, so that effective nutrient management of the QLB can be facilitated. Three main conclusions can be drawn.

First, optimization of localized model parameters suitable for hilly areas with monsoon climate enables the SWAT model to provide reliable estimates of N loads in the basin. The model confirms that, owing to the shallow soil layer and intensive rainfall, N loss from subsurface flow in the period before rainy season is the main driver of the N transport process. Complex terrain and severe rainfall heterogeneity enhance the complexity of the problem.

Second, non-point sources are the major contributors to N loads across the QLB, with cropland, tea land, and the basin's extensive tracts of highly erodible forest land accounting for 18%, 15%, and 10% of the TN loads, respectively. Major point sources include domestic sewage (21%) and livestock production (20%). The basin is characterized by the existence of hotspot sub-basins of high TN yield, non-uniform loading of TN along the reach of the Xin'anjiang River, and spatially determined variability in the relative contributions of different transport processes to TN movement.

Third, physical-based modeling suggests that N pollution control in the QLB can be made more comprehensive and efficient. Upstream to downstream analysis of the N transport pathway highlights the need not only to construct more wastewater treatment plants in areas identified here as sites of high domestic and industrial pollution, but also to increase their operational efficiency. In addition, soil and water conservation practices should be targeted to areas identified here with intensive agriculture and high TN yields. Insights from this study thus are likely to aid regional management of the QLB and provide a framework for similar monsoon basins elsewhere.

Supplementary Materials: The following are available online at http://www.mdpi.com/2073-4441/12/4/1075/s1. Figure S1: Land use classes used for SWAT analysis of the Qiandao Lake basin, Figure S2: The soil types map of the Qiandao Lake basin that was generated for use in SWAT analysis, Figure S3: Temporal and spatial heterogeneity of precipitation as recorded by individual rainfall stations (dots) compared to that recorded at two meteorological stations (dashed lines), Table S1: Fertilization and other agricultural management operations, Table S2: Main localized crop parameters of tea land.

Author Contributions: Conceptualization: D.C. and H.L.; methodology: D.C., W.Z., and J.P.; software, validation, and formal analysis: D.C.; investigation, resources, and data curation: D.C. and Y.D.; writing—original draft preparation: D.C.; writing, review, and editing: D.C., H.L., W.Z. and S.G.P. All authors have read and agreed to the published version of the manuscript.

Funding: This research was funded by an Agricultural and Social Development Project of Hangzhou, Zhejiang Province, grant number [20180417A06], the National Natural Science Foundation, grant number [41877513], the National Key R&D Program of China [2018YFD1100102], and the Thirteenth Five-Year Plan of the Nanjing Institute of Geography and Limnology [NIGLAS2018GH06].

Acknowledgments: The authors wish to express their gratitude to the Environmental Protection Agency of Chun'an and the Hangzhou Environmental Protection Research Institute for funding the collaboration that generated this paper. S.G.P. acknowledges the Nanjing Agricultural University-Michigan State University Asia Hub Program for supporting his participation in the investigation. We thank Weimin Chen, Zhixu Wu, and Yicai Han for insights on aquatic management, and Pengcheng Li, Jianwei Geng, and Wenyan Lu for advice, encouragement, and support.

Conflicts of Interest: The authors declare no conflict of interest.

References

1. Markogianni, V.; Mentzafou, A.; Dimitriou, E. Assessing the impacts of human activities and soil erosion on the water quality of Plastira mountainous Mediterranean Lake, Greece. *Environ. Earth Sci.* **2016**, *75*, 915. [CrossRef]
2. Wang, X.; Wang, Q.; Wu, C.; Liang, T.; Zheng, D.; Wei, X. A method coupled with remote sensing data to evaluate non-point source pollution in the Xin'anjiang catchment of China. *Sci. Total Environ.* **2012**, *430*, 132–143. [CrossRef] [PubMed]
3. Nayyeri, H.; Zandi, S. Evaluation of the effect of river style framework on water quality: Application of geomorphological factors. *Environ. Earth Sci.* **2018**, *77*, 343. [CrossRef]
4. Wang, H.; Wu, Z.; Hu, C. A Comprehensive Study of the Effect of Input Data on Hydrology and non-point Source Pollution Modeling. *Water Resour. Manag.* **2015**, *29*, 1505–1521. [CrossRef]
5. Zhou, P.; Huang, J.; Hong, H. Modeling nutrient sources, transport and management strategies in a coastal watershed, Southeast China. *Sci. Total Environ.* **2018**, *610–611*, 1298–1309. [CrossRef] [PubMed]
6. Hu, M.; Liu, Y.; Wang, J.; Dahlgren, R.A.; Chen, D. A modification of the Regional Nutrient Management model (ReNuMa) to identify long-term changes in riverine nitrogen sources. *J. Hydrol.* **2018**, *561*, 31–42. [CrossRef]
7. Que, Z.; Seidou, O.; Droste, R.L.; Wilkes, G.; Sunohara, M.; Topp, E.; Lapen, D.R. Using AnnAGNPS to Predict the Effects of Tile Drainage Control on Nutrient and Sediment Loads for a River Basin. *J. Environ. Qual.* **2015**, *44*, 629–641. [CrossRef]
8. Mishra, B.K.; Regmi, R.K.; Masago, Y.; Fukushi, K.; Kumar, P.; Saraswat, C. Assessment of Bagmati river pollution in Kathmandu Valley: Scenario-based modeling and analysis for sustainable urban development. *Sustain. Water Qual. Ecol.* **2017**, *9–10*, 67–77. [CrossRef]
9. Wang, Q.; Xu, Y.; Xu, Y.; Wu, L.; Wang, Y.; Han, L. Spatial hydrological responses to land use and land cover changes in a typical catchment of the Yangtze River Delta region. *Catena* **2018**, *170*, 305–315. [CrossRef]
10. Zhang, Y.; Lan, J.; Li, H.; Liu, F.; Luo, L.; Wu, Z.; Yu, Z.; Liu, M. Estimation of external nutrient loadings from the main tributary (Xin'anjiang) into Lake Qiandao, 2006–2016. *J. Lake Sci.* **2019**, *31*, 1534–1546. (In Chinese)
11. Arnold, J.G.; Fohrer, N. SWAT2000: Current capabilities and research opportunities in applied watershed modelling. *Hydrol. Process.* **2005**, *19*, 563–572. [CrossRef]
12. Luzio, M.D.; Srinivasan, R.; Arnold, J.G. A GIS-Coupled Hydrological Model System for the Watershed Assessment of Agricultural Nonpoint and Point Sources of Pollution. *Trans. GIS* **2010**, *8*, 113–116. [CrossRef]
13. Cerro, I.; Antiguedad, I.; Srinavasan, R.; Sauvage, S.; Volk, M.; Sanchez-Perez, J.M. Simulating land management options to reduce nitrate pollution in an agricultural watershed dominated by an alluvial aquifer. *J. Environ. Qual.* **2014**, *43*, 67–74. [CrossRef] [PubMed]
14. Withers, P.J.A.; May, L.; Jarvie, H.P.; Jordan, P.; Doody, D.; Foy, R.H.; Bechmann, M.; Cooksley, S.; Dils, R.; Deal, N. Nutrient emissions to water from septic tank systems in rural catchments: Uncertainties and implications for policy. *Environ. Sci. Policy* **2012**, *24*, 71–82. [CrossRef]
15. Malago, A.; Bouraoui, F.; Vigiak, O.; Grizzetti, B.; Pastori, M. Modelling water and nutrient fluxes in the Danube River Basin with SWAT. *Sci. Total Environ.* **2017**, *603–604*, 196–218. [CrossRef] [PubMed]
16. Nguyen, H.H.; Recknagel, F.; Meyer, W.; Frizenschaf, J.; Shrestha, M.K. Modelling the impacts of altered management practices, land use and climate changes on the water quality of the Millbrook catchment-reservoir system in South Australia. *J. Environ. Manag.* **2017**, *202*, 1–11. [CrossRef]
17. Maharjan, G.R.; Park, Y.S.; Kim, N.; Shin, D.S.; Choi, J.; Hyun, G.; Jeon, J.-H.; Ok, Y.S.; Lim, K.J. Evaluation of SWAT sub-daily runoff estimation at small agricultural watershed in Korea. *Front. Environ. Sci. Eng.* **2013**, *7*, 109–119. [CrossRef]
18. Glavan, M.; White, S.; Holman, I.P. Evaluation of River Water Quality Simulations at a Daily Time Step—Experience with SWAT in the Axe Catchment, UK. *Clean-Soil AirWater* **2011**, *39*, 43–54. [CrossRef]
19. Tuppad, P.; Douglas-Mankin, K.R.; Koelliker, J.K.; Hutchinson, J.M.S. SWAT Discharge Response to Spatial Rainfall Variability in a Kansas Watershed. *Trans. ASABE* **2010**, *53*, 65–74. [CrossRef]
20. Yang, X.; Warren, R.; He, Y.; Ye, J.; Li, Q.; Wang, G. Impacts of climate change on TN load and its control in a River Basin with complex pollution sources. *Sci. Total Environ.* **2018**, *615*, 1155–1163. [CrossRef]

21. Shao, G.; Guan, Y.; Zhang, D.; Yu, B.; Zhu, J. The Impacts of Climate Variability and Land Use Change on Streamflow in the Hailiutu River Basin. *Water* **2018**, *10*, 814. [CrossRef]
22. Zhang, H.; Peng, S.; Zhou, Y.; Yuan, H.; Chen, J. Analysis of current pollutant loads and investigation of total pollutant discharge limits in Qiandao Lake. *Water Resour. Prot.* **2014**, *30*, 53–56. (In Chinese)
23. MEP. *Chinese Environmental Quality Standards for Surface Water (GB 3838-2002)*; Ministry of Environmental Protection of the People's Republic of China: Beijing, China, 2002. (In Chinese)
24. Da, W.; Li, Y.; Zhu, G.; Xu, H.; Liu, M.; Lan, J.; Wang, Y.; WU, Z.; Zheng, W. Influence of Hydrometeorogical Process on Nutrient Dynamics in Qiandao Lake. *J. Hydroecol.* **2019**, *40*, 9–19. (In Chinese)
25. Han, X.; Zhu, G.; Wu, Z.; Chen, W.; Zhu, M. Spatial-temporal variations of water quality parameters in Xin'anjiang Reservoir (Lake Qiandao) and the water protection strategy. *J. Lake Sci.* **2013**, *25*, 836–845. (In Chinese)
26. Wu, Z.; Liu, M.; Lan, J.; He, J.; Yu, Z. Vertical distribution of phytoplankton and physico-chemical characteristics in the lacustrine zone of Xin'anjiang Reservoir(Lake Qiandao) in subtropic China during summer stratification. *J. Lake Sci.* **2012**, *24*, 460–465. (In Chinese)
27. Li, Z.; Liu, M.; Zhao, Y.; Liang, T.; Sha, J.; Wang, Y. Application of Regional Nutrient Management Model in Tunxi Catchment: In Support of the Trans-boundary Eco-compensation in Eastern China. *Clean-Soil Air Water* **2014**, *42*, 1729–1739. [CrossRef]
28. FAO. *The World Fertilizer Outlook*; Food and Agriculture Organization of the United Nations: Rome, Italy, 2015.
29. Gao, T.; Xie, L.; Liu, B. Association of extreme precipitation over the Yangtze River Basin with global air-sea heat fluxes and moisture transport. *Int. J. Climatol.* **2016**, *36*, 3020–3038. [CrossRef]
30. Wang, R.; Chen, J.; Chen, X.; Wang, Y. Variability of precipitation extremes and dryness/wetness over the southeast coastal region of China, 1960–2014. *Int. J. Climatol.* **2017**, *37*, 4656–4669. [CrossRef]
31. Pueppke, S.G.; Zhang, W.; Li, H.; Chen, D.; Ou, W. An Integrative Framework to Control Nutrient Loss: Insights from Two Hilly Basins in China's Yangtze River Delta. *Water* **2019**, *11*, 2036. [CrossRef]
32. MWR. *Hydrological Data of River Basins in China*; Ministry of Water Resources of China: Beijing, China, 2016. (In Chinese)
33. CBS. *Chun'an Statistical Yearbook*; Chun'an Bureau of Statistic, China: Hangzhou, China, 2016. (In Chinese)
34. HBS. *Huangshan Statistical Yearbook*; Huangshan Bureau of Statistic, China: Huangshan, China, 2016. (In Chinese)
35. Uniyal, B.; Jha, M.K.; Verma, A.K. Assessing Climate Change Impact on Water Balance Components of a River Basin Using SWAT Model. *Water Resour. Manag.* **2015**, *29*, 4767–4785. [CrossRef]
36. Abbaspour, K.C.; Rouholahnejad, E.; Vaghefi, S.; Srinivasan, R.; Yang, H.; Kløve, B. A continental-scale hydrology and water quality model for Europe: Calibration and uncertainty of a high-resolution large-scale SWAT model. *J. Hydrol.* **2015**, *524*, 733–752. [CrossRef]
37. Luo, P.; Zhou, M.; Deng, H.; Lyu, J.; Cao, W.; Takara, K.; Nover, D.; Geoffrey Schladow, S. Impact of forest maintenance on water shortages: Hydrologic modeling and effects of climate change. *Sci. Total Environ.* **2018**, *615*, 1355–1363. [CrossRef] [PubMed]
38. Nash, J.E.; Sutcliffe, J.V. River flow forecasting through conceptual models, part I. A discussion of principles. *J. Hydrol.* **1970**, *10*, 282–290. [CrossRef]
39. Moriasi, D.N.; Arnold, J.G.; Liew, M.W.V.; Bingner, R.L.; Harmel, R.D.; Veith, T.L. Model Evaluation Guidelines for Systematic Quantification of Accuracy in Watershed Simulations. *Trans. ASABE* **2007**, *50*, 885–900. [CrossRef]
40. Du, X.; Shrestha, N.K.; Wang, J. Assessing climate change impacts on stream temperature in the Athabasca River Basin using SWAT equilibrium temperature model and its potential impacts on stream ecosystem. *Sci. Total Environ.* **2019**, *650*, 1872–1881. [CrossRef]
41. Shen, Z.Y.; Gong, Y.W.; Li, Y.H.; Hong, Q.; Xu, L.; Liu, R.M. A comparison of WEPP and SWAT for modeling soil erosion of the Zhangjiachong Watershed in the Three Gorges Reservoir Area. *Agric. Water Manag.* **2009**, *96*, 1435–1442. [CrossRef]
42. Han, L.; Huang, M.; Ma, M.; Wei, J.; Hu, W.; Chouhan, S. Evaluating sources and processing of nonpoint source nitrate in a small suburban watershed in China. *J. Hydrol.* **2018**, *559*, 661–668. [CrossRef]
43. Zhang, M.K.; Xu, J.M. Restoration of surface soil fertility of an eroded red soil in southern China. *Soil Tillage Res.* **2005**, *80*, 13–21. [CrossRef]

44. Zhai, X.; Zhang, Y.; Wang, X.; Xia, J.; Liang, T. Non-point source pollution modelling using Soil and Water Assessment Tool and its parameter sensitivity analysis in Xin'anjiang catchment, China. *Hydrol. Process.* **2014**, *28*, 1627–1640. [CrossRef]
45. Tuo, Y.; Duan, Z.; Disse, M.; Chiogna, G. Evaluation of precipitation input for SWAT modeling in Alpine catchment: A case study in the Adige river basin (Italy). *Sci. Total Environ.* **2016**, *573*, 66–82. [CrossRef]
46. Campling, P.; Gobin, A.; Feyen, J. Temporal and spatial rainfall analysis across a humid tropical catchment. *Hydrol. Process.* **2001**, *15*, 359–375. [CrossRef]
47. Lan, J.; Wang, Y.; Chen, S.; Yu, C.; Wu, Z.; Zhang, Y. Time Variation Characteristics of the Main Inflow Nutrient at Xin'anjiang Reservoir and Its Impact factors in 2007–2016. *Environ. Monit. China* **2019**, *35*, 95–101. (In Chinese)
48. Sonnenborg, T.O.; Hinsby, K.; van Roosmalen, L.; Stisen, S. Assessment of climate change impacts on the quantity and quality of a coastal catchment using a coupled groundwater–surface water model. *Clim. Chang.* **2011**, *113*, 1025–1048. [CrossRef]
49. Wen, J.; Luo, D.; Luo, X.; Tang, D.; Chen, S. Agriculture non-point source pollution control measures of Qiandao lake area. *J. Soil Water Conserv.* **2004**, *18*, 126–129. (In Chinese)
50. FAO. *Case Studies on Remuneration of Positive Externalities (RPE)*; Payments for Environmental Services (PES): Rome, Italy, 2013.
51. Wang, J.; Wang, Y.; Liu, G.; Zhao, Y. The first eco-compensation for crossing provinces of downstream and upstream in China: A model of Xinanjiang River. *Environ. Prot.* **2016**, *14*, 38–40. (In Chinese) [CrossRef]
52. Chen, P.; Mao, Z.; Zhang, Z.; Liu, H.; Pan, D. Detecting subsurface phytoplankton layer in Qiandao Lake using shipborne lidar. *Opt. Express* **2020**, *28*, 558. [CrossRef]
53. Peterson, B.J.; Wollheim, W.M.; Mulholland, P.J.; Webster, J.R.; Meyer, J.L.; Tank, J.L.; Marti, E.; Bowden, W.B.; Valett, H.M.; Hershey, A.E.; et al. Control of nitrogen export from watersheds by headwater streams. *Science* **2001**, *292*, 86–90. [CrossRef]
54. Xia, J.; Zhang, Y.-Y.; Zhan, C.; Ye, A.Z. Water Quality Management in China: The Case of the Huai River Basin. *Int. J. Water Resour. Dev.* **2011**, *27*, 167–180. [CrossRef]
55. Yu, S.; Xu, Z.; Wu, W.; Zuo, D. Effect of land use types on stream water quality under seasonal variation and topographic characteristics in the Wei River basin, China. *Ecol. Indic.* **2016**, *60*, 202–212. [CrossRef]
56. Wan, R.; Cai, S.; Li, H.; Yang, G.; Li, Z.; Nie, X. Inferring land use and land cover impact on stream water quality using a Bayesian hierarchical modeling approach in the Xitiaoxi River Watershed, China. *J. Environ. Manag.* **2014**, *133*, 1–11. [CrossRef]
57. Jia, X.; Luo, W.; Wu, X.; Wei, H.; Wang, B.; Phyoe, W.; Wang, F. Historical record of nutrients inputs into the Xin'an Reservoir and its potential environmental implication. *Environ. Sci. Pollut. Res. Int.* **2017**, *24*, 20330–20341. [CrossRef] [PubMed]
58. Yang, G. Land use and land cover change and regional economic development: The revelation of the change in cropland area in the Yangtze River delta during the past 50 years. *Acta Geogr. Sin.* **2004**, *59*, 41–46. (In Chinese)
59. Li, H.; Chen, W.; Yang, G.; Nie, X. Reduction of nitrogen and phosphorus emission and zoning management targeting at water quality of lake or reservoir systems: A case study of Shahe Reservoir within Tianmuhu Reservoir area. *J. Lake Sci.* **2013**, *25*, 785–798. (In Chinese)
60. Zhang, W.; Li, H.; Pueppke, S.; Diao, Y.; Nie, X.; Geng, J.; Chen, D.; Pang, J. Nutrient loss is sensitive to land cover changes and slope gradients of agricultural hillsides: Evidence from four contrasting pond systems in a hilly catchment. *Agric. Water Manag.* **2020**, in press.
61. Xu, Q. The Study of Agricultural Non-Point Source Pollution Control Policy System. Master's Thesis, Michigan Technological University, Houghton, MI, USA, 2014.
62. Li, X.; Zhang, L.; Yang, G.; Li, H.; He, B.; Chen, Y.; Tang, X. Impacts of human activities and climate change on the water environment of Lake Poyang Basin, China. *Geoenviron. Disasters* **2015**, 2. [CrossRef]
63. Han, Y.; Li, H.; Nie, X.; Xu, X. Nitrogen and phosphorus budget of different land use types in hilly area of Lake Taihu upper river basin. *J. Lake Sci.* **2012**, *24*, 829–837. (In Chinese)
64. Beven, K.; Heathwaite, L.; Haygarth, P.; Walling, D.; Brazier, R.; Withers, P. On the concept of delivery of sediment and nutrients to stream channels. *Hydrol. Process.* **2005**, *19*, 551–556. [CrossRef]
65. Gao, W.; Howarth, R.W.; Hong, B.; Swaney, D.P.; Guo, H.C. Estimating net anthropogenic nitrogen inputs (NANI) in the Lake Dianchi basin of China. *Biogeosciences* **2014**, *11*, 4577–4586. [CrossRef]

66. Tomer, M.D.; Moorman, T.B.; Kovar, J.L.; Cole, K.J.; Nichols, D.J. Eleven years of runoff and phosphorus losses from two fields with and without manure application, Iowa, USA. *Agric. Water Manag.* **2016**, *168*, 104–111. [CrossRef]
67. Vadas, P.A.; Powell, J.M. Monitoring nutrient loss in runoff from dairy cattle lots. *Agric. Ecosyst. Environ.* **2013**, *181*, 127–133. [CrossRef]
68. Potter, T.L.; Bosch, D.D.; Strickland, T.C. Tillage impact on herbicide loss by surface runoff and lateral subsurface flow. *Sci. Total Environ.* **2015**, *530–531*, 357–366. [CrossRef] [PubMed]
69. Alexander, R.B.; Boyer, E.W.; Smith, R.A.; Schwarz, G.E.; Moore, R.B. The Role of Headwater Streams in Downstream Water Quality. *J. Am. Water Resour. Assoc.* **2007**, *43*, 41–59. [CrossRef] [PubMed]
70. Suescún, D.; Villegas, J.C.; León, J.D.; Flórez, C.P.; García-Leoz, V.; Correa-Londoño, G.A. Vegetation cover and rainfall seasonality impact nutrient loss via runoff and erosion in the Colombian Andes. *Reg. Environ. Chang.* **2016**, *17*, 827–839. [CrossRef]
71. Lai, X.; Zhu, Q.; Zhou, Z.; Liao, K.; Lv, L. Optimizing the spatial pattern of land use types in a mountainous area to minimize non-point nitrogen losses. *Geoderma* **2020**, *360*, 114016. [CrossRef]
72. Zhang, W.; Swaney, D.P.; Hong, B.; Howarth, R.W.; Han, H.; Li, X. Net anthropogenic phosphorus inputs and riverine phosphorus fluxes in highly populated headwater watersheds in China. *Biogeochemistry* **2015**, *126*, 269–283. [CrossRef]
73. ZPMF. *Horizontal Ecological Compensation Agreement for Upstream and Downstream of Xinanjiang River Basin*; Zhejiang Provincial Ministry of Finance: Hangzhou, China, 2018. (In Chinese)
74. Zheng, H.; Liu, Z.; Zuo, J.; Wang, L.; Nie, X. Characteristics of Nitrogen Loss through Surface-Subsurface Flow on Red Soil Slopes of Southeast China. *Eurasian Soil Sci.* **2018**, *50*, 1506–1514. [CrossRef]
75. Wu, X.; Zhang, L.; Zhang, M.; Ni, H.; Wang, H. Research on characteristics of nitrogen loss in sloping land under different rainfall intensities. *Acta Geogr. Sin.* **2007**, *27*, 4576–4582. (In Chinese)
76. Chen, X.; Yang, J.; Zheng, T.; Zhang, J. Sediment, runoff, nitrogen and phosphorus losses of sloping cropland of quaternary red soil in Northern Jiangxi. *Trans. Chin. Soc. Agric. Eng.* **2015**, *31*, 162–167. (In Chinese) [CrossRef]
77. Lane, S.N.; Reaney, S.M.; Heathwaite, A.L. Representation of landscape hydrological connectivity using a topographically driven surface flow index. *Water Resour. Res.* **2009**, *45*. [CrossRef]
78. Wang, P.; Sadeghi, A.; Linker, L.; Arnold, J.; Shenk, G.; Wu, J. Simulated soil water content effect on plant nitrogen uptake and export in watershed management. In *Quantifying and Understanding Plant Nitrogen Uptake for Systems Modeling*; Ma, L., Ahuja, L.R., Bruulsema, T., Eds.; CRC Press: Boca Raton, FL, USA, 2009; pp. 277–304.
79. Zhang, W.; Li, H.; Li, Y. Spatio-temporal dynamics of nitrogen and phosphorus input budgets in a global hotspot of anthropogenic inputs. *Sci. Total Environ.* **2019**, *656*, 1108–1120. [CrossRef]
80. Freeze, R.A. Role of subsurface flow in generating surface runoff. 2. Upstream source areas. *Water Resour. Res.* **1972**, *8*, 1272–1283. [CrossRef]
81. Zhang, W.; Li, H.; Kendall, A.D.; Hyndman, D.W.; Diao, Y.; Geng, J.; Pang, J. Nitrogen transport and retention in a headwater catchment with dense distributions of lowland ponds. *Sci. Total Environ.* **2019**, *683*, 37–48. [CrossRef] [PubMed]
82. Weigelhofer, G.; Hein, T.; Bondar-Kunze, E. Phosphorus and nitrogen dynamics in riverine systems: Human impacts and management options. In *Riverine Ecosystem Management*; Scmutz, S., Sendzimir, J., Eds.; Springer: Cham, Switzerland, 2018; pp. 187–202.
83. Yang, X.; Fang, S. Practices, perceptions, and implications of fertilizer use in East-Central China. *Ambio* **2015**, *44*, 647–652. [CrossRef] [PubMed]
84. Xu, H.; Xu, Z.; Liu, P. Estimation of Nonpoint Source Pollutant Loads and Optimization of the Best Management Practices (BMPs) in the Zhangweinan River Basin. *Environ. Sci.* **2013**, *34*, 882–891. (In Chinese) [CrossRef]
85. Kang, M.S.; Park, S.W.; Lee, J.J.; Yoo, K.H. Applying SWAT for TMDL programs to a small watershed containing rice paddy fields. *Agric. Water Manag.* **2006**, *79*, 72–92. [CrossRef]
86. Lü, Y.; Ma, Z.; Zhang, L.; Fu, B.; Gao, G. Redlines for the greening of China. *Environ. Sci. Policy* **2013**, *33*, 346–353. [CrossRef]
87. Xu, X.; Tan, Y.; Yang, G.; Barnett, J. China's ambitious ecological red lines. *Land Use Policy* **2018**, *79*, 447–451. [CrossRef]

88. Diyabalanage, S.; Samarakoon, K.K.; Adikari, S.B.; Hewawasam, T. Impact of soil and water conservation measures on soil erosion rate and sediment yields in a tropical watershed in the Central Highlands of Sri Lanka. *Appl. Geogr.* **2017**, *79*, 103–114. [CrossRef]
89. Wenger, A.S.; Atkinson, S.; Santini, T.; Falinski, K.; Hutley, N.; Albert, S.; Horning, N.; Watson, J.E.M.; Mumby, P.J.; Jupiter, S.D. Predicting the impact of logging activities on soil erosion and water quality in steep, forested tropical islands. *Environ. Res. Lett.* **2018**, *13*, 044035. [CrossRef]
90. Gonzalez, S.O.; Almeida, C.A.; Calderon, M.; Mallea, M.A.; Gonzalez, P. Assessment of the water self-purification capacity on a river affected by organic pollution: Application of chemometrics in spatial and temporal variations. *Environ. Sci. Pollut. Res.* **2014**, *21*, 10583–10593. [CrossRef]

Article

Implementing a Statewide Deficit Analysis for Inland Surface Waters According to the Water Framework Directive—An Exemplary Application on Phosphorus Pollution in Schleswig-Holstein (Northern Germany)

Phuong Ta [1,*], Björn Tetzlaff [1], Michael Trepel [2] and Frank Wendland [1]

1 Institute of Bio- and Geosciences—Agrosphere (IBG-3), Forschungszentrum Jülich GmbH, 52425 Jülich, Germany; b.tetzlaff@fz-juelich.de (B.T.); f.wendland@fz-juelich.de (F.W.)
2 Department for Water Management, Marine and Coastal Protection; Ministry of Energy, Agriculture, the Environment, Nature and Digitalization Schleswig-Holstein, Mercatorstraße 3, 24106 Kiel, Germany; michael.trepel@melund.landsh.de
* Correspondence: p.ta@fz-juelich.de; Tel.: +49-2461-61-96915

Received: 2 April 2020; Accepted: 9 May 2020; Published: 12 May 2020

Abstract: Deficit analysis—which principally deals with the question "how big are the gaps between current water status and good ecological status?"—has become an essential element of the river basin management plans prescribed by the European Water Framework Directive (WFD). In a research project on behalf of the Ministry of Energy, Agriculture, the Environment, Nature and Digitalization Schleswig-Holstein (MELUND), a deficit analysis based on distributed results from the water balance and phosphorus emission model system GROWA-MEPhos at high spatial resolution was performed. The aim was, inter alia, to identify absolute and relative required reduction in total phosphorus at any river segment or lake within the state territory as well as to highlight significant emission sources. The results of the deficit analysis were successfully validated and show an exceedance of the phosphorus target concentrations in 60% of the analyzed subcatchments. Statewide, 269 tons of phosphorus needs to be reduced yearly, which corresponds to approximately 31% of the total emission. Detailed data as well as maps generated by the deficit analysis benefit the planning and implementation of regionally efficient measures, which are indispensable with regard to meeting the environmental quality objectives set by the WFD.

Keywords: deficit analysis; phosphorus; inland surface waters; Water Framework Directive; LAWA; Schleswig-Holstein

1. Introduction

With the European Water Framework Directive (WFD) 2000/60/EC, the European Union is pursuing a holistic protection and utilization concept for European waters. The fundamental aim is to establish good ecological status in natural waters and good ecological potential in "heavily modified" and "artificial" waters. To achieve this objective, the EU Member States are obliged to draw up river basin management plans at regular intervals. The management plans must contain, inter alia, a summary of significant pressures and impact of human activity on water status, a map of established monitoring networks and results of monitoring programs carried out as well as a summary of binding measures required [1].

In order to properly target their programs of measures, Member States need to identify how measures can be combined in the most cost-effective way to close the gaps between current water status and good ecological status [2]. A deficit analysis—which principally deals with the question "how big are the gaps between current water status and good ecological status?"—needs to be carried out to determine what has to be done to meet the objective, how much time it will take and who will bear the expenses [2,3]. Furthermore, exceptions due to technical impracticability or disproportionate costs can only be duly justified on the basis of this analysis [2]. Further, even where derogations are justified, Member States must ensure that measures are taken to achieve the closest possible approximation to the objective [2]. Acknowledging the importance of the deficit analysis in measure planning processes, the German Working Group on water issues of the Federal States and the Federal Government (LAWA) developed a guidance document for a harmonized approach in the context of nutrient management [3]. The LAWA proposes that the deficit analysis should be carried out for river basin districts as well as their hydrological subareas and not only on basis of nutrient concentrations, but also based on nutrient loads. Monitoring, modeled data or a combination of both can be used in a deficit analysis. Furthermore, for each deficit identified, significant emission sources should be highlighted.

Due to the federal political structure in Germany, the 16 German states are responsible for the implementation of the EU Directives in the water sector [4]. In Schleswig-Holstein, the input of phosphorus into surface waters presents a substantial problem [5]. The physicochemical conditions of surface waters in the state were assessed in 2017 on the basis of monitoring data. The assessment revealed that approximately two-thirds of the river water bodies in Schleswig-Holstein do not comply with the orientation values for total phosphorus, which are stipulated in Annex 7, Ordinance on the Protection of Surface Waters 2016 (OGewV 2016). This result exerts a great pressure on the responsible water management institutions.

Although the results of the monitoring can be used to calculate the distance to good ecological status as a concentration, mg/L, information as the load in tons per year or information about the most significant emission paths are not feasible. It should also be noted that, due to the lack of measurements, some river segments had to be assessed on the basis of neighboring water bodies. Since monitoring data are not available in sufficient quantities for every water body in Schleswig-Holstein, it is advantageous to take model results into account. A prerequisite is that the models are implemented consistently and comprehensively at the state level and all the relevant diffuse and point source inputs are considered. The high-resolution model results by Tetzlaff et al. [6] fulfill these requirements and therefore can be taken into consideration.

Against this background, the main objective of this paper is to develop a method to perform a deficit analysis according to the LAWA for the entire federal state of Schleswig-Holstein based on available data and model results. Inter alia, the following results are to be achieved:

- Absolute and area-specific phosphorus loads for each subcatchment;
- Maximum and second-highest input path from a comparison of all paths;
- Percentage shares of point and diffuse sources for each subcatchment;
- Expected phosphorus concentration in each subcatchment;
- Absolute and relative required reduction in mg/L, t/yr or % for each subcatchment in order to meet the orientation values according to OGewV;
- Number of subcatchments which do not achieve good ecological status, statewide and for each river basin district;
- Required reduction statewide and for each river basin district. Results of the deficit analysis are also available for more detailed units such as planning units (Planungseinheiten) and subbasin areas (Bearbeitungsgebiete). In order to keep this paper short and easily readable, these results will not be included but can be provided if necessary.

The research works have been carried out by the authors in a project on behalf of the Ministry of Energy, Agriculture, the Environment, Nature and Digitalization Schleswig-Holstein (MELUND).

2. Methods

2.1. Study Area

The Federal State of Schleswig-Holstein is located in Northern Germany and comprises a total area of 15,800 km^2, with a population of 2.9 million and a population density of 183.7 inhabitants per km^2 (2019).

Schleswig-Holstein is characterized by a dense network of streams and rivers, which sum up to a total length of approximately 30,000 km, as well as a large number of natural lakes. Typical for Schleswig-Holstein are streams with small catchment areas, low gradients and a short flow path towards the nearest lake or sea. Most of the streams are no wider than 2 m. Larger river systems are Treene, Eider, Stör, Trave, Füsinger Au and Schwentine [7]. Approximately 300 natural [8] and a few artificial lakes located along the North Sea Coast encompass an area of more than 350 km^2, which makes up more than 2% of the state area.

Schleswig-Holstein is divided into three river basin districts, Elbe, Eider and Schlei/Trave [9], as well as three natural regions, "Marsh" (on the North Sea coast), "Geest" (in the state interior) and "Morainic Uplands" (in the east) [10] (p. 8). Approximately 69% of the state area is used for agriculture. Arable land (46%) is widely distributed in all parts of the country, while grassland (23%) is strongly limited to the "Marsh" as well as lowlands and floodplains. Special crops play a minor role in comparison to other agricultural land use types.

2.2. Methods of the Deficit Analysis

The deficit analysis makes use of available spatially distributed results on mean annual discharges and mean phosphorus emissions in tons per year for ten different pathways, which were modeled in a previous research project on behalf of MELUND (2010–2014) [6]. The main idea is to determine long-term phosphorus concentrations at any river segment or lake in Schleswig-Holstein and compare them with orientation concentrations for achieving good ecological status according to OGewV 2016. In a first step, modeled phosphorus loads and long-term modeled annual discharges were assigned for each of the LAWA subcatchments, which represent the smallest existing subdivision of a river basin according to the rules written by the LAWA [11]. This was performed in GIS based on spatial relationships. Subsequently, the modeled phosphorus loads and discharges were summed up from upstream to downstream subcatchments based on their hierarchy, which can be decoded from the hydrological area register Schleswig-Holstein. The modeled phosphorus concentration in the river segment of each subcatchment was then derived as the quotient of the total load and the total discharge. If the calculated concentration is higher than the orientation value, the required reduction in mg/L is calculated as the difference between these two values. In order to obtain the absolute required reductions in tons per year, the required reductions in mg/L are multiplied by the total discharges. The reduction amount can also be expressed as a relative percentage. The methods described are illustrated in the example below.

Figure 1 shows a total of three subcatchments (CA). Following the river flow directions, subcatchments 1 and 2 are upper basins and both drain into subcatchment 3. The phosphorus loads in t/yr arising from individual subcatchments are marked as P_1, P_2 and P_3. Q_1 to Q_3 characterize the average annual discharges in mm/yr (or m^3/yr) originating from the respective subcatchments 1 to 3. A specific phosphorus orientation value in mg/L based on the characteristics of the water body was taken from OGewV 2016 and assigned for each subcatchment. Since subcatchments 1 and 2 are also basins without loads from upstream, the total loads and total discharges (subscript k, considered all upstream subcatchments) correspond to their own loads and discharges: $P_{1,k} = P_1$ and $Q_{1,k} = Q_1$ as well as $P_{2,k} = P_2$ and $Q_{2,k} = Q_2$. For subcatchment 3, the total phosphorus load is calculated as:

$$P_{3,k} = P_1 + P_2 + P_3 \tag{1}$$

and the total discharge, respectively:

$$Q_{3,k} = Q_1 + Q_2 + Q_3 \tag{2}$$

The phosphorus concentration in the river segment of the subcatchment numbered i can be determined as follows:

$$C_i = \frac{P_{i,k}}{Q_{i,k}} \tag{3}$$

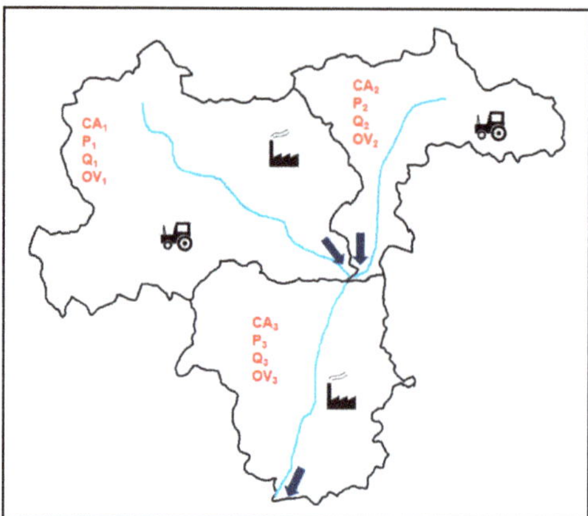

Figure 1. Example for the deficit analysis (CA: subcatchment, P: phosphorus load, Q: annual discharge, and OV: phosphorus orientation value).

The LAWA recommends that phosphorus retention and release in lakes should be taken into account, as neglecting such processes can lead to incorrect calculation of the concentration in the outflows and therefore the required reduction amount. However, the phosphorus cycle in lakes is complex and usually requires a large number of temporally and spatially high-resolution parameters, e.g., sedimentation rates, kinetic coefficients or diffusion coefficients of different layers of the water column, etc. Due to the availability of data, only phosphorus concentrations in lakes could be estimated based on inflow phosphorus loads. In a lake subcatchment numbered j, the phosphorus concentration is estimated using the formula according to Vollenweider and Kerekes [12] (Equation (4)). The retention and release processes, which possibly affect the lake outflows were neglected in this study due to the lack of suitable formulae for German, or rather Schleswig-Holstein, conditions as well as the lack of measurements.

$$C_j = \frac{L_{j,k} \cdot TW_j}{z_j \cdot \left(1 + \sqrt{TW_j}\right)} \tag{4}$$

C_j = modeled phosphorus concentration in lake $\left[\frac{mg}{L}\right]$,
$L_{j,k}$ = annual phosphorus load per lake area $\left[\frac{g}{yr \cdot m^2}\right]$,
TW_j = theoretical water retention time [yr], and
z_j = average depth of the lake [m].

If the determined concentration C_i or C_j is higher than the orientation value OV_i or OV_j, respectively, there is a need to reduce the phosphorus loads (Figure 2).

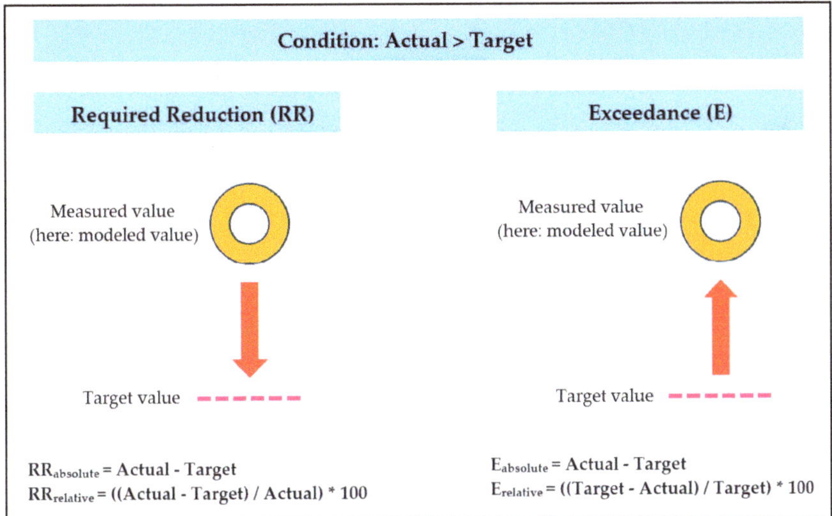

Figure 2. Methodological approaches of deficit analysis in the context of nutrients, according to the guidance of the German Working Group on water issues of the Federal States and the Federal Government (LAWA) [3]. Instead of measured values, modeled values are used as discussed in the introduction chapter.

An important indicator in terms of nutrient loads according to the LAWA [3] is the number of water bodies that are not in a good ecological status. The deficit analysis carried out in this paper indicates the number of subcatchments in which the river segments or lakes do not achieve good ecological status, statewide as well as for each of the river basin districts.

A further important question is how much of the phosphorus load each river basin district has to reduce. This is also subject of the deficit analysis. The calculation of the total reduction for each unit is explained using the example below. Assuming that there are n river subcatchments and m lake subcatchments in a river basin district, the required reduction to be met from this river basin district is calculated as follows:

$$\sum_{I=1}^{n}(C_i - OV_i)\cdot Q_i + \sum_{J=1}^{m}\left(C_j - OV_j\right)\cdot z_j\cdot\left(1+\sqrt{TW_j}\right)\cdot\frac{A_j}{TW_j} \tag{5}$$

for $C_i > OV_i$ and $C_j > OV_j$ with A_j is the area of the lake.

Equation (5) applies under the assumption that each individual subcatchment complies with the orientation value on the basis of its own phosphorus load and its own discharge. As a result, neither negative reductions nor "dilution effects" are taken into consideration.

In addition to the phosphorus concentrations and the amount of reductions that may be necessary, further results are provided by the deficit analysis. Information and maps of the absolute and area-specific phosphorus loads for each subcatchment—from individual subcatchments as well as from all upstream areas inclusive (as a sum value or broken down by paths)—can serve to identify hotspots. Furthermore, the maximum and second-highest phosphorus input path for each subcatchment can be highlighted. In the context of an integrated and holistic river basin management approach followed by the WFD, these results enable a more targeted planning of measures.

2.3. Input Data for the Deficit Analysis

Table 1 provides an overview of the input data required for the deficit analysis. These include, among others, the phosphorus emission over all input paths, the average annual discharge, the information on the hierarchy between subcatchments as well as the orientation values according to OGewV 2016. All data must be fully available for each subcatchment.

Table 1. Input data for the deficit analysis.

Input Data	Data Source	Processing Method
hierarchy between subcatchments	hydrological area register Schleswig-Holstein (GFV)	decoding based on [11]
total and agricultural area of subcatchments	GFV, Integrated Administration and Control System (InVeKoS 2011)	derivation in GIS
information relating to river basin districts	State Agency for Agriculture, Environment and Rural Areas Schleswig-Holstein (LLUR)	transfer to subcatchments
lake information	expert information system for water management: lakes	transfer to lake subcatchments
average annual discharge	water balance model GROWA [6]	spatially intersecting in GIS
phosphorus emission	phosphorus emission model MEPhos [6]	spatially intersecting in GIS
type of surface water body	GFV, database of surface water body characteristics	link column SH_CD_WB from GFV to surface water body database
orientation value	OGewV 2016, LLUR (Table 2)	transfer to subcatchments based on the type of surface water body

2.3.1. The German Catchment Coding System and the Hydrological Area Register Schleswig-Holstein

The German catchment coding system was firstly established in 1970 by the LAWA and updated two times, in 1993 and 2005 [11]. The numbering is executed from the source of the river to its mouth. Thus, the system is hydrology based and highly hierarchical. Every catchment can be subdivided into 9 subcatchments. Every subdivision of a catchment into subcatchments introduces a new digit, where an odd digit indicates the intercatchment areas along the main river and an even digit implies tributary catchments. The greater the number of digits, the more detailed the subcatchments. The latest version of the hydrological area register Schleswig-Holstein comprises 6428 subcatchments with a maximum digit number of 9, which is equivalent to the 9th level of subdivision. The average size of each subcatchment is approximately 2.5 km^2. A few subcatchments, which have the same coding numbers and further attributes, e.g., orientation value, are merged together.

For each of the subcatchments, the direct downstream basin was determined based on the described LAWA regulations. In order to check the correctness of the derived drain directions, a further routine was developed and implemented, which can identify and graphically illustrate codings, which do not comply with the guideline written by the LAWA [11]. These issues were discussed in several stages with the State Agency for Agriculture, Environment and Rural Areas Schleswig-Holstein (LLUR) and corrected where necessary.

Within the framework of this study, the focus lies on inland waters or, to be more precise, on inland rivers and lakes. Coastal and transitional water bodies are not part of the work and therefore were removed from their subcatchments (Figure 3). The remaining parts of the cut-off subcatchments are treated as outflows into the sea and are assigned river orientation values. Consequently, the state area of Schleswig-Holstein can be completely taken into account.

After all the above-mentioned processing steps, instead of 6428, 6407 subcatchments are available for the state of Schleswig-Holstein.

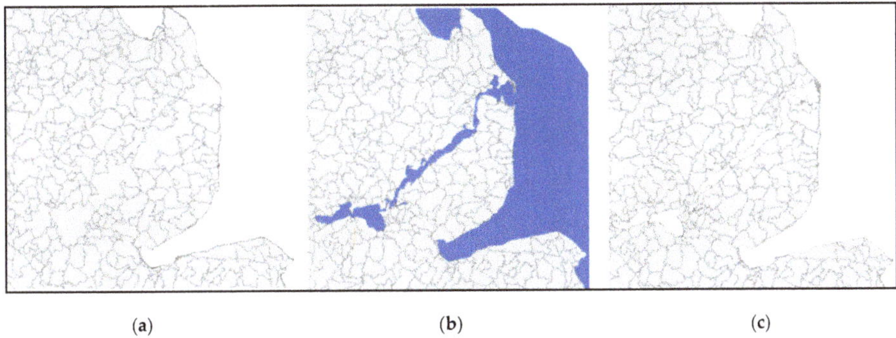

(a) (b) (c)

Figure 3. Removal of the coastal water bodies from the subcatchments using the example of the river Schlei: (**a**) original subcatchments; (**b**) coastal water bodies (blue); (**c**) cut-off subcatchments.

2.3.2. Average Annual Discharge Modeled with GROWA

GROWA [13] is a grid-based empirical water-balance model that was developed, calibrated and validated for central-European site conditions. GROWA conceptually combines distributed meteorological data (precipitation and potential evapotranspiration) with distributed site parameters (land use, soil properties, slope gradient and exposure, mean depth to groundwater) to calculate long-term annual averages of water-balance components [14]. During the last 17 years, GROWA has been applied and successfully used for various water management issues at different scales, from national, i.e., in Lower Saxony [15], Hamburg Metropolitan Region [16], North Rhine-Westphalia [14], etc., to international, i.e., Greece [17] and Slovenia [18].

The statewide average annual discharge was determined with the GROWA model for the period 1971–2000 in the framework of the research project "Regional differentiated modeling of nutrient inputs into groundwater and surface waters in Schleswig-Holstein" on behalf of MELUND [6]. The result was successfully validated [6] (p. 120) and is available as a high-resolution raster of 25 m × 25 m [6] (p. 112). The highest values partly exceeded 400 mm/yr and can be observed in the "Geest" region. As the annual precipitation decreases towards the east, the total discharge falls to values of 100 to 200 mm/yr at a maximum.

2.3.3. Phosphorus Emission Modeled with MEPhos

The MEPhos model is based on an area- and pathway-differentiated emissions approach [19]. The model uses a modular structure to quantify the mean long-term phosphorus inputs separately for different pathways. While entries from municipal sewage treatment plants and industrial effluents are considered as site specific using point data, entries by rainwater sewers and combined sewer overflows are calculated integratively for subcatchments of 10–520 km^2 in size [20]. Within MEPhos, diffuse phosphorus entries via artificial drainage, groundwater outflow, interflow, erosion and wash-off are modeled based on phosphotopes [20]. Phosphotopes are regarded as homogenous types of subareas representing discontinuous source areas for non-point phosphate inputs. Analogous to hydrotopes, phosphotopes include a set of parameters that control the phosphorus emissions, e.g., soil types, land use or hydraulic connectivity to surface waters, etc. By means of MEPhos, differentiated phosphorus emissions in Schleswig-Holstein were modeled and successfully validated [6] (p. 168). The highest phosphorus input into surface waters originates from artificial drainage (371 t/yr), followed by municipal sewage treatment plants (145 t/yr), groundwater outflow (136 t/yr), rainwater sewers (82 t/yr), erosion (70 t/yr), deposition (38 t/yr), small sewage treatment plants (37 t/yr) and industrial effluents (19 t/yr). Inputs via interflow and wash-off are considerably lower and amount to 5 and 2 t/yr, respectively. All the results are available as high-resolution datasets.

2.3.4. Phosphorus Orientation Values for Surface Waters

Phosphorus orientation values for achieving good ecological status vary depending on the type of water body and are defined in OGewV 2016 (Table 2).

Table 2. Orientation values for phosphorus according to Ordinance on the Protection of Surface Waters 2016 (OGewV 2016).

Type		Explanation	Phosphorus Orientation Value [mg/L]
Rivers	14	Small sand-dominated lowland rivers	0.1
	15	Mid-sized and large sand and loam-dominated lowland rivers	0.1
	16	Small gravel-dominated lowland rivers	0.1
	17	Mid-sized and large gravel-dominated lowland rivers	0.1
	19	Small streams in riverine floodplains	0.15
	20	Very large sand-dominated rivers	0.1
	21_N	Lake outflows in the North German lowlands	0.1
	22.1	Marshland streams with catchments almost completely inside the marshes, which flow directly into the North Sea or lower reaches of large rivers	0.3
	22.2	Marshland streams with catchments in ground moraines of young and old moraine landscapes	0.3
	22.3	Very large marshland rivers (only Elbe and Weser)	0.3
	77	Kiel Canal	0.15 (in agreement with LLUR)
Lakes	10.1	Lowland layered lakes with relatively large catchment area	0.0325
	10.2	Lowland layered lakes with relatively large catchment area	0.037
	11.1	Lowland polymictic lakes with relatively large catchment area	0.04
	11.2	Lowland polymictic lakes with relatively large catchment area	0.045
	12	Lowland river-like lakes	0.075
	13	Lowland layered lakes with relatively small catchment area	0.03
	14	Lowland polymictic lakes with relatively small catchment area	0.037
	99	Artificial lakes located along the North Sea Coast	Not considered as in good condition

By means of the processing method presented in Table 1, the orientation values can be assigned to each of the subcatchments. Their spatial distribution is illustrated in Figure 4.

It should be noted at this point that a wide range of methods are currently being used by Member States of the European Union for deriving nutrient thresholds to support good ecological status in surface waters [21]. As a consequence, thresholds vary greatly among countries, even for similar water body types and in some cases show more than a 10-fold difference in concentrations [21,22]. Poikane et al. suggest that nutrient criteria should be derived from biological responses to nutrients and a harmonization of methods between countries should be established [21].

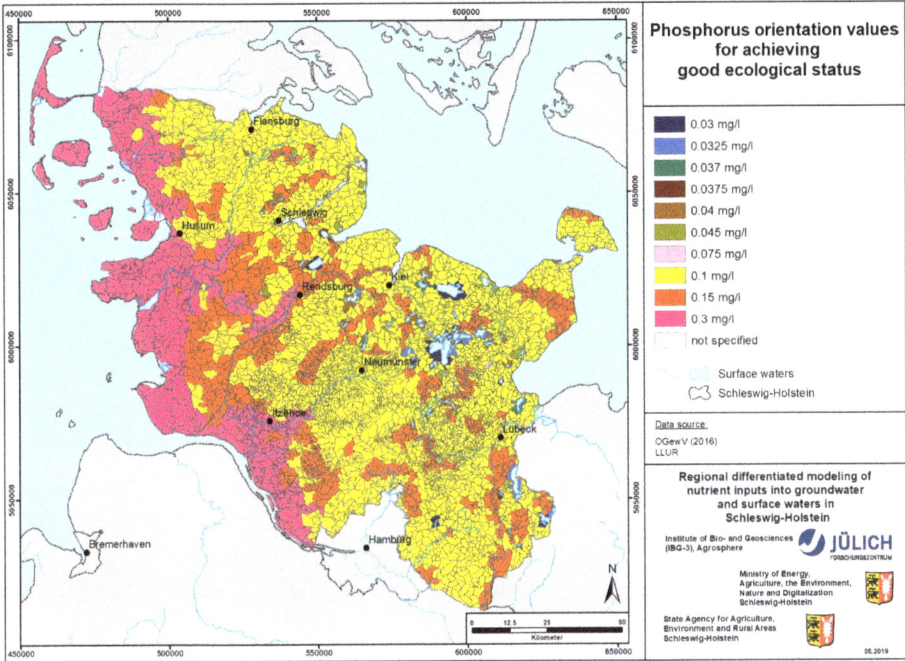

Figure 4. Phosphorus orientation values for achieving good ecological status.

3. Results

The results of the deficit analysis using the methods described above and based on the water balance and phosphorus emission model combination GROWA-MEPhos are shown in Figures 5–9. Since these figures consider the phosphorus emissions originating from individual subcatchments as well as from their upstream areas, they support the integrated and holistic river basin management approach followed by the WFD.

The maximum phosphorus emission path is illustrated in Figure 5a. It is obvious that, in the landscapes "Marsh" and "Geest", artificial drainage represents the dominant input path. In the landscape "Schleswig-Holstein Morainic Uplands" in the east, erosion characterizes the maximum phosphorus emission. As upstream areas are considered, site-specific sewage treatment plants become more noticeable in the map. In the large cities of Kiel and Lübeck and the medium-sized towns of Flensburg, Rendsburg, etc., phosphorus mostly emits to the surface waters through rainwater sewers. Deposition plays a major role in the subcatchment areas of the lakes.

Figure 5b provides information on which emission path represents the second-highest phosphorus input into surface waters. It can be seen that input via groundwater, which is not particularly noticeable in Figure 5a, dominates in almost the entire state. Phosphorus input through artificial drainage, sewage treatment plants and erosion can still be observed. The emission from small sewage treatment plants emerges spatially aggregated in the areas southeast of Flensburg, southwest of Itzehoe and in the northern surroundings of Lübeck.

The percentage shares of point sources (comprised of wastewater treatment plants, industry, rainwater sewers and small sewage treatment plants) in the total modelled phosphorus loads are shown in Figure 6a. The map shows a high proportion (over 70%) in the cities and surrounding areas. Figure 6b showing the percentage shares of the diffuse sources due to agriculture (artificial drainage, erosion and wash-off) indicates almost the opposite of Figure 6a. A very high proportion of agricultural sources of over 70% can be observed in the "Marsh", which is strongly caused by artificial drainage.

Except in the cities, at least 30% and mostly more than 50% of phosphorus emissions originate from agricultural sources.

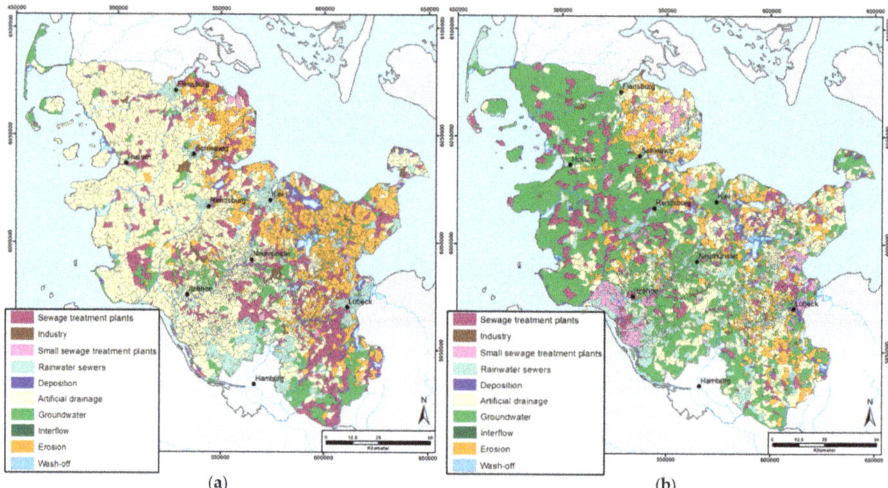

Figure 5. (**a**) Maximum and (**b**) second-highest phosphorus emission path from individual subcatchments and upstream areas.

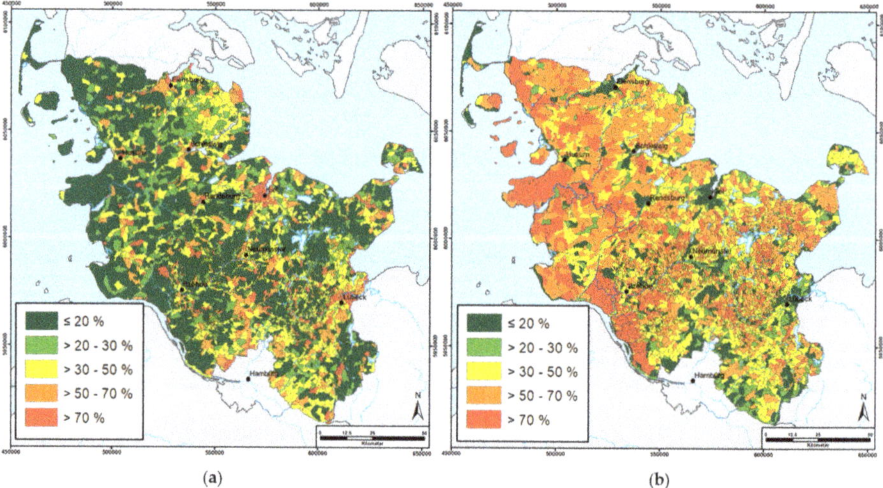

Figure 6. (**a**) Point sources and (**b**) agricultural diffuse sources as percentage shares of the total modeled phosphorus loads from individual subcatchments and upstream areas.

Figure 7a provides an overview of the modeled absolute phosphorus loads. Along large rivers, the phosphorus loads accumulate, which is clearly shown in the map, for example, along the Treene, Schwentine, Stör and Trave. The map of absolute phosphorus loads is strongly dependent on the size of the catchments. In order to assess and compare the spatial distribution of phosphorus emissions, it is of advantage to relate the loads to the catchment area sizes. Figure 7b shows the so-called area-specific phosphorus loads in kg/(ha·yr). Compared to the rest of the state, higher values in the "Marsh" can be observed. Alternatively, phosphorus loads due to agriculture can also be related to the agricultural area of catchments.

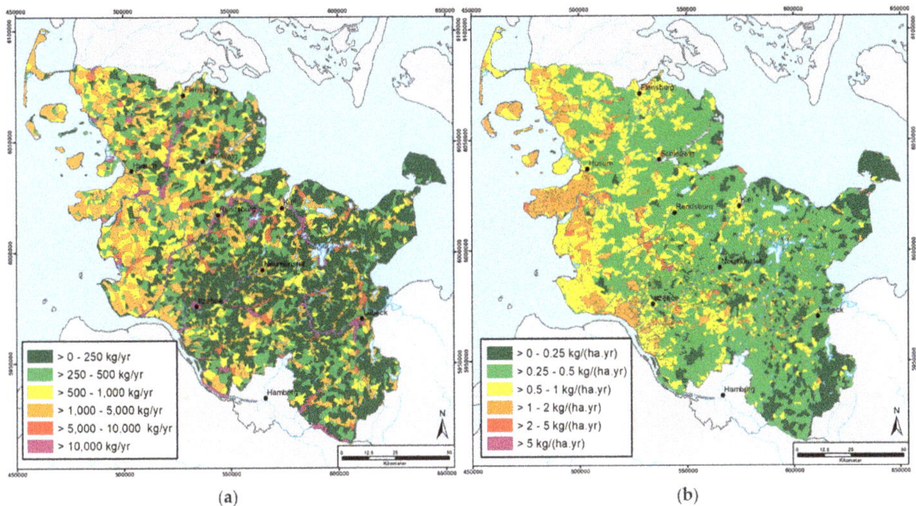

Figure 7. (**a**) Absolute and (**b**) area-specific phosphorus loads from individual subcatchments and upstream areas.

The modeled phosphorus concentrations in the surface waters of the subcatchments are shown in Figure 8a. Due to the high loads (Figure 7a,b), high phosphorus concentrations were calculated for the "Marsh", which, to a large extent, exceed the orientation value of 0.3 mg/L that applies in this region. In the rest of the state, with the exception of cities and their surroundings, modeled phosphorus concentrations are predominantly below 0.15 mg/L.

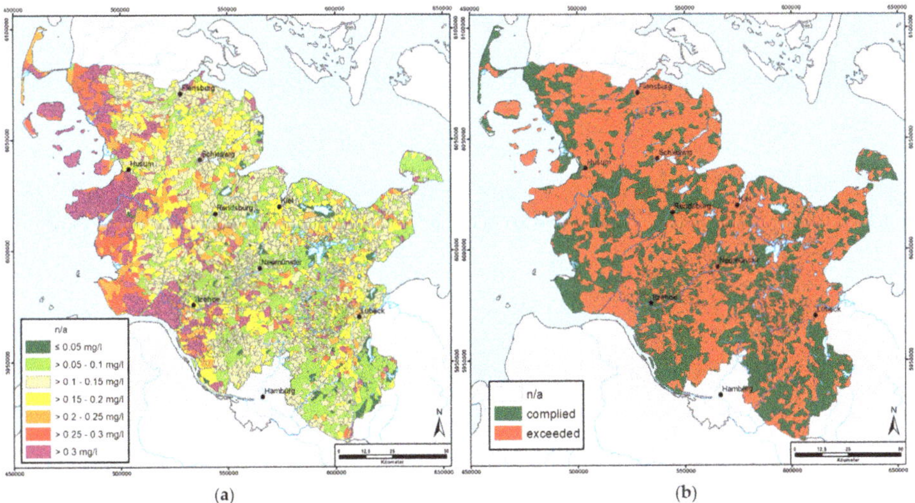

Figure 8. (**a**) Modeled phosphorus concentration; (**b**) compliance or exceedance of the phosphorus orientation values based on model results.

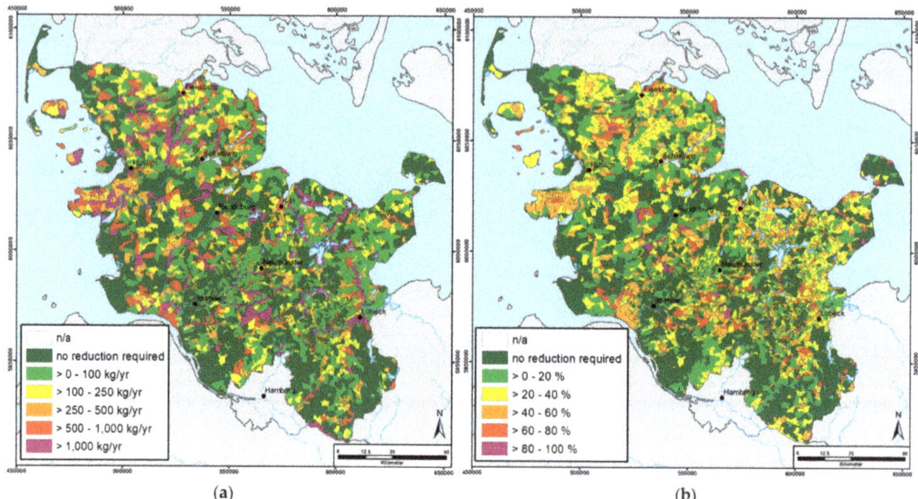

Figure 9. (**a**) Absolute and (**b**) relative required reduction in phosphorus emissions from individual subcatchments and upstream areas.

If the modeled phosphorus concentrations are compared with the orientation values from Figure 4, problematic areas and required reductions can be identified, which are shown in Figures 8b and 9a,b. It is obvious that for a large part of the state, phosphorus reduction is urgently necessary. In some isolated subcatchments, over 80% of total phosphorus emissions needs to be reduced. In such cases, information about the maximum emission path as well as the percentage shares of point and diffuse sources offers great help in developing targeted measures and assessing their feasibility.

The need for action regarding phosphorus emissions into inland surface waters in Schleswig-Holstein is summarized in Table 3. Accordingly, 60% of the subcatchments exceed the orientation values for phosphorus. Approximately 31% of the total phosphorus emissions should be reduced in order to achieve good ecological according to the WFD. This corresponds to approximately 269 t/yr. It should be noted that the total phosphorus load for Schleswig-Holstein in Table 3 differs from the sum given in Section 2.3.3 (905 t/yr). This is due to the neglect of coastal and transitional water bodies, whereby large water areas with high phosphorus inputs through deposition are not taken into consideration. However, the difference of 25 t/yr or 2.7% of the total emission is to be regarded as minor.

Table 3. Modeled phosphorus loads and required reductions.

River Basin District	Number of Subcatchments	Number of Subcatchments Exceeding Orientation Values	Phosphorus Loads [t/yr]	Required Reductions [t/yr]
Elbe	2815	1609 (57%)	313	101 (32%)
Eider	1170	718 (61%)	359	100 (28%)
Schlei/Trave	2422	1542 (64%)	208	68 (33%)
Schleswig-Holstein	6407	3869 (60%)	880	269 (31%)

4. Validation of Results and Discussion

In order to assess the accuracy and reliability of the implemented deficit analysis, possibilities for validating the results were sought. In the "Waterbody and Nutrient Information System Schleswig-Holstein", measured phosphorus concentrations in rivers for the period 1991–2018 are available. Additionally, seasonal mean concentrations of phosphorus in the lakes were provided by the LLUR's lake department.

These data underwent a detailed analysis and subsequently were processed and then compared with the modeled results. If there is a satisfactory agreement for a sufficiently large number of quality monitoring stations, it can be assumed that representative statements can be made with the underlying deficit analysis. In order to keep unavoidable uncertainties as small as possible, only monitoring stations with at least 50 concentration measurements were used.

4.1. Validation Based on Phosphorus Concentrations Measured in Rivers (at Least 50 Measurements per Measuring Station)

In total, there are 138 monitoring stations that meet the quality criterion of at least 50 measurements. With approximately 12 measurements per year, the criterion corresponds to approximately four years of monitoring. The 138 monitoring stations are distributed homogeneously throughout the state and cover all river basin districts and the state's major rivers. Figure 10 shows the comparison between the mean measured and modeled phosphorus concentrations. Most of the points are located in the zone of ±30% deviation. The mean absolute percentage error amounts to 29%. However, larger deviations, which are over 100% in extreme cases, can be observed. A more detailed analysis revealed that monitoring stations with the largest deviations are located near coastal areas, for example in the island of Fehmarn, as the tidal influence could not be considered in the models. Furthermore, for some of the stations, the monitoring time was not continuous, so that their average values are not fully comparable with model results, which generally characterize mean long-term conditions. It should also be noted that the input parameters for the GROWA-MEPhos model combination was derived from statewide available databases. As a consequence, local obviously deviating model results may have been due to databases being insufficiently accurate for local issues [23].

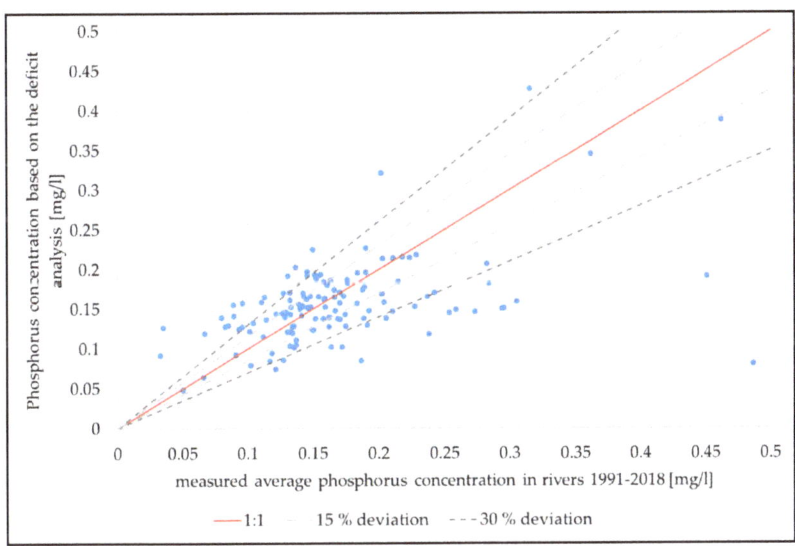

Figure 10. Validation based on phosphorus concentrations measured in rivers.

Figure 11 shows a comparison of measured and modeled relative required reduction for the 138 monitoring stations used for validation as well as their agreement based on classes defined in the upper panels. In general, good agreement can be considered, as there are mainly no, one or two class differences between measured and modeled results. A perfect match between them is also not realistic, since on the one hand the available data used for validation and the modeled results do not cover the exact same period of time, on the other hand there are fundamental uncertainties in the modeling, but also in the handling of measured values below the determination limits or in the estimation method used to determine the phosphorus loads at the monitoring stations.

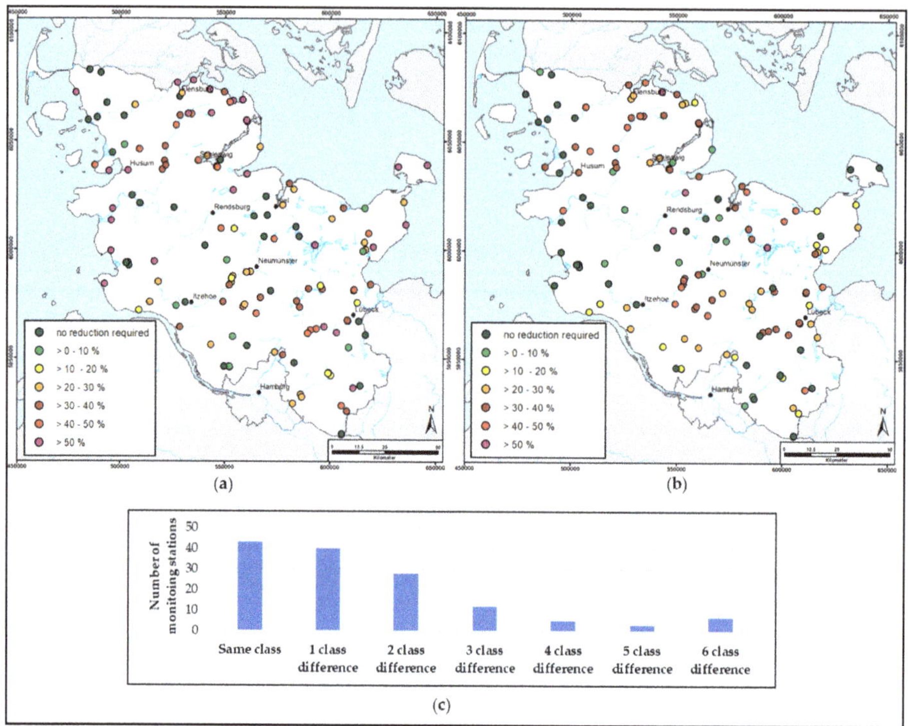

Figure 11. (**a**) Measured and (**b**) modeled relative required reduction for 138 monitoring stations used for validation; (**c**) agreement between panels (**a**) and (**b**) based on their classes.

4.2. Validation Based on Phosphorus Concentrations Measured in Lakes

Figure 12 compares the phosphorus concentrations estimated according to Vollenweider and Kerekes [12] with the seasonal mean values provided by the LLUR's Lake Department for a total of 59 lakes. The mean absolute percentage error amounts to 39% and is relatively high at first sight. However, some aspects have to be considered. First of all, the formula according to Vollenweider and Kerekes only provides an estimate of the lake concentrations. Uncertainties can reach up to a factor of 2. Furthermore, some lakes (e.g., Großer Ratzeburger See or Schaalsee) receive loads from Mecklenburg-Vorpommern, which could not be taken into account. Finally, the measured seasonal mean values are based on the period from March to October and only one seasonal mean value was provided for each lake. In view of the uncertainties listed above, which could add up in unfavorable cases, the result can be classified as satisfactory.

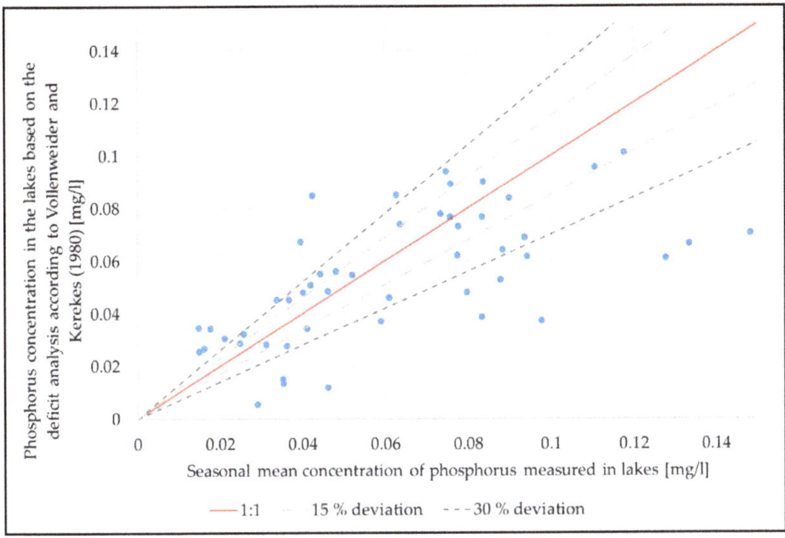

Figure 12. Validation based on phosphorus concentrations measured in lakes.

5. Conclusions and Outlook

A deficit analysis according to [3] was performed for all inland surface waters in the Federal State of Schleswig-Holstein. The results could be considered as successfully validated. According to the analysis, 60% of the subcatchments exceed the orientation values for total phosphorus. The phosphorus reduction required for Schleswig-Holstein amounts to approximately 31% on average, which corresponds to approximately 269 t/yr. These model results are consistent with the investigation carried out by Obernolte and Trepel [24], in which the type-specific orientation values for total phosphorus are not met in 68% of the water bodies in Schleswig-Holstein and a required reduction for phosphorus emissions into rivers and lakes of approximately 30% is necessary [25]. Since a large part of phosphorus inputs into surface waters in Schleswig-Holstein comes from agricultural diffuse sources, an implementation of measures with regard to fertilization management as well as erosion control is conceivable. Nevertheless, an increase in clarification efficiency in sewage treatment plants must not be out of consideration, especially in urban areas and surroundings, where they are playing the major role as maximum phosphorus emission path.

Generally, there is a need for further refinements and developments of the model combination, which generated the results used in this paper, so that the deviations can be minimized for better representation of the reality. This includes the use of actual and more accurate data, e.g., soil map and emissions of sewage treatment plants as well as calculation at higher time resolutions, such as on a daily or monthly basis. In the context of the deficit analysis, technically sound and plausible methods for describing phosphorus retention processes, especially in lakes as well as deep and slow rivers, should be developed. Furthermore, a new module for studying the impacts of possibly applied measures should be implemented. This could benefit the planning of regional efficient measures, which are indispensable with regard to achieving the quality objectives set by the WFD.

The methods of the deficit analysis presented in this paper are not limited to phosphorus or to the Federal State of Schleswig-Holstein. Other substances, such as nitrogen, which was also modeled statewide for Schleswig-Holstein [6], can be the next subject of the deficit analysis. Such a nitrogen analysis is advantageous, especially in the context of the EU policy on marine protection. The described deficit analysis can also be implemented for other federal states in Germany, e.g., North Rhine-Westfalia, Mecklenburg-Vorpommern, etc., where the model combination GROWA-MEPhos was successfully applied [26,27]. In other European countries, such as Austria and Switzerland, other models [28,29] were used to quantify nitrogen and phosphorus emissions to surface waters. With appropriate modification, these results can also be used as input data for the deficit analysis.

Author Contributions: Conceptualization, P.T., B.T., M.T. and F.W.; methodology, P.T., B.T. and M.T.; software, P.T. and B.T.; validation, P.T., B.T., M.T. and F.W.; resources, B.T. and M.T.; data curation, P.T.; writing—original draft preparation, P.T.; writing—review and editing, P.T., B.T., M.T. and F.W.; visualization, P.T.; supervision, B.T., M.T. and F.W.; project administration, B.T. and M.T.; funding acquisition, B.T. All authors have read and agreed to the published version of the manuscript.

Funding: This research was funded by the Ministry of Energy, Agriculture, the Environment, Nature and Digitalization Schleswig-Holstein (MELUND) through the project: "Further development of nutrient model Schleswig-Holstein"; project duration: 2017–2019; grant number: T/Z 1015.07.17.

Acknowledgments: The authors would like to thank the State Agency for Agriculture, Environment and Rural Areas Schleswig-Holstein for providing important data and valuable insights as well as the editors and the anonymous reviewers for their crucial comments, which helped improve the quality of this paper.

Conflicts of Interest: The authors declare no conflict of interest.

References

1. European Parliament; Council of the European Union. Directive 2000/60/EC of the European Parliament and of the Council of 23 October 2000 Establishing a Framework for Community Action in the Field of Water Policy. *Off. J. Eur. Communities* **2000**, *L327*, 1–73. Available online: https://eur-lex.europa.eu/eli/dir/2000/60/oj (accessed on 10 January 2020).
2. European Union: European Commission. Wasserrahmenrichtlinie und Hochwasserrichtlinie-Maßnahmen zum Erreichen eines guten Gewässerzustands in der EU und zur Verringerung der Hochwasserrisiken, Mitteilung der Kommission an das Europäische Parlament und den Rat, COM(2015) 120 Final: Brussels, 9 March 2015. Available online: https://ec.europa.eu/environment/water/water-framework/pdf/4th_report/COM_2015_120_de.pdf (accessed on 10 January 2020).
3. The German Working Group on water issues: LAWA. *Empfehlungen für eine harmonisierte Vorgehensweise zum Nährstoffmanagement (Defizitanalyse, Nährstoffbilanzen, Wirksamkeit landwirtschaftlicher Maßnahmen) in Flussgebietseinheiten*; Ständiger Ausschuss Oberirdische Gewässer und Küstengewässer LAWA-AO: Magdeburg, Germany, 14 July 2017. Available online: https://www.wasserblick.net/servlet/is/142651/WRRL_AO_35-36-37_Vorgehensweise_Naehrstoffmanagement_20170714.pdf?command=downloadContent&filename=WRRL_AO_35-36-37_Vorgehensweise_Naehrstoffmanagement_20170714.pdf (accessed on 10 January 2020).
4. Tetzlaff, B.; Haider, J.; Kreins, P.; Kuhr, P.; Kunkel, R.; Wendland, F. Grid-based modelling of nutrient inputs from diffuse and point sources for the state of North Rhine-Westphalia (Germany) as a tool for river basin management according to EU-WFD. *River Syst.* **2013**, *20*, 213–229. [CrossRef]
5. Haustein, V.; Janson, P.; Petenati, T.; Plambeck, G.; Schrey, J.; Steinmann, F.; Trepel, M. *Nährstoffe in Gewässern Schleswig-Holsteins—Entwicklung und Bewirtschaftungsziele*; Landesamt für Landwirtschaft, Umwelt und ländliche Räume Schleswig-Holstein: Flintbek, Germany, 2014. Available online: https://www.umweltdaten.landsh.de/nuis/wafis/fliess/gewaessernaehrstoffe.pdf (accessed on 10 January 2020).
6. Tetzlaff, B.; Keller, L.; Kuhr, P.; Kreins, P.; Kunkel, R.; Wendland, F. Räumlich differenzierte Quantifizierung der Nährstoffeinträge ins Grundwasser und die Oberflächengewässer Schleswig-Holsteins unter Anwendung der Modellkombination RAUMIS-GROWA-WEKU-MEPhos; Final Report: Jülich, Germany, 2017. Available online: https://www.schleswig-holstein.de/DE/Fachinhalte/W/wasserrahmenrichtlinie/Downloads/Bewirtschaftungszeitraum2/endberichtNaehrstoffmodellierung.pdf;jsessionid=7A97800F5A6E2E67340B24574A4C1BF1.delivery2-master?__blob=publicationFile&v=1 (accessed on 10 January 2020).
7. Official Website of the Federal State of Schleswig-Holstein: Rivers and Streams. Available online: https://www.schleswig-holstein.de/DE/Themen/F/fluesse_baeche.html (accessed on 10 January 2020).
8. Official Website of the Federal State of Schleswig-Holstein: Lakes. Available online: https://www.schleswig-holstein.de/DE/Themen/S/seen.html (accessed on 10 January 2020).

9. Official Website of the Federal State of Schleswig-Holstein: Flussgebietseinheiten (FGE), Bearbeitungsgebiete und Wasserkörper. Available online: https://www.schleswig-holstein.de/DE/Fachinhalte/F/fluesse_baeche/flussgebietseinheiten.html (accessed on 22 April 2020).
10. Statistisches Amt für Hamburg und Schleswig-Holstein. Kartenatlas zur Landwirtschaft in Hamburg und Schleswig-Holstein 2013, Endgültige Ergebnisse der Agrarstrukturerhebung 2013. Available online: https://www.statistik-nord.de/fileadmin/Dokumente/Statistische_Berichte/landwirtschaft/C_IV_Teil_1_S_Bodennutzung_Agrarstruktur/Kartenatlas_ASE_2013_Teil_1.pdf (accessed on 10 January 2020).
11. The German Working Group on water issues: LAWA. Richtlinie für die Gebiets- und Gewässerverschlüsselung, ausgearbeitet vom LAWA-Unterausschuss Gewässerverschlüsselung in der Wasserrahmenrichtlinie beim Ständigen Ausschuss Daten, Herausgegeben von der Länderarbeitsgemeinschaft Wasser (LAWA), Ministerium für Umwelt und Naturschutz, Landwirtschaft und Verbraucherschutz des Landes Nordrhein-Westfalen, April 2005.
12. Vollenweider, R.; Kerekes, J. The loading concept as a basis for controlling eutrophication: Philosophy and preliminary results of the OECD programme on eutrophication. *Progress Water Technol.* **1980**, *12*, 5–38.
13. Kunkel, R.; Wendland, F. The GROWA98 model for water balance analysis in large river basins—The river Elbe case study. *J. Hydrol.* **2002**, *259*, 152–162. [CrossRef]
14. Bogena, H.; Kunkel, R.; Schobel, T.; Schrey, H.P.; Wendland, F. Distributed modeling of groundwater recharge at the macroscale. *Ecol. Model.* **2005**, *187*, 15–26. [CrossRef]
15. Wendland, F.; Kunkel, R.; Tetzlaff, B.; Dorhofer, G. GIS-based determination of the mean long-term groundwater recharge in Lower Saxony. *Environ. Geol.* **2003**, *45*, 273–278. [CrossRef]
16. Tetzlaff, B.; Kunkel, R.; Taugs, R.; Wendland, F. *Grundlagen für eine nachhaltige Bewirtschaftung von Grundwasserressourcen in der Metropolregion Hamburg*; Umwelt/Environment; Forschungszentrum Jülich: Jülich, Germany, 2004; Volume 46.
17. Panagopoulos, A.; Arampatzis, G.; Kuhr, P.; Kunkel, R.; Tziritis, E.; Wendland, F. Area-differentiated modeling of water balance in Pinios Basin, central Greece. *Glob. Nest* **2015**, *17*, 221–235.
18. Tetzlaff, B.; Andjelov, M.; Kuhr, P.; Uhan, J.; Wendland, F. Model-based assessment of groundwater recharge in Slovenia. *Environ. Earth Sci.* **2015**, *74*, 6177–6192. [CrossRef]
19. Tetzlaff, B.; Vereecken, H.; Kunkel, R.; Wendland, F. Modelling phosphorus inputs from agricultural sources and urban areas in river basins. *Environ. Geol.* **2008**, *57*, 183–193. [CrossRef]
20. Tetzlaff, B.; Kreins, P.; Kunkel, R.; Wendland, F. Area-differentiated modelling of P-fluxes in heterogeneous macroscale river basins. *Water Sci. Technol.* **2007**, *55*, 123–131. [CrossRef] [PubMed]
21. Poikane, S.; Kelly, M.; Salas Herrero, F.; Pitt, J.; Jarvie, H.; Claussen, U.; Leujak, W.; Lyche Solheim, A.; Teixeira, H.; Phillips, G. Nutrient criteria for surface waters under the European Water Framework Directive: Current state-of-the-art, challenges and future outlook. *Sci. Total Environ.* **2019**, *695*, 133888. [CrossRef] [PubMed]
22. Phillips, G.; Kelly, M.; Teixeira, H.; Salas, F.; Free, G.; Leujak, W.; Solheim, A.L.; Várbíró, G.; Poikane, S. *Best Practice for Establishing Nutrient Concentrations to Support Good Ecological Status*; Technical Report EUR 29329 EN.; Publications Office of the European Union: Luxembourg, 2018; p. 67. Available online: https://publications.jrc.ec.europa.eu/repository/bitstream/JRC112667/bpg_online_corrected.pdf (accessed on 22 April 2020).
23. Wendland, F.; Bergmann, S.; Eisele, M.; Gömann, H.; Herrmann, F.; Kreins, P.; Kunkel, R. Model-Based Analysis of Nitrate Concentration in the Leachate—The North Rhine-Westfalia Case Study, Germany. *Water* **2020**, *12*, 550. [CrossRef]
24. Obernolte, M.; Trepel, M. Beurteilung der physikalisch-chemischen Bedingungen der Fließgewässer Schleswig-holsteins und Maßnahmen zur Verringerung der Nährstoffbelastung. 2017. Available online: http://www.umweltdaten.landsh.de/public/wrrl/massnahmen_db/download/FG_pc_sh_171222.pdf (accessed on 10 January 2020).
25. Trepel, M. Entwicklung von Nährstoffkonzentrationen- und -frachten in Fließgewässern Schleswig-Holsteins. Presentation, 20 October 2015. Available online: https://www.lung.mv-regierung.de/dateien/20151020_trepel.pdf (accessed on 10 January 2020).

26. Wendland, F.; Kreins, P.; Kuhr, P.; Kunkel, R.; Tetzlaff, B.; Vereecken, H. *Räumlich differenzierte Quantifizierung der N- und P-Einträge in Grundwasser und Oberflächengewässer in Nordrhein-Westfalen unter besonderer Berücksichtigung diffuser landwirtschaftlicher Quellen*; Schriften des Forschungszentrums Jülich, Reihe Energie & Umwelt/Energy & Environment, Band/Volume 88; Forschungszentrums Jülich: Jülich, Germany, 2010. Available online: https://www.flussgebiete.nrw.de/system/files/atoms/files/raeumlich_differenzierte_quantifizierung_eintraege_n_p_grundwasser_ow_wendland_etal_2010.pdf (accessed on 10 January 2020).
27. Wendland, F.; Keller, L.; Kuhr, P.; Kunkel, R.; Tetzlaff, B. Regional differenzierte Quantifizierung der Nährstoffeinträge in das Grundwasser und in die Oberflächengewässer Mecklenburg-Vorpommerns unter Anwendung der Modellkombination GROWA-DENUZ-WEKU-MEPhos; Final Report. 2015. Available online: https://www.wrrl-mv.de/doku/hintergrund/modellierung_naehrstoffeintraege_mv.pdf (accessed on 10 January 2020).
28. Schilling, C.; Zessner, M.; Kovacs, A.; Hochedlinger, G.; Windhofer, G.; Gabriel, O.; Thaler, S.; Parajka, J.; Natho, S. Stickstoff- und Phosphorbelastungen der Fließgewässer Österreichs und Möglichkeiten zu deren Reduktion. *Österreichische Wasser- und Abfallwirtschaft* **2011**, *63*, 105–116. [CrossRef]
29. Hürdler, J.; Prasuhn, V.; Spiess, E. *Abschätzung diffuser Stickstoff- und Phosphoreinträge in die Gewässer der Schweiz MODIFFUS 3.0*; Bericht im Auftrag des Bundesamtes für Umwelt (BAFU); Publ. Agroscope INH: Zürich, Switzerland, 2015; pp. 1–117. Available online: https://www.agroscope.admin.ch/agroscope/de/home/themen/nutztiere/bienen/bienenprodukte/honig/honigqualitaet/_jcr_content/par/columncontrols_1833305568/items/0/column/externalcontent_1350153509.external.exturl.html/aHR0cHM6Ly9pcmEuYWdyb3Njb3BlLmNoL2VuLVVTL3B1YmxpY2F0aW9uLzM1MDUy.html (accessed on 10 January 2020).

Article

Effects of Land Use on Stream Water Quality in the Rapidly Urbanized Areas: A Multiscale Analysis

Yu Song [1,2], Xiaodong Song [3,*], Guofan Shao [4] and Tangao Hu [1,2]

1 Institute of Remote Sensing and Earth Sciences (IRSE), College of Science, Hangzhou Normal University, Hangzhou 311121, China; songyu@hznu.edu.cn (Y.S.); hutangao@hznu.edu.cn (T.H.)
2 Zhejiang Provincial Key Laboratory of Urban Wetlands and Regional Change, Hangzhou 311121, China
3 College of Geomatics & Municipal Engineering, Zhejiang University of Water Resources and Electric Power, Hangzhou 310018, China
4 Department of Forestry and Natural Resources, Purdue University, West Lafayette, IN 47907, USA; shao@purdue.edu
* Correspondence: xdsong@zjweu.edu.cn; Tel.: +86-133-3614-9126

Received: 22 March 2020; Accepted: 8 April 2020; Published: 15 April 2020

Abstract: The land use and land cover changes in rapidly urbanized regions is one of the main causes of water quality deterioration. However, due to the heterogeneity of urban land use patterns and spatial scale effects, a clear understanding of the relationships between land use and water quality remains elusive. The primary purpose of this study is to investigate the effects of land use on water quality across multi scales in a rapidly urbanized region in Hangzhou City, China. The results showed that the response characteristics of stream water quality to land use were spatial scale-dependent. The total nitrogen (TN) was more closely related with land use at the circular buffer scale, whilst stronger correlations could be found between land use and algae biomass at the riparian buffer scales. Under the circular buffer scale, the forest and urban greenspace were more influential to the TN at small buffer scales, whilst significant positive or negative correlations could be found between the TN and the areas of industrial land or the wetland and river as the buffer scales increased. The redundancy analysis (RDA) showed that more than 40% variations in water quality could be explained by the landscape metrics at all circular and riparian buffer scales, and this suggests that land use pattern was an important factor influencing water quality. The variation in water quality explained by landscape metrics increased with the increase of buffer size, and this implies that land use pattern could have a closer correlation with water quality at larger spatial scales.

Keywords: stream water pollution; land use pattern; scale effect; redundancy analysis; urbanization

---

1. Introduction

Human activities have greatly affected the physical and chemical properties of water quality and the stability of aquatic ecosystems at regional and even global scales [1]. The rapidly urbanized areas are the places where human activities are the most concentrated and land use is changing drastically and, not surprisingly, the most typical areas for water quality degradation [2,3]. The level of urbanization in the world had reached 50% in 2008 and is expected to increase to 60% in 2030, and future urban population growth will mainly occur in developing countries. For China, urbanization has entered a period of rapid growth, and the percentages of the urban population to the total population were 19%, 26% and 57% in 1980, 1990 and 2016, respectively. During the process of urbanization, the rapidly expanded built-up areas together with the gathered population and industries have huge impacts on the original natural vegetation, soil environment and aquatic ecosystem. For example, the increased impervious surfaces and land use transition during urbanization may have direct impacts on the pollution concentrations in urban streams [4,5].

The spatial distribution of land use could well represent human activity and intensity and is a key factor affecting the quality of stream water in urbanized regions. Generally, industrial, mining and arable land types have relative high pollution risks compared with the other land use types; in contrast, forests and wetlands are sinks of potential pollutants in water bodies, while the riparian vegetation buffer zone usually has a filtering and barrier effect on pollutants [6,7]. The significant changes in regional land use and land cover caused by urbanization are mainly reflected in the increase of impermeable surfaces such as buildings; asphalt or concrete roads; squares and the reduction of natural underlying surfaces, e.g., woodlands and wetlands, with degraded capacities against pollutants [8,9]. Various types of impervious surfaces together with pervious surfaces form heterogeneous spatial structures of land use in urbanized regions. The negative hydrological effects caused by this particular land use form are mainly manifested by reduced soil infiltration and increased surface runoff, as well as increased sediment and sources of pollutants [10,11]. From a long-term perspective, the regional urbanization process is usually accompanied by the degradation of surface water quality and aquatic ecosystems [12–14]. For streams, the common pollutants, e.g., nitrogen, phosphorus and heavy metals, usually increased significantly with rapid urbanization, and it has been suggested that stream water quality and aquatic ecosystems might be damaged if the proportion of impervious surfaces in watersheds reach 10%–15% [15–18]. However, it is of limited reference value, especially for urban planning and management, if the spatial pattern and scale are missing for a particular study on the response mechanism of water quality to land use with only the statistical information provided [19,20].

The "disturbance-response" process of a stream system generally occurs in the regional context, while the impact of land use on stream water quality depends on the spatial scale of the disturbance. The stream system could be divided into four spatial scales, i.e., watershed, sub-catchment, riparian zone and local area. It is generally suggested that water quality of a high-order stream is mainly affected by a land use pattern in the upstream region, while water quality of a low-order stream is primarily determined by the local land use patterns [21]. Numerous multiscale related studies focused on the following categories—for example, the buffer zone versus the whole watershed [22–24], comparisons of multiscale watersheds [25–27] and comparisons at different buffer widths. For regions featured by a plain stream network or urban areas with complex underlying surfaces, varied buffer scales are often applied to study the relationship between land use and water quality [28–30]. However, the optimal spatial scale (in a correlation perspective) between land use and water quality still varies on a case-by-case basis. The characteristics of stream watersheds, intensity of human interference and data accuracy all have varying degrees of influences on multiscale studies about the relationship between land use and water quality. The biomass and composition of algae is another important indicator of stream water quality. At present, most of the studies about the feedback of biomass or the community structure of algae on environmental factors are based on laboratory cultivation, ecological simulation or field sampling [31,32]. However, studies about the impact of land use change-induced water quality disturbances on the biomass and community structures of algae are still not adequate at the regional scale [33,34].

Intense human activities in urbanized regions have led to changes in land use patterns, which can greatly affect the physical and chemical properties of water bodies and the health of aquatic ecosystems. Under the background of the rapid urbanization in China in recent decades, this study takes a typical area in Hangzhou City, one of the most rapidly expanded megacities in China, as the study area. We extracted the fine-resolution land use data of the study area using high-resolution remote-sensing images. The relationships between water quality, including the algae biomass, and land use across multiple spatial scales were analyzed with the support of field survey data. Our objectives were to quantitatively address the following questions: (1) For typical urbanized areas, how does the land use affect the water quality of urban streams? (2) What is the most influential landscape metrics correlated with water quality? Additionally, (3) is there a spatial scale dependence between land use and water quality in urbanized areas?

2. Materials and Methods

2.1. Study Area

Hangzhou City is a rapidly urbanizing city in the Yangtze River Delta in China (Figure 1). The study area is located in the west of Hangzhou, Zhejiang Province, which is a geographical unit that transitions from the low mountains and hills in the south to the northern plains. There are two large wetlands—the Hemu and Wuchang wetlands, in the study area, with intertwined stream networks. The study area has been experiencing intense urbanization during the past decade, and there have been increased contradictions between urban development and wetland protection in recent years. The study area is in the subtropical humid monsoon climate zone. The mountainous area is mainly composed by bedrock and residual slope deposits, and the plain area is mainly formed by Holocene sediments. The study area is about 120.19 km^2, with most of the areas as plain and low-lying lands. The soils have been affected by human activities for centuries and can be categorized as anthrosols (FAO–UNESCO Soil Map of The World, 1988). The background concentrations of the heavy metals Cd, Pb, As, Ni, Cr, Zn and Cu in the plain area of the study area were 0.206, 38.2, 10.0, 41.1, 92.1, 110.0 and 40.8 (unit: mg/kg), respectively [35]. Except As, the concentrations of the other heavy metals were slightly higher than the corresponding values of the 1st class natural background given in the Soil Environment Quality Standard of China (GB15618-1995).

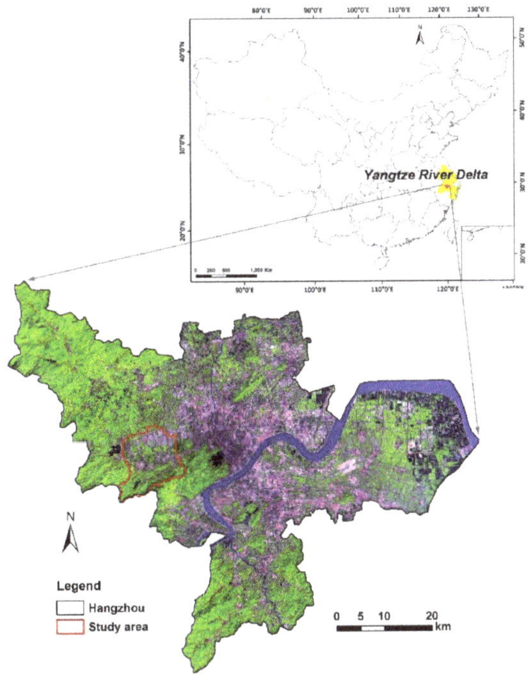

Figure 1. Locations of the study area.

2.2. Land Use Information Extraction

One cloudless SPOT 6 image obtained on 26 October 2014 was used to extract the land use data over the study area. The SPOT 6 image had four multispectral bands and one panchromatic band with 1.5-m resolution. The ancillary data included ASTER DEM and field survey data.

The land use data was extracted using the following steps: (1) projection conversion and geometric correction of SPOT 6 image; (2) fusing the multispectral and panchromatic data using the Gram-Schmidt method (the fused image had a spatial resolution of 1.5 m with blue, green, red and near-infrared bands) and then (3) extracting the land use data by visual interpretation (Figure 2). The classification accuracy was evaluated using a stratified random sampling with reference to the ground-truth data collected in the field survey. The assessment results showed that the overall accuracy of the land use classification was 93%. We adopted a two-level classification schema from the Chinese Standard of Land Use Classification [36] and classified the lands into five Level I classes and 12 Level II classes (Table 1).

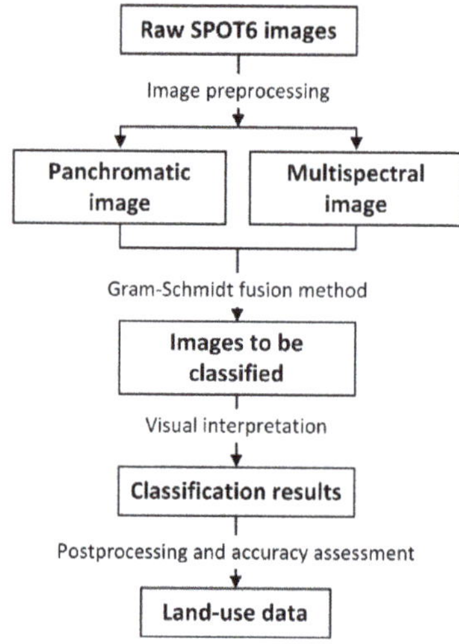

Figure 2. Technical schema in the land use data extracted from SPOT 6 images.

Table 1. Land use classification scheme.

Level I	Level II	Description
1 Cropland	11 Cropland	Cultivated land
2 Forest	21 Forest	A large area dominated by trees
3 Water	31 River	Natural-flowing watercourse and artificial canal
	32 Lake and pond	lake, reservoir and pond
4 Construction land	41 Residential	Houses and apartment buildings
	42 Industrial	Land and buildings used for manufacturing, logistics, warehouse and mining
	43 Commercial	Houses and buildings for businessman working, commercial retails, restaurants, lodging and entertainments
	44 Public management and service	Lands used for administration, public services and municipal utilities, education and research
	45 Urban greenspace	Parks and greenspace lands used for entertainments and environmental conservations
	46 Road	Paved roads including freeways, major and minor city roads
	47 Under construction	Under construction but not yet finished
5 Bareland	51 Bareland	Bare soil and bare rock

2.3. Water Sampling and Buffer Zone Delineation

The water quality of streams during the high water period is affected by the dual effects of land use, especially the construction land and cropland nonpoint sources, which could effectively reflect the content and intensity of human activities in the regional context [37–40]. The water quality sampling work at 25 stream sections was finished within one week when the stream flow was relatively stable during the local high-water period in June 2014 (Figure 3). The water quality parameters selected were total nitrogen (TN); total phosphorus (TP) and soluble heavy metals arsenic (As), copper (Cu), lead (Pb), chromium (Cr), cadmium (Cd), zinc (Zn), manganese (Mn) and nickel (Ni). The heavy metals selected in this study are the major heavy metals concentrated in the western plain areas of Zhejiang Province. Water samples were collected 0.3–0.5 m below the water surface from the middle of the stream using a 250-mL organic glass hydrophore. At each sampling site, three parallel samples were collected to avoid accidental error. The samples were stored in iceboxes during transport before water quality measurements were carried out in the laboratory. TP and TN were determined using the alkaline potassium persulfate digestion UV spectrophotometric method (GB11893-89, China National Standards) and the ammonium molybdate spectrophotometric method (GB11894-89, China National Standards), respectively. The heavy metal concentrations were determined using inductively coupled plasma optical emission spectrometry (ICP-OES, Thermo Fisher Scientific, Waltman, MA, USA). In addition, concentrations of chlorophyll a (TChla), including a chlorophyll a concentration of Cyanophyta ($Chla_{Cyan}$), a chlorophyll a concentration of Chlorophyta ($Chla_{Chlo}$) and a chlorophyll a concentration of Bacillariophyta and Dinophyta ($Chla_{Baci\text{-}Dino}$), were measured using a four-wavelength-excitation chlorophyll fluorometer (PHYTO-PAM Fa. Walz, Effeltrich, Germany), and each sample was sampled three times.

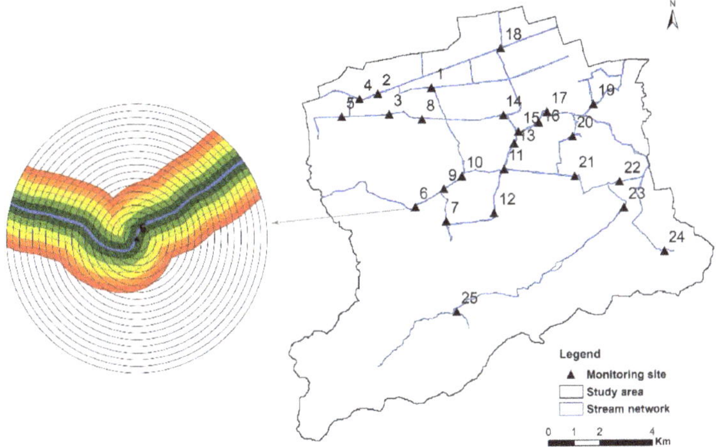

Figure 3. Spatial distribution of surface water quality monitoring sites. The left subfigure shows the buffering schema, i.e., the circular buffer and riparian buffer, used in this study. The interval of the neighboring buffers is 50 m. The circular buffer is a series of buffers around the sampling point with radiuses ranging from 50 to 1000 m, and the riparian buffer is along the particular stream bank with distances ranging from 50 to 300 m.

To explore the spatial scale effects of land use on water quality, two kinds of buffer scales were utilized, i.e., the riparian buffer zone along a specific stream and the circular buffer zone around the sampling point. In the literature, the buffer extent (width or radius) generally ranges from 50 to 1000 m, even 2000 m, according to the topography or sampling density in a case-specific manner, and the intervals between neighboring buffer zones are usually 50 to 100 m [28,41,42]. In this study, a series of buffers with varying widths/radiuses were tested to screen the optimal buffering schema. For the circular buffer scale, 20 buffering zones with radiuses ranging from 50 to 1000 m and an interval of 50 m were created around each sampling point. For the riparian buffer scale, six total buffering zones with widths ranging from 50 to 300 m with an interval of 50 m were created for each specific stream with a sampling point (Figure 3).

2.4. Landscape Metrics

It is acknowledged that there are correlations between regional landscape patterns and the hydrological, physical, chemical and biological indicators of a specific stream system [20,43–45]. We selected four categories of landscape metrics, i.e., fragmentation, dominance, connectedness and aggregation and shape complexity, to study the sensitivity of water quality on land use patterns at the class levels (Table 2).

Table 2. Description of the landscape metrics selected at the class levels.

Attributes	Name (Abbreviation)	Unit	Description
Fragmentation	Edge density (ED)	m/ha	Total length of all edge segments divided by total area for the corresponding patch type
	Patch density (PD)	n/km^2	Number of patches of the corresponding patch type per unit area
Dominance	Percentage of landscape (PLAND)	%	Percentage of the landscape comprised of the corresponding patch type
	Largest patch index (LPI)	%	Proportion of total area occupied by the largest patch of a patch type
Shape complexity	Mean shape index (SHMN)	-	Mean patch perimeter divided by the minimum perimeter of the corresponding land use area
	Mean fractal dimension index (FDMN)	-	Sum of 2 times the logarithm of the patch perimeter divided by the logarithm of the total area for the corresponding patch type divided by the number of patches
	Landscape shape index (LSI)	-	Perimeter-to-area ratio for the corresponding class, increasing with irregular shapes
Connectedness and aggregation	Contiguity index (CONTIG)	-	Assessing patch shapes based on the spatial connectedness of cells within a patch
	Cohesion index (COHE)	-	Indicates the physical connectedness of the corresponding patch type
	Aggregation index (AI)	%	Number of like adjacencies involving the corresponding class, divided by the maximum possible number of like adjacencies involving the corresponding land use type

2.5. Statistical Analysis

The correlations between the areas of each land use type and water quality at different spatial scales were analyzed. The Shapiro-Wilk test showed that some land uses and water quality data at certain buffering scales/sampling points did not follow the normal distribution, so we used the Spearman rank correlation, a nonparametric rank-based method which makes no assumption about the distribution of data, in the following analysis.

We also analyzed the relationships between water quality and land use patterns with a multivariate approach. The detrended correspondence analysis (DCA) was applied to evaluate the gradient length of the water quality data and revealed the existence of short gradients less than 3 standard deviations, so the redundancy analysis (RDA) was utilized to explore the relationships among water quality and land use pattern parameters, i.e., the species and environmental variables at the RDA context, respectively. A Monte Carlo permutation test (499 permutations) was used to determine the statistical validity of the RDA, and the significant explanatory variables were selected [46]. The RDAs were performed using the CANOCO 4.5 program.

3. Results

3.1. Land Use Patterns

As shown in the land use map of the study area (Figure 4), the forest land is concentrated in the southern mountain area; the lake and pond scattered in the central part of the study area (mainly concentrated in the Hemu and Wuchang wetlands) featured by a dense stream network, ponds and broken forest plots; the cropland concentrated in the north-central area; the industrial land mainly distributed in the north, west and south areas in a state of agglomeration and the large chunks of residential land mainly scattered along the major traffic roads at the eastern and southern areas, as well as the western area.

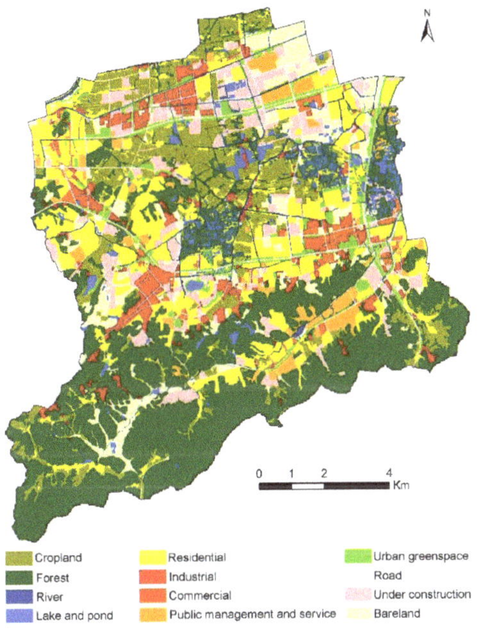

Figure 4. Land use map of the study area.

Figure 5 shows the average composition of land uses at each buffer scale. Overall, the ratios of the pervious surface and impervious surface were roughly equal. The agricultural, forest and residential lands were the dominant land uses. For each buffer zone, the buffer area and the areas of each land use were calculated, respectively. We summed up the areas of each land use in all buffer zones and calculated the ratios of the total areas of each land use to the total area of all the buffers. The following are the results for each land use. The ratios of the total area of the agricultural, forest and residential lands to the total area of all the buffers were 45% and 52% for the circular and riparian buffers, respectively; the ratios of the three land uses were 16.2%, 13.3% and 19.6%, respectively, at the riparian buffer scale, slightly higher than the corresponding ratios, i.e., 15.8%, 12.1% and 16.8% at the circular buffer scale. Taking the road, under construction and industrial land uses as a whole, the ratios were 25.5% and 20.8% for the circular and riparian buffers, respectively. The ratios of river, urban greenspace, lake and pond and bareland were all about 25% at the circular and riparian buffer scales. The commercial and public management and service had the smallest land areas, and the ratios at both buffering schemas were less than 5%.

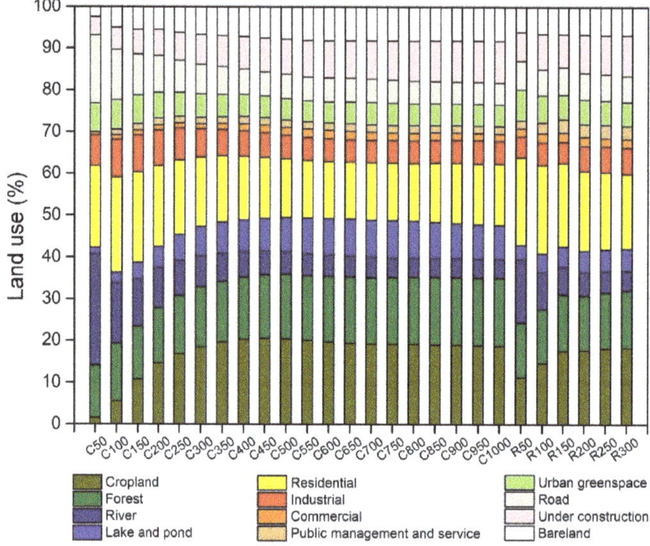

Figure 5. Proportions of each land use at different buffer scales. The labels at the x-axis indicate different buffer zones for the circular and riparian buffer scales, respectively. The initial character "C" or "R" represents the circular or riparian buffer types, and the numerical values after that character means the corresponding buffer radius. For example, "C50" means a circular buffer with a radius of 50 m around a sampling point. Hereinafter, the same.

3.2. Characteristics of Water Quality

The water quality analysis results of the sampling points are presented in Table S1 in the Supplementary Materials. The statistical analysis of the water quality parameters of the 25 river sections is shown in Table 3. The mean values of these parameters were compared with two Chinese national standards of water quality, i.e., the Environmental Quality Standards for Surface Water (EQSSW) (GB3838-2002, State Environmental Protection Administration of China) and the Sanitary Standards for Drinking Water Quality (SSDWQ) (GB 5749-2006). According to the EQSSW, the water quality is classified into five classes; the higher the class, the worse the water quality. Since Mn and Ni are not listed in the EQSSW, we referred the corresponding thresholds listed in the SSDWQ.

The average concentrations of the TN and TP exceeded the Class V surface water concentration limits, i.e., 2.0 and 0.4 mg/L for the TN and TP, respectively. For the TN, there were 16% and 80% sampling points, except No. 19 (Class IV), that were Class V and beyond; for the TP, there were 12%, 24%, 8% and 52% sampling points in Class III, IV, V and beyond, respectively, except No. 25 (Class II). The heavy metals at most of the sampling points were far below the surface water concentration limit of Class I or the limits in the SSDWQ, except No. 16 with Cr in Class II and No. 3 with Mn over the limits in the SSDWQ. Additionally, there were 40%, 84% and 92% sampling points that detected no Ni, Cd and Zn, respectively; all sampling points detected no Cu.

The mean concentrations of $Chla_{Cyan}$, $Chla_{Chlo}$ and $Chla_{Baci\text{-}Dino}$ were 8.39, 63.56 and 25.04 μg/L, respectively. The $Chla_{Baci\text{-}Dino}$ in the Hemu and Wuchang wetlands were higher than the other areas. The mean value of TChla was 96.99 μg/L, a typical eutrophic water quality according to the Organization for Economic Co-Operation and Development (OECD) eutrophication evaluation criteria [47]. It was noted that there were 64% sampling points that detected no $Chla_{Cyan}$.

Table 3. Statistics of the stream water quality parameters. TN: total nitrogen and TP: total phosphorus.

Indicator	Maximum	Minimum	Mean (Std.)
TN (mg/L)	31.69	1.34	6.46 (6.73)
TP (mg/L)	2.80	0.09	0.61 (0.62)
As (μg/L)	8.20	0.00	4.48 (2.28)
Cd (μg/L)	1.00	0.00	0.10 (0.27)
Cr (μg/L)	12.30	0.00	3.66(3.36)
Cu (μg/L)	0.00	0.00	0.00 (0.00)
Mn (μg/L)	676.40	0.00	30.40 (134.80)
Ni (μg/L)	1.70	0.00	0.32 (0.43)
Pb (μg/L)	5.40	0.00	1.98 (1.43)
Zn (μg/L)	2.00	0.00	0.13 (0.46)
$Chla_{Cyan}$ (μg/L) [1]	130.43	0.00	8.39 (26.52)
$Chla_{Chlo}$ (μg/L) [2]	233.29	0.00	63.56 (69.46)
$Chla_{Baci-Dino}$ (ug/L) [3]	138.74	0.00	25.04 (31.18)
TChla (μg/L) [4]	328.84	2.08	96.99 (83.75)

[1] $Chla_{Cyan}$: chlorophyll a concentration of Cyanophyta. [2] $Chla_{Chlo}$: chlorophyll a concentration of Chlorophyta. [3] $Chla_{Baci-Dino}$: chlorophyll a concentration of Bacillariophyta and Dinophyta. [4] TChla: total chlorophyll a concentration.

3.3. Land Use Types and Water Quality

The correlation analysis showed that the nutrient concentration parameters, i.e., TN and TP, had certain correlations with the land use area at different buffer scales (Table S3 in the Supplementary Materials). Under the circular buffer scale, the TN had significant positive correlations with land areas of industrial, road, urban greenspace and commercial land uses, respectively. Specifically, the TN had significant positive correlations with commercial and urban greenspace land use areas at relatively small buffer scales ranging from 50–100 m and 50–400 m, respectively. The TN also had significant positive correlations with the industrial land use areas at relatively large buffer scales (450–1000 m), whilst the TN significantly positively correlated with the road areas at all buffer scales (50–1000 m). On the contrary, the TN had negative significant correlations with the areas of forest, lake and pond (hereafter referred to as wetland) and river land uses. For forests, the significant correlations could be observed at buffer scales ranging from 50–250 m, while, for the river as well as wetland, the ranges were 450–1000 m and 700–1000 m, respectively (Figure 6). Compared with the TN, there were less land uses that significantly correlated with the TP. At relatively large buffer scales, e.g., 650–750 m and 550–1000 m, the TP showed significant positive correlations with the areas of residential and industrial land uses, respectively, while, at small buffer scales (e.g. 50–100 m), the TP had significant positive correlations with commercial land areas (Figure 7). Under the riparian buffer scales, the TN was significantly negatively correlated with the forest areas, while the TP was significantly positively correlated with the areas of urban greenspaces.

The correlation analysis between the algae biomass and land use areas showed that the TChla and $Chla_{Baci-Dino}$ had significant correlations with land uses at certain buffer scales (Table S4 in the Supplementary Materials). Under the riparian buffer scales, the TChla had significant positive correlations with the areas of river, wetland, industrial, urban greenspace, road and under construction land uses (Figure 8); the $Chla_{Baci-Dino}$ had significant positive correlations with the areas of river, wetland and residential land uses (Figure 9). Under the circular buffer scale, the $Chla_{Baci-Dino}$ showed significant positive correlations with the areas of forest, river and wetland at the ranges of 100–300 m, 50–1000 m and 150–1000 m, respectively, as well as negative correlations with the urban greenspace and road areas at the ranges of 650–1000 m and 150–1000 m, respectively. In addition, we found that the areas of wetland and road both had significant positive correlations with the $Chla_{Chlo}$. Since most of the sampling points detected no Cyanophyta, no further analysis was carried out for this parameter.

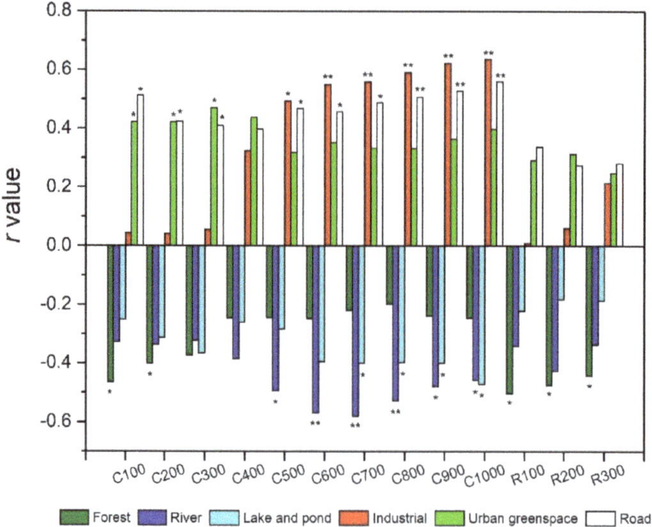

Figure 6. Correlations between TN and the areas of particular land uses under the circular and riparian buffer scales at different spatial scales based on a Spearman's rank correlation analysis. The one-asterisk (two-asterisk) superscripts denote above 90% (95%) confidence levels. Note that only the land uses that significantly correlated with the TN were shown.

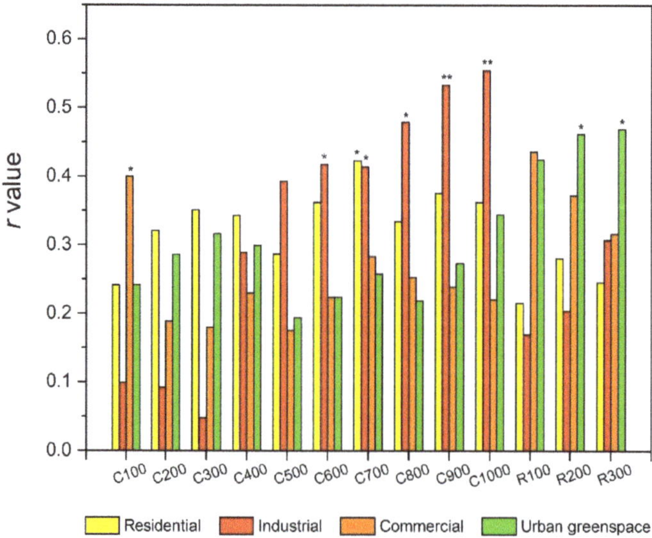

Figure 7. Correlations between TP and the areas of particular land uses under the circular and riparian buffer scales at different spatial scales based on a Spearman's rank correlation analysis. The one-asterisk (two-asterisk) superscripts denote above 90% (95%) confidence levels. Note that only the land uses that significantly correlated with the TP were shown.

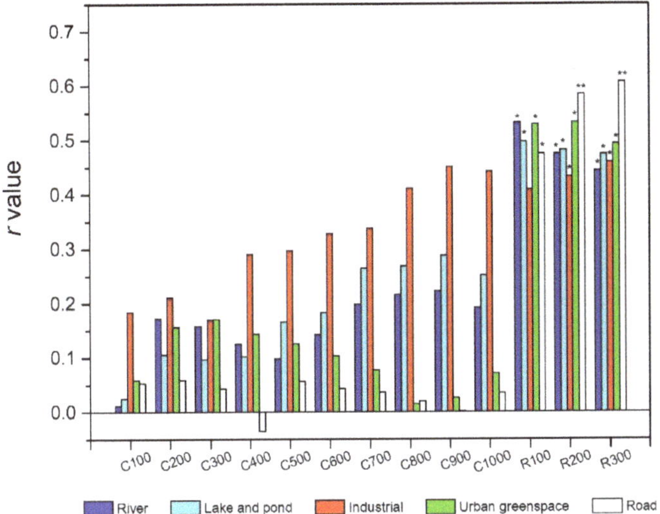

Figure 8. Correlations between TChla and the areas of particular land uses under the circular and riparian buffering scales at different spatial scales based on a Spearman's rank correlation analysis. The one-asterisk (two-asterisk) superscripts denote above 90% (95%) confidence levels. Note that only the land uses that significantly correlated with TChla were shown.

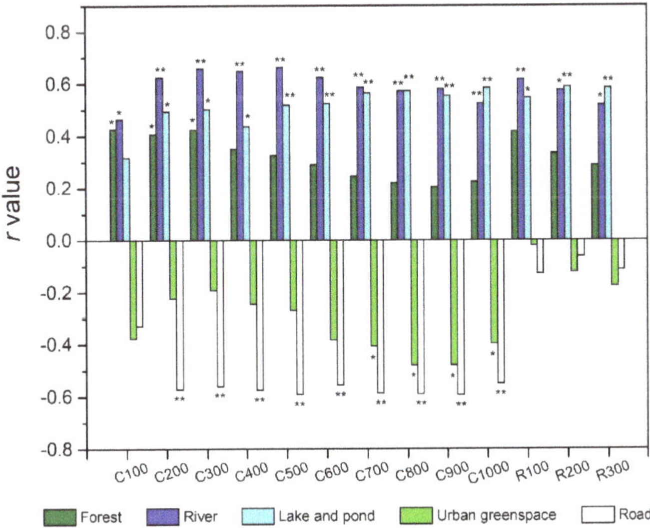

Figure 9. Correlations between $Chla_{Baci\text{-}Dino}$ and the areas of particular land use types under the circular and riparian buffering scales at different spatial scales based on a Spearman's rank correlation analysis. The one-asterisk (two-asterisk) superscripts denote above 90% (95%) confidence levels. Note that only the land uses that significantly correlated with the $Chla_{Baci\text{-}Dino}$ were shown.

The heavy metal concentrations also showed significant correlations with the areas of different kinds of land uses (Table S5 in the Supplementary Materials). For example, Cr and Ni had significant positive correlations with the area of industrial land, Mn and Ni both showed significant positive correlations with the areas of road and urban greenspace, As and Mn had significant positive and negative correlations with the agricultural land area, respectively and Cr and Pb had similar correlations with residential land areas; in addition, Pb also had significant positive correlations with the areas of wetland as well as the residential land uses and a significant negative correlation with the under construction land area. Since most of the sampling points did not detect Cd, Cu and Zn, no further analysis on these heavy metals was carried out.

3.4. Land Use Patterns and Water Quality

We did a RDA at several typical buffer scales, i.e., 100, 500 and 1000 m, for the circular buffer and 200 m for the riparian buffer, respectively. The percentages of variance explained by the ordination axis are shown in Table 4. The RDA ordinations showed that there were statistically significant relationships between water quality and different numbers of explanatory variables, i.e., 5, 9 and 15 at 100, 500 and 1000-m circular buffer scales, respectively; at each scale, the landscape metrics could explain 39.5%, 71.7% and 80.9% of the variations in water quality, respectively. For the 200-m riparian buffer scale, there were statistically significant relationships between water quality and nine explanatory variables, and 60.1% of the variation in water quality could be explained by these variables. More than 40% water quality variation was explained by the first two axes, in which the first axis explained about twice as much as the second axis.

Table 4. Variations in water quality explained by land uses and landscape metrics at different buffer scales in the redundancy analysis (RDA).

Scales	Explained Variation (%)			Explanatory Variables Selected ($p < 0.05$)
	Axis 1	Axis 2	All Axes	
100-m circular buffer	26.0	12.2	39.5	31COHE, 31PLAND, 46LPI, 46PLAND and 46COHE
500-m circular buffer	43.9	25.2	71.6	21LSI, 21PLAND, 31LPI, 31PLAND, 42COHE, 42CONTIG, 42PLAND, 46LPI and 46PLAND
1000-m circular buffer	47.4	23.7	80.9	21AI, 21CONTIG, 21FRMN, 21LSI, 21SHMN, 31LPI, 32ED, 32LSI, 32PLAND, 42COHE, 42CONTIG, 42PLAND, 46ED, 46LPI and 46PLAND
200-m riparian buffer	38.2	20.3	60.1	32ED, 32PD, 32LSI, 32PLAND, 42AI, 42CONTIG, 46AI, 46CONTIG and 46SHMN

Notes: Landscape metrics include the edge density (ED), patch density (PD), percentage of landscape (PLAND), largest patch index (LPI), mean shape index (SHMN), mean fractal dimension index (FDMN), landscape shape index (LSI), contiguity index (CONTIG), cohesion index (COHE) and aggregation index (AI). The land use metrics refer to the following land use types, i.e., 11 (cropland), 21 (forest), 31 (river), 32 (wetland), 42 (industrial) and 46 (road). The p-values were derived from Monte Carlo permutation tests (499 permutations) of all of the canonical axes.

Figure 10 shows ordination diagrams derived from the RDA using water quality variables and significant explanatory variables for the four typical buffer scales. Under all scales, the first axis represents variables related with eutrophication, which was mainly related to the industrial land use. The second axis consistently displayed a gradient of nutrients, which was mainly positively correlated with roads and negatively correlated with wetlands. Particularly, the LPI and PLAND of roads were positively correlated with the TN and TP, whilst negatively correlated with $Chla_{Baci\text{-}Dino}$ at 100, 500 and 1000-m circular buffer scales. Under 500 and 1000-m circular buffer scales, the PLAND of industrial land use was positively correlated with the TN and TP; except the TN and TP, the landscape metrics COHE and CONTIG were also positively correlated with the TChla and $Chla_{Chlo}$. The CONTIG and AI of the road and industrial land uses were positively correlated with the TP, TN, TChla and $Chla_{Chlo}$, whilst negatively correlated with the $Chla_{Baci\text{-}Dino}$ at 200-m the riparian buffer scale. The SHMN of the roads was negatively correlated with the TP, TN, TChla and $Chla_{Chlo}$.

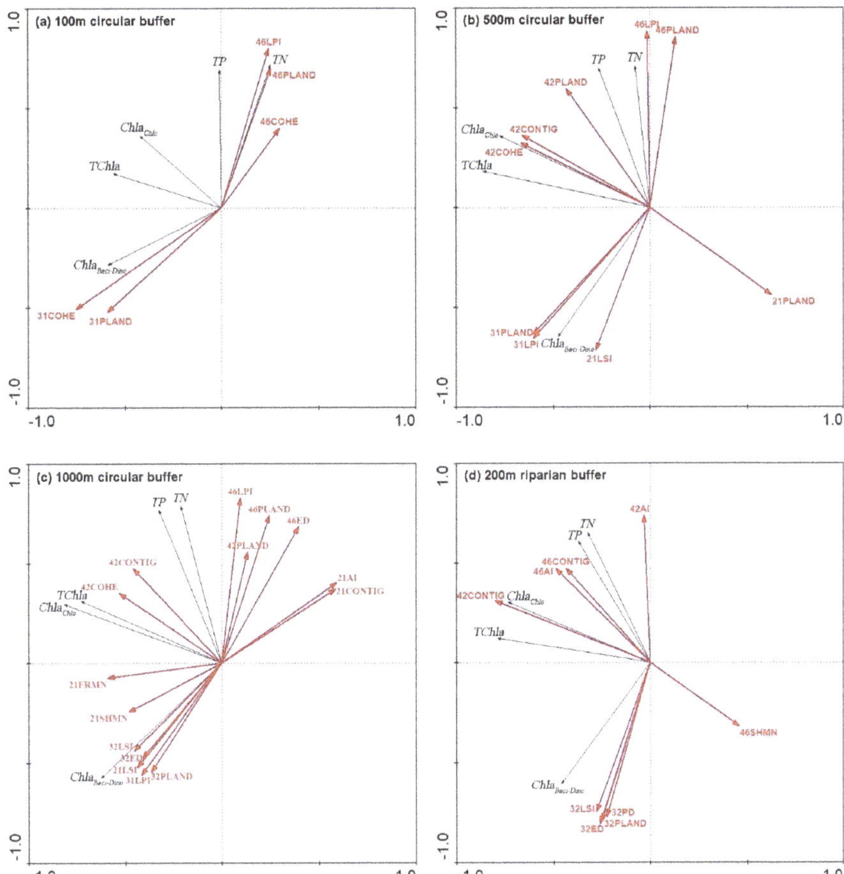

Figure 10. Biplots of water quality parameters (represented by dark blue lines) and landscape metrics (represented by red lines) within the circular buffers of 100 m (**a**), 500 m (**b**) and 1000 m (**c**) and the riparian buffer of 200m (**d**) according to the redundancy analysis (RDA). The length of the arrow represents the standard deviation of variables in the sorting space, and the direction indicates the change of gradient direction. The cosine values of the arrow and axis represent the correlation.

The PLAND and LPI of the rivers were positively correlated with the $Chla_{Baci\text{-}Dino}$, whilst negatively correlated with the TN and TP at 100, 500 and 1000-m circular buffer scales. The PLAND, LSI and ED metrics of the wetlands were positively correlated with the $Chla_{Baci\text{-}Dino}$, while negatively correlated with the TN and TP at the 1000-m circular buffer scale and 200-m riparian buffer scale, respectively.

The PLAND of the forest land use was negatively correlated with the TN, TP, TChla and $Chla_{Chlo}$, and the LSI was positively correlated with the $Chla_{Baci\text{-}Dino}$ but negatively correlated with the TN and TP at the 500-m circular buffer scale. The FRMN, SHMN and LSI metrics of the forest land use were positively correlated with the $Chla_{Baci\text{-}Dino}$, $Chla_{Chlo}$ and TChla, but the AI and CONTIG were negatively correlated with these three water quality parameters at the 1000-m circular buffer scale.

4. Discussion

4.1. Water Pollution in Urbanized Areas

The land use pattern in the study area was mainly dominated by the urbanization process. The proportion of the impervious surface area was over 50% if adding up all the buffer zones used in this study—of which, the proportions of the residential, under construction, road, industrial, commercial and public management and service land uses decreased in turn (Figure 4). The less proportion of the commercial, as well as the public management and service land use, areas, compared with that of the residential land, suggested that the study area was still in the suburban urbanization process. It has been acknowledged that the early stages of rapid urbanization are often the beginning of the accelerated deterioration of urban surface water environments [48]. In the context of rapid urbanization, the TN and TP in stream water usually originated from industrial, living and agricultural activities [49,50]. The sampling data showed that the variations of water quality data were relatively large, and the concentrations of TN and TP were much higher than the Class V thresholds defined in the EQSSW (Table 3), a typical heavily polluted stream water environment.

4.2. Effects of Land Uses Types on Stream Water Quality

4.2.1. Nutrient and Heavy Metal Concentrations

The construction land area was an important factor to explain the change of TN concentration in stream waters (Figure 6). Under the circular buffer scale, the industrial, commercial, road and urban greenspace land uses were the primary sources of nitrogen pollution, whilst the areas of forest, wetland and river had significant negative correlations with the TN, showing effective degradation and purification effects on nitrogen pollution [23,51–54]. If only considering the TN, more than 80% sampled streams were beyond Class V, and this could be closely related with the densely distributed industrial land in this region, and further, the associated pollution might have exceeded the pollution load capacity of local aquatic ecosystems.

Compared with the TN, the TP had relative weak correlations with particular land uses. We found the TP had significant correlation with the industrial land area but no significant correlations with the area of forest, as well as wetland (Figure 7). This suggests that the TN and TP might have varied sensitivities to land use patterns, and the TN might be a more sensitive indicator of land use changes. We suggested that the TN might be more heavily influenced by human activities coupled with specific land use patterns, while the TP might be more controlled by strong point source emissions and had relative weak correlations with land use changes [55,56].

There was uncertainty in the relationships between the cropland area and concentrations of the TN and TP. The cropland area negatively correlated with the TN and TP, but neither was statistically significant, so it was not an appropriate factor that can be closely related to the eutrophication of water quality in this study area, despite in some cases where the cropland area was proved to be a promising predictor [57,58]. More specifically, this uncertainty could be closely related with local topography, cultivation practices and fertilizer usages [59–61]. The densely distributed stream network, including numerous ponds, made this study area more like a wetland from the functional perspective, and this kind of special geographical environment may have greatly weakened the pollution concentration coming from agriculture.

Generally, the construction land is the main source of heavy metal pollution, while the forest and wetland could absorb and promote the transformation of heavy metals [58,62]. In this study, we found significant positive correlations between heavy metals (e.g., Ni) and the areas of industrial, road and urban greenspace land uses, including significant negative correlations with the forest area at certain buffer scales. Similar to the cases of the TN and TP, there were also uncertainties in correlations between heavy metals and the cropland area, and both positive and negative correlations could be found (e.g., As and Mn). It was noted that there were relationships hard to interpret between certain heavy metals and some specific land uses (e.g., wetland and residential land uses); we suggested that this might be connected with the regional background sources [63,64].

The urban greenspace is a kind of urban construction land with special ecological significance, but we found that the area of urban greenspace, similar to the road, usually had significant positive correlations with the TN, TP and Mn at multiple buffer scales. Considering that the study area was still in the process of rapid urbanization, large parts of the greenspace distributed along with the road and traffic facilities, so it could be a main sink of vehicle pollutant emissions.

4.2.2. Algae Biomass

The TN and TP showed close relationships with the algae biomass (Table S2 in the Supplementary Materials), as shown by the positive correlations with the TChla and $Chla_{Chlo}$, respectively. However, the relationship between the algae biomass and land use pattern was still not straightforward. For example, it was not surprising that the TChla positively correlated with the areas of construction lands, i.e., the industrial, road, urban greenspace and under construction land uses, at the riparian buffer scales (Figure 8). However, our results showed that the TChla positively correlated with the wetland areas, which was against the common consensus that the wetlands, e.g., stream, as well as lake and pond, could reduce the concentration of the TN and TP. It was suggested that, except for the concentrations of the TN and TP, the structure and growth of algae can also be affected by multiple factors, e.g., the ratio between the TN and TP, water pH and dissolved oxygen [65,66].

It was noted that the $Chla_{Baci\text{-}Dino}$ was not sensitive to the TN and TP (Table S2 in the Supplementary Materials), whilst it showed significant negative correlations with the areas of construction lands, i.e., road and urban greenspace, and significant positive correlations with the forest and wetlands areas (Figure 9). In terms of spatial distribution, the $Chla_{Baci\text{-}Dino}$ in the central part of the study area (Hemu and Wuchang wetlands) was relatively higher (Figure S1 in the Supplementary Materials). We suggested that this anomaly might be caused by the specific community structure of the $Chla_{Baci\text{-}Dino}$ in the study area [67]. The wetlands in the study area were mainly composed of the stream network and cropland with closed ponds at the interior, and the impact of this complex land use composition on the structure and stability of the algae community still needs further study.

4.3. Effects of Land Use Patterns on Stream Water Quality

The land use patterns could explain more than 40% variation in water quality at all spatial scales (Table 4). This confirmed that land use patterns had a strong impact on water quality in the study area. For example, we found that the TN and TP were positively correlated with landscape metrics of the PLAND, LPI, CONTIG, COHE and AI of the road and industrial land uses, whilst negatively correlated with the PLAND and LPI of the wetland and river land uses in the RDA (Figure 10). The literature also suggested that the PLAND, LPI and AI are influential metrics on water quality [38,43,68,69].

We found that the larger ED, or higher LSI of the wetland, both correlated with a larger $Chla_{Baci\text{-}Dino}$ at the 1000-m circular buffer scale. In addition, the higher LSI, SHMN and FDMN or lower AI and CONTIG of the forest were also positively correlated with the $Chla_{Baci\text{-}Dino}$ (Figure 10). This suggested that the $Chla_{Baci\text{-}Dino}$ biomass may have close relationships with the complexity and fragmentation characteristics of land use composition, and it emphasized the importance of the spatial structure of land use in this kind of study.

It was noted that the percentages of the explained variations of water quality by the CONTIG and COHE (connectedness); PLAND and LPI (dominance) and SHMN, FAMN and LSI (complexity) were very close. This finding might be helpful in water quality management practices, in that different kinds of compositions of landscape metrics could have similar explanation capabilities in water quality variations.

4.4. Influence of Spatial Scale on Land Use-Water Quality Relationships

Our results showed that the response characteristics of stream water quality to land use were spatial scale-dependent. Under the circular buffer scale, the land use was more closely related with the TN, but the corresponding buffer scales varied with particular land uses (Figure 6). The forest and urban greenspace were more influential to the TN at small buffer scales, and there was a tendency that the correlations (positive) between the TN and industrial land area increased, whilst the correlations (negative) generally decreased with the areas of wetland and river as the buffer scales increased. This suggests that the TN and TP could be partially mitigated by a more rational spatial arrangement of the lands of industry and forest or wetlands [70,71]. We also found strong correlations between land use and algae biomass at the riparian buffer scales (Figure 8), and this implies that the riparian buffer zone could play a key role in the conservation of aquatic ecosystems [72].

The land use patterns also showed scale effects on the water quality in the study area. The variation in water quality explained by the landscape metrics increased with the increasing of the spatial scale, and the total explained variation was up to 80% at the 1000-m circular buffer scale (Table 4). It means that the land use pattern could have continuously improving explanatory capabilities on the water quality, and this finding is consistent with other studies [25,29,68]. In this study, the maximum buffer width was 1000 m to avoid overlap with the neighboring buffer zones, but it would be worth it to explore what would be the relationship if the buffer scale further increased if data allowed.

5. Conclusions

The construction land area was an important factor to explain the variation of the TN concentrations in stream water. At the circular buffer scales, the industrial, commercial, road and urban greenspace land use areas significantly positively correlated with the TN; whilst the forest, wetland and river land use areas had significant negative correlations with the TN. The response characteristics of the stream water quality to land use were spatial scale-dependent. Under the circular buffer scale, the TN was more closely related with the land use status, but the corresponding buffer scales varied with the particular land use types. The forest and urban greenspace were more influential to the TN at small buffer scales, whilst significant positive or negative correlations could be found between the TN and the areas of industrial land or the wetland and stream as the buffer scales increased. Therefore, it might be possible to more quantitatively manage the water environment by identifying the particular buffer scales that closely relate with specific water quality parameters, for example, a more rational spatial arrangement of the industry land at larger spatial scales and the forest at smaller spatial scales to mitigate the TN and TP. In addition, there were more relevant correlations between the land use and algae biomass at the riparian buffer scales, and this suggests that the riparian buffer zone could play a key role in the conservation of aquatic ecosystems.

The land use pattern was an important factor influencing the water quality. The variations in water quality explained by landscape metrics were greater than 40% at all circular and riparian buffer scales. The increased landscape metrics of dominance and connectedness and aggregation of the industrial and road land uses, as well as the decreased landscape dominance metrics of the wetland and river, were correlated with the increased TN and TP. The land use pattern showed a scale effect on the water quality. The variation in water quality explained by the landscape metrics increased with the increasing of the buffer size, and it implies that the land use pattern could have a closer correlation with the water quality at larger spatial scales. This suggests that more attention should be paid to the landscape patterns at relatively larger spatial scales in water environment management practices.

Supplementary Materials: The following are available online at http://www.mdpi.com/2073-4441/12/4/1123/s1, Figure S1: Spatial variation of the (a) TN, (b) TP, (c) TChla, (d) $Chla_{Cyan}$, (e) $Chla_{Chlo}$ and (f) $Chla_{Baci\text{-}Dino}$. The range of the corresponding concentration values were classified into six classes based on the natural breaks method. Table S1: Water quality analysis results of the sampling points. Table S2: Correlation coefficients between the stream nutrient concentration, heavy metals and algal biomass based on a Spearman's rank order correlation analysis. Table S3: Correlations between the TN/TP and land uses at different spatial scales based on a Spearman's rank correlation coefficient. Table S4: Correlations between the TChla, $Chla_{Chlo}$, $Chla_{Baci\text{-}Dino}$ and land use at different spatial scales based on a Spearman's rank correlation coefficient. Table S5: Correlations between heavy metal concentrations and land uses at different spatial scales based on a Spearman's rank correlation coefficient.

Author Contributions: Conceptualization, Y.S., X.S., G.S. and T.H.; data curation, Y.S., X.S. and T.H.; formal analysis, Y.S. and X.S.; investigation, Y.S. and X.S.; methodology, Y.S., X.S., G.S. and T.H.; software, Y.S.; supervision, X.S. and G.S.; validation, G.S.; writing—original draft, Y.S. and writing—review and editing, X.S., G.S. and T.H. All authors have read and agreed to the published version of the manuscript.

Funding: This research was funded by the Natural Science Foundation of Zhejiang Province, China (grant number: LQ13C030007), Entrepreneurship and Innovation Project for High-level Overseas Returnees in Hangzhou City in 2019 and the Science and Technology Project of the Water Resources Department of Zhejiang Province (grant number: RC1810).

Conflicts of Interest: The authors declare no conflicts of interest.

References

1. Vörösmarty, C.J.; McIntyre, P.B.; Gessner, M.O.; Dudgeon, D.; Prusevich, A.; Green, P.; Glidden, S.; Bunn, S.E.; Sullivan, C.A.; Liermann, C.R. Global threats to human water security and river biodiversity. *Nature* **2010**, *467*, 555–561. [CrossRef] [PubMed]
2. Arnold, C.L., Jr.; Gibbons, C.J. Impervious surface coverage: The emergence of a key environmental indicator. *J. Am. Plan. Assoc.* **1996**, *62*, 243–258. [CrossRef]
3. Giri, S.; Qiu, Z. Understanding the relationship of land uses and water quality in Twenty First Century: A review. *J. Environ. Manag.* **2016**, *173*, 41–48. [CrossRef] [PubMed]
4. Dewan, A.M.; Yamaguchi, Y. Land use and land cover change in Greater Dhaka, Bangladesh: Using remote sensing to promote sustainable urbanization. *Appl. Geogr.* **2009**, *29*, 390–401. [CrossRef]
5. Ren, W.; Zhong, Y.; Meligrana, J.; Anderson, B.; Watt, W.E.; Chen, J.; Leung, H.-L. Urbanization, land use, and water quality in Shanghai: 1947–1996. *Environ. Int.* **2003**, *29*, 649–659. [CrossRef]
6. Chen, Q.; Mei, K.; Dahlgren, R.A.; Wang, T.; Gong, J.; Zhang, M. Impacts of land use and population density on seasonal surface water quality using a modified geographically weighted regression. *Sci. Total Environ.* **2016**, *572*, 450–466. [CrossRef]
7. Tong, S.T.; Chen, W. Modeling the relationship between land use and surface water quality. *J. Environ. Manag.* **2002**, *66*, 377–393. [CrossRef]
8. Deng, J.S.; Wang, K.; Hong, Y.; Qi, J.G. Spatio-temporal dynamics and evolution of land use change and landscape pattern in response to rapid urbanization. *Landsc. Urban Plan.* **2009**, *92*, 187–198. [CrossRef]
9. Delphin, S.; Escobedo, F.; Abd-Elrahman, A.; Cropper, W. Urbanization as a land use change driver of forest ecosystem services. *Land Use Policy* **2016**, *54*, 188–199. [CrossRef]
10. Göbel, P.; Dierkes, C.; Coldewey, W. Storm water runoff concentration matrix for urban areas. *J. Contam. Hydrol.* **2007**, *91*, 26–42. [CrossRef]
11. McGrane, S.J. Impacts of urbanisation on hydrological and water quality dynamics, and urban water management: A review. *Hydrol. Sci. J.* **2016**, *61*, 2295–2311. [CrossRef]
12. Kannel, P.R.; Lee, S.; Kanel, S.R.; Khan, S.P.; Lee, Y.-S. Spatial–temporal variation and comparative assessment of water qualities of urban river system: A case study of the river Bagmati (Nepal). *Environ. Monit. Assess.* **2007**, *129*, 433–459. [CrossRef] [PubMed]
13. Booth, D.B.; Roy, A.H.; Smith, B.; Capps, K.A. Global perspectives on the urban stream syndrome. *Freshw. Sci.* **2016**, *35*, 412–420. [CrossRef]
14. Zhang, Y.; Wang, X.-N.; Ding, H.-Y.; Dai, Y.; Ding, S.; Gao, X. Threshold Responses in the Taxonomic and Functional Structure of Fish Assemblages to Land Use and Water Quality: A Case Study from the Taizi River. *Water* **2019**, *11*, 661. [CrossRef]
15. Brabec, E.; Schulte, S.; Richards, P.L. Impervious surfaces and water quality: A review of current literature and its implications for watershed planning. *J. Plan. Lit.* **2002**, *16*, 499–514. [CrossRef]
16. Schiff, R.; Benoit, G. Effects of Impervious Cover at Multiple Spatial Scales on Coastal Watershed Streams 1. *J. Am. Water Resour. Assoc.* **2007**, *43*, 712–730. [CrossRef]
17. Zampella, R.A.; Procopio, N.A.; Lathrop, R.G.; Dow, C.L. Relationship of Land-Use/Land-Cover Patterns and Surface-Water Quality in The Mullica River Basin. *J. Am. Water Resour. Assoc.* **2007**, *43*, 594–604. [CrossRef]

18. Klein, R.D. Urbanization and stream quality impairment. *J. Am. Water Resour. Assoc.* **1979**, *15*, 948–963. [CrossRef]
19. Dai, X.; Zhou, Y.; Ma, W.; Zhou, L. Influence of spatial variation in land-use patterns and topography on water quality of the rivers inflowing to Fuxian Lake, a large deep lake in the plateau of southwestern China. *Ecol. Eng.* **2017**, *99*, 417–428. [CrossRef]
20. Alberti, M.; Booth, D.; Hill, K.; Coburn, B.; Avolio, C.; Coe, S.; Spirandelli, D. The impact of urban patterns on aquatic ecosystems: An empirical analysis in Puget lowland sub-basins. *Landsc. Urban Plan.* **2007**, *80*, 345–361. [CrossRef]
21. Buck, O.; Niyogi, D.K.; Townsend, C.R. Scale-dependence of land use effects on water quality of streams in agricultural catchments. *Environ. Pollut.* **2004**, *130*, 287–299. [CrossRef]
22. Johnson, L.; Richards, C.; Host, G.; Arthur, J. Landscape influences on water chemistry in Midwestern stream ecosystems. *Freshw. Biol.* **1997**, *37*, 193–208. [CrossRef]
23. Sliva, L.; Williams, D.D. Buffer zone versus whole catchment approaches to studying land use impact on river water quality. *Water Res.* **2001**, *35*, 3462–3472. [CrossRef]
24. de Mello, K.; Valente, R.A.; Randhir, T.O.; dos Santos, A.C.A.; Vettorazzi, C.A. Effects of land use and land cover on water quality of low-order streams in Southeastern Brazil: Watershed versus riparian zone. *Catena* **2018**, *167*, 130–138. [CrossRef]
25. Zhou, T.; Wu, J.; Peng, S. Assessing the effects of landscape pattern on river water quality at multiple scales: A case study of the Dongjiang River watershed, China. *Ecol. Indic.* **2012**, *23*, 166–175. [CrossRef]
26. Hurley, T.; Mazumder, A. Spatial scale of land-use impacts on riverine drinking source water quality. *Water Resour. Res.* **2013**, *49*, 1591–1601. [CrossRef]
27. Zhang, W.; Li, H.; Sun, D.; Zhou, L. A statistical assessment of the impact of agricultural land use intensity on regional surface water quality at multiple scales. *Int. J. Environ. Res. Public Health* **2012**, *9*, 4170–4186. [CrossRef]
28. Kolpin, D.W. Agricultural chemicals in groundwater of the midwestern United States: Relations to land use. *J. Environ. Qual.* **1997**, *26*, 1025–1037. [CrossRef]
29. Zhao, J.; Lin, L.; Yang, K.; Liu, Q.; Qian, G. Influences of land use on water quality in a reticular river network area: A case study in Shanghai, China. *Landsc. Urban Plan.* **2015**, *137*, 20–29. [CrossRef]
30. Shen, Z.; Hou, X.; Li, W.; Aini, G.; Chen, L.; Gong, Y. Impact of landscape pattern at multiple spatial scales on water quality: A case study in a typical urbanised watershed in China. *Ecol. Indic.* **2015**, *48*, 417–427. [CrossRef]
31. Lewis, W.M., Jr.; Mccutchan, J.R., Jr. Ecological responses to nutrients in streams and rivers of the Colorado mountains and foothills. *Freshw. Biol.* **2010**, *55*, 1973–1983. [CrossRef]
32. Elliott, J.; Jones, I.; Thackeray, S. Testing the sensitivity of phytoplankton communities to changes in water temperature and nutrient load, in a temperate lake. *Hydrobiologia* **2006**, *559*, 401–411. [CrossRef]
33. Rimet, F. Recent views on river pollution and diatoms. *Hydrobiologia* **2012**, *683*, 1–24. [CrossRef]
34. An, K.-J.; Lee, S.-W.; Hwang, S.-J.; Park, S.-R.; Hwang, S.-A. Exploring the non-stationary effects of forests and developed land within watersheds on biological indicators of streams using geographically-weighted regression. *Water* **2016**, *8*, 120. [CrossRef]
35. Wang, Q.H.; Dong, Y.X.; Zhou, G.H.; Zheng, W. Soil geochemical baseline and environmental background values of agricultural regions in Zhejiang province. *J. Ecol. Rural. Environ.* **2007**, *23*, 81–88. (In Chinese)
36. National Bureau of Standards. *Current Land Use Classification (GB/T 21010-2017)*; China Quality and Standards Publishing & Media Co., Ltd.: Beijing, China, 2017. (In Chinese)
37. Bu, H.; Wei, M.; Yuan, Z.; Wan, J. Relationships between land use patterns and water quality in the Taizi River basin, China. *Ecol. Indic.* **2014**, *41*, 187–197. [CrossRef]
38. Ai, L.; Shi, Z.H.; Yin, W.; Huang, X. Spatial and seasonal patterns in stream water contamination across mountainous watersheds: Linkage with landscape characteristics. *J. Hydrol.* **2015**, *523*, 398–408. [CrossRef]
39. Pan, Y.; Herlihy, A.; Kaufmann, P.; Wigington, J.; Van Sickle, J.; Moser, T. Linkages among land-use, water quality, physical habitat conditions and lotic diatom assemblages: A multi-spatial scale assessment. *Hydrobiologia* **2004**, *515*, 59–73. [CrossRef]
40. Xu, J.; Jin, G.; Tang, H.; Mo, Y.; Wang, Y.-G.; Li, L. Response of water quality to land use and sewage outfalls in different seasons. *Sci. Total Environ.* **2019**, *696*, 134014. [CrossRef]

41. Allan, J.D. Landscapes and Riverscapes: The Influence of Land Use on Stream Ecosystems. *Annu. Rev. Ecol. Evol. Syst.* **2004**, *35*, 257–284. [CrossRef]
42. Damanik-Ambarita, M.N.; Everaert, G.; Goethals, P.L. Ecological models to infer the quantitative relationship between land use and the aquatic macroinvertebrate community. *Water* **2018**, *10*, 184. [CrossRef]
43. Lee, S.-W.; Hwang, S.-J.; Lee, S.-B.; Hwang, H.-S.; Sung, H.-C. Landscape ecological approach to the relationships of land use patterns in watersheds to water quality characteristics. *Landsc. Urban Plan.* **2009**, *92*, 80–89. [CrossRef]
44. Zhang, G.; Guhathakurta, S.; Dai, G.; Wu, L.; Yan, L. The control of land-use patterns for stormwater management at multiple spatial scales. *Environ. Manag.* **2013**, *51*, 555–570. [CrossRef] [PubMed]
45. Zhang, X.; Liu, Y.; Zhou, L. Correlation analysis between landscape metrics and water quality under multiple scales. *Int. J. Environ. Res. Public Health* **2018**, *15*, 1606. [CrossRef] [PubMed]
46. Ter Braak, C.J.; Smilauer, P. *Canoco Reference Manual and User's Guide: Software for Ordination, Version 5.0*; Microcomputer Power: Ithaca, NY, USA, 2012.
47. Vollenweider, R.A.; Kerekes, J.J. *Eutrophication of Waters: Monitoring, Assessment and Control*; OECD: Paris, France, 1982.
48. Hwang, S.-A.; Hwang, S.-J.; Park, S.-R.; Lee, S.-W. Examining the relationships between watershed urban land use and stream water quality using linear and generalized additive models. *Water* **2016**, *8*, 155. [CrossRef]
49. Tromboni, F.; Dodds, W. Relationships between land use and stream nutrient concentrations in a highly urbanized tropical region of Brazil: Thresholds and riparian zones. *Environ. Manag.* **2017**, *60*, 30–40. [CrossRef] [PubMed]
50. Beaulac, M.N.; Reckhow, K.H. An examination of land use—Nutrient export relationships. *J. Am. Water Resour. Assoc.* **1982**, *18*, 1013–1024. [CrossRef]
51. Paul, M.J.; Meyer, J.L. Streams in the urban landscape. *Annu. Rev. Ecol. Syst.* **2001**, *32*, 333–365. [CrossRef]
52. Rodríguez-Romero, A.J.; Rico-Sánchez, A.E.; Mendoza-Martínez, E.; Gómez-Ruiz, A.; Sedeño-Díaz, J.E.; López-López, E. Impact of changes of land use on water quality, from tropical forest to anthropogenic occupation: A multivariate approach. *Water* **2018**, *10*, 1518.
53. Matteo, M.; Randhir, T.; Bloniarz, D. Watershed-scale impacts of forest buffers on water quality and runoff in urbanizing environment. *J. Water Resour. Plan. Manag.* **2006**, *132*, 144–152. [CrossRef]
54. Brogna, D.; Dufrêne, M.; Michez, A.; Latli, A.; Jacobs, S.; Vincke, C.; Dendoncker, N. Forest cover correlates with good biological water quality. Insights from a regional study (Wallonia, Belgium). *J. Environ. Manag.* **2018**, *211*, 9–21. [CrossRef] [PubMed]
55. Caccia, V.G.; Boyer, J.N. Spatial patterning of water quality in Biscayne Bay, Florida as a function of land use and water management. *Mar. Pollut. Bull.* **2005**, *50*, 1416–1429. [CrossRef] [PubMed]
56. Puckett, L.J. Identifying the major sources of nutrient water pollution. *Environ. Sci. Technol.* **1995**, *29*, 408A–414A. [CrossRef]
57. Meador, M.R.; Goldstein, R.M. Assessing water quality at large geographic scales: Relations among land use, water physicochemistry, riparian condition, and fish community structure. *Environ. Manag.* **2003**, *31*, 0504–0517. [CrossRef]
58. Lenat, D.R.; Crawford, J.K. Effects of land use on water quality and aquatic biota of three North Carolina Piedmont streams. *Hydrobiologia* **1994**, *294*, 185–199. [CrossRef]
59. Casalí, J.; Gastesi, R.; Álvarez-Mozos, J.; de Santisteban, L.; de Lersundi, J.D.V.; Giménez, R.; Larrañaga, A.; Goñi, M.; Agirre, U.; Campo, M. Runoff, erosion, and water quality of agricultural watersheds in central Navarre (Spain). *Agric. Water Manag.* **2008**, *95*, 1111–1128. [CrossRef]
60. Ding, J.; Jiang, Y.; Fu, L.; Liu, Q.; Peng, Q.; Kang, M. Impacts of land use on surface water quality in a subtropical River Basin: A case study of the Dongjiang River Basin, Southeastern China. *Water* **2015**, *7*, 4427–4445. [CrossRef]
61. Skaggs, R.W.; Breve, M.; Gilliam, J. Hydrologic and water quality impacts of agricultural drainage. *Crit. Rev. Environ. Sci. Technol.* **1994**, *24*, 1–32. [CrossRef]
62. Bai, J.; Yang, Z.; Cui, B.; Gao, H.; Ding, Q. Some heavy metals distribution in wetland soils under different land use types along a typical plateau lake, China. *Soil Tillage Res.* **2010**, *106*, 344–348. [CrossRef]
63. Ouyang, W.; Wang, Y.; Lin, C.; He, M.; Hao, F.; Liu, H.; Zhu, W. Heavy metal loss from agricultural watershed to aquatic system: A scientometrics review. *Sci. Total Environ.* **2018**, *637*, 208–220. [CrossRef]

64. Lindström, M. Urban land use influences on heavy metal fluxes and surface sediment concentrations of small lakes. *Water Air Soil Pollut.* **2001**, *126*, 363–383. [CrossRef]
65. Smith, V.H. Low nitrogen to phosphorus ratios favor dominance by blue-green algae in lake phytoplankton. *Science* **1983**, *221*, 669–671. [CrossRef]
66. Zelnik, I.; Balanč, T.; Toman, M.J. Diversity and Structure of the tychoplankton diatom community in the limnocrene spring Zelenci (Slovenia) in relation to environmental factors. *Water* **2018**, *10*, 361. [CrossRef]
67. Teittinen, A.; Taka, M.; Ruth, O.; Soininen, J. Variation in stream diatom communities in relation to water quality and catchment variables in a boreal, urbanized region. *Sci. Total Environ.* **2015**, *530*, 279–289. [CrossRef] [PubMed]
68. Ding, J.; Jiang, Y.; Liu, Q.; Hou, Z.; Liao, J.; Fu, L.; Peng, Q. Influences of the land use pattern on water quality in low-order streams of the Dongjiang River basin, China: A multi-scale analysis. *Sci. Total Environ.* **2016**, *551*, 205–216. [CrossRef]
69. Uuemaa, E.; Roosaare, J.; Mander, Ü. Landscape metrics as indicators of river water quality at catchment scale. *Hydrol. Res.* **2007**, *38*, 125–138. [CrossRef]
70. Wang, X. Integrating water-quality management and land-use planning in a watershed context. *J. Environ. Manag.* **2001**, *61*, 25–36. [CrossRef]
71. Mander, Ü.; Tournebize, J.; Tonderski, K.; Verhoeven, J.T.; Mitsch, W.J. Planning and establishment principles for constructed wetlands and riparian buffer zones in agricultural catchments. *Ecol. Eng.* **2017**, *103*, 296–300. [CrossRef]
72. Junior, R.F.V.; Varandas, S.G.; Pacheco, F.A.; Pereira, V.R.; Santos, C.F.; Cortes, R.M.; Fernandes, L.F.S. Impacts of land use conflicts on riverine ecosystems. *Land Use Policy* **2015**, *43*, 48–62. [CrossRef]

Article

Predicting the Trend of Dissolved Oxygen Based on the kPCA-RNN Model

Yi-Fan Zhang [1,*], Peter Fitch [2] and Peter J. Thorburn [1]

[1] Agriculture & Food, CSIRO, Brisbane, QLD 4067, Australia; Peter.Thorburn@csiro.au
[2] Land & Water, CSIRO, Canberra, ACT 2601, Australia; peter.fitch@csiro.au
* Correspondence: yi-fan.zhang@csiro.au; Tel.: +61-452-208-587

Received: 19 December 2019; Accepted: 14 February 2020; Published: 20 February 2020

Abstract: Water quality forecasting is increasingly significant for agricultural management and environmental protection. Enormous amounts of water quality data are collected by advanced sensors, which leads to an interest in using data-driven models for predicting trends in water quality. However, the unpredictable background noises introduced during water quality monitoring seriously degrade the performance of those models. Meanwhile, artificial neural networks (ANN) with feed-forward architecture lack the capability of maintaining and utilizing the accumulated temporal information, which leads to biased predictions in processing time series data. Hence, we propose a water quality predictive model based on a combination of Kernal Principal Component Analysis (kPCA) and Recurrent Neural Network (RNN) to forecast the trend of dissolved oxygen. Water quality variables are reconstructed based on the kPCA method, which aims to reduce the noise from the raw sensory data and preserve actionable information. With the RNN's recurrent connections, our model can make use of the previous information in predicting the trend in the future. Data collected from Burnett River, Australia was applied to evaluate our kPCA-RNN model. The kPCA-RNN model achieved R^2 scores up to 0.908, 0.823, and 0.671 for predicting the concentration of dissolved oxygen in the upcoming 1, 2 and 3 hours, respectively. Compared to current data-driven methods like Feed-forward neural network (FFNN), support vector regression (SVR) and general regression neural network (GRNN), the predictive accuracy of the kPCA-RNN model was at least 8%, 17% and 12% better than the comparative models in these three cases. The study demonstrates the effectiveness of the kPAC-RNN modeling technique in predicting water quality variables with noisy sensory data.

Keywords: water quality; machine learning; recurrent neural network; PCA

---

1. Introduction

Surface water quality has a strong dependence on the nature and extent of agricultural, industrial and other anthropogenic activities within a region's catchments [1]. The reliable prediction of water quality is crucial in order for decision-makers to improve water quality management and protection activities [2]. However, forecasting the temporal variation of water quality parameters for surface river system can be a significantly challenging task owing to rapidly changing environmental conditions and insufficiently historical data records [3].

Dissolved oxygen (DO) content is one of the most vital water quality variables as it directly indicates the status of the aquatic ecosystem and its ability to sustain aquatic life [4]. Rapid decomposition of organic materials, including manure or wastewater sources, can quickly take the DO out of water in few hours, resulting in deficient DO levels that can lead to stress and death of aquatic fauna [5]. For example, DO levels that remain below 1–2 mg/L for a few hours can result in large fish kills. In pond management, an aeration system can quickly increase dissolved oxygen levels if the decreasing of dissolved oxygen in the water can be predicted. Hence, short-term predictions of DO are critical in delivering good water quality management [6].

Various mechanism models have been applied for predicting the concentration of DO [7]. The mechanism model considers many factors such as physical, chemical, and biological factors affecting the change of water quality. The common mechanism models include the BASINS model system [8], the MIKE model system [9], and the QUAL2K model system [10]. However, it is often challenging to simulate the target water quality systems when lacking adequate monitoring data or background information [11]. Consequently, those models are not likely to be able to be generalized without significant parameter adjustment [12].

Data-driven models have received increasing attention in predicting the concentration of DO based on the sensory data. For example, in the study proposed by Zhang [13], a multi-layer feedforward neural network (FFNN) is designed for predicting the trend of dissolved oxygen of the Baffle Creek in Australia. In their approach, a mutual information-based feature selection strategy is introduced to pick up the relevant water quality variables for DO forecasting. Antanasijević et al. [14] tested the effectiveness of applying general regression neural network (GRNN) models for the forecasting of DO in the Danube River, Europe. In their experiments, 19 water quality parameters, five different data normalization methods, and three input selection techniques were tested to find the best combination. In addition, Li et al. [15] evaluated the performance of support vector regression (SVR) for the prediction of DO concentration based on multiple water quality parameters. The SVR was optimized by the particle swarm optimization algorithm and achieved superior performance than linear regression models. Though various data-driven models have been tested in predicting the trend of DO, most existing models lack the mechanisms in processing temporal data. Under these circumstances, seasonal or diurnal patterns within the water quality data are hard to be captured [16].

Apart from model architectures, the quality of input data also has an enormous influence on the data-driven model's performance [17]. The high-frequency data collected by sensors are prevalent in building water quality forecasting models. However, random errors generated by the environment, instruments or network transmission are unavoidable when monitoring water quality variables [18,19]. Though techniques such as z-score and min-max are used in preprocessing input data for data-driven models [14], those techniques aim to rescale the numeric range of water quality variables instead of reducing sensor noise. Accordingly, the unwanted noise would be accepted by the data-driven models, which increases the challenges for generating accurate predictions for water quality variables.

In this paper, we propose a water quality predictive model based on Kernel Principal Component Analysis (kPCA) and Recurrent Neural Network (RNN) to solve the above issues. Our work differs from other comparative approaches in the following two aspects:

- Kernel Principal Component Analysis (kPCA) is implemented to reconstruct the input water quality data. Instead of feeding the water quality sensor data into the data-driven models directly, we pick up the top-ranked principal components as the new inputs. Meanwhile, the dropped principal components are expected to contain background noise. In this way, the reconstructed inputs only have useful information included.
- A recurrent neural network (RNN) is designed to capture the temporal variations within water quality variables and utilize the historical changing patterns as a guide for predicting water quality in the future.

This study aims to evaluate the predictive accuracy of the kPCA-RNN model by comparing it with three data-driven methods discussed above. The evaluation is undertaken on a case study of DO concentrations in Burnett River, Australia.

2. Material and Methods

2.1. Study Area and Monitoring Data

2.1.1. Overview

The Burnett River is located on the southern Queensland coast and flows into the coral sea of the South Pacific Ocean. Cultivation of sugar cane and small crops are important land uses in this region. The total area of the catchment is about 33,000 km^2. Figure 1 illustrates the location and extent of the catchment. Time series physiochemical water quality variables analysed in this study were obtained by a YSI 6 Series sonde sensor near the Bundaberg Co-op Wharf (Figure 1) [20]. Water quality variables such as temperature, electric conductivity (EC), pH, dissolved oxygen (DO), turbidity, and chlorophyll-a (Chl-a) are recorded with 1 h time interval for 5 months in 2015 (Table 1).

Figure 1. Burnett River catchment area and the monitoring site. This monitoring site is part of the Queensland Government's water quality monitoring network [21].

Table 1. Water quality data from 1 June 2015 to 31 October 2015.

Variables	No. of Data	Unit	Min	Max	Median	Mean	SD [1]	CV [2] (%)
Temperature	3672	°C	16.1	27.9	21.0	21.4	2.3	11
EC	3672	uS·cm^{-1}	613.0	49,150.0	45,750.0	44,712.1	3566.8	8
pH	3672		7.5	8.4	7.9	7.8	0.1	2
DO	3672	mg·L^{-1}	5.2	13.0	6.8	6.9	0.9	13
Turbidity	3672	NTU	2.6	63.0	8.2	9.5	4.7	50
Chl-a	3672	μg·L^{-1}	0.1	137.6	2.6	3.5	3.6	102

[1] standard deviation; [2] coefficient of variation.

2.1.2. Water Quality Statistical Analysis

As demonstrated in Table 1, Chl-a and turbidity have larger variability than other water quality variables (CV > 50%). In the case of turbidity, this is due to extreme weather events [22]. The variability of Chl-a concentration can be affected by the discharge of river, temperature, and salinity variation. The high variability in turbidity and Chl-a are caused by a small number of observations with high values (Figure 2). Additionally, outliers of EC tend to have lower measurement values. These outliers can be caused by variations in river flow of other characteristics of the catchment. Ignoring those variations may cause serious information loss.

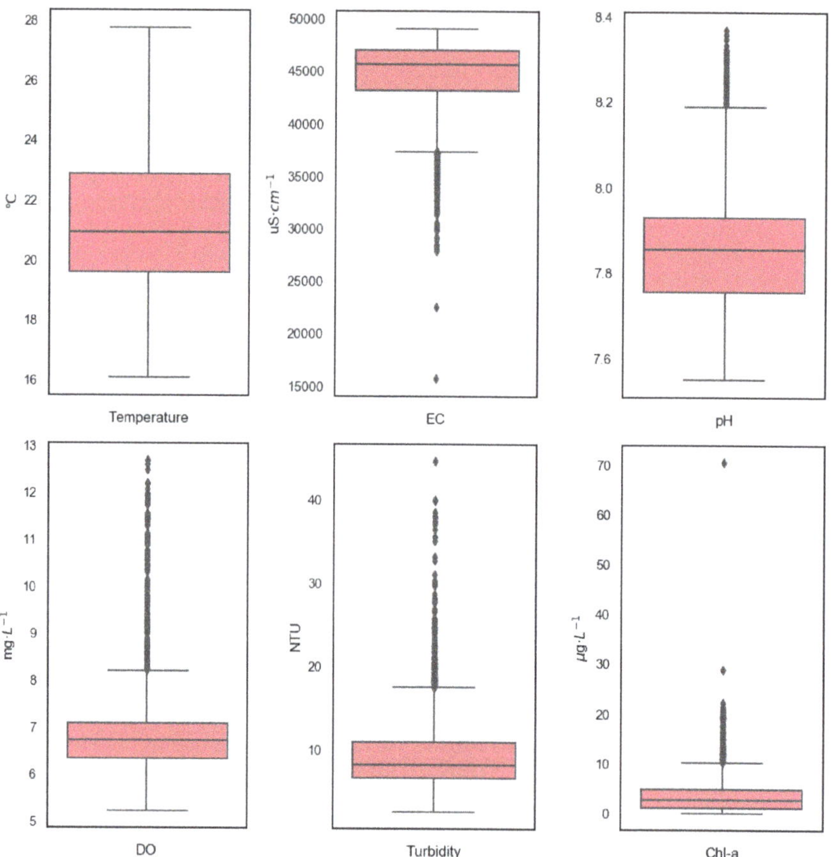

Figure 2. Data distribution for six water quality variables.

Figure 3 illustrates the changing patterns of DO both within a day and over a consecutive number of days. It is obvious that the concentration of DO follows a similar daily pattern, which makes it possible to predict the changing of DO. However, when tracking the concentration of DO in a larger time scale, it is plain to see that the mean value of the concentration of DO is increasing incrementally in the first half of the month and reach the peak value around 21 September. After keeping the high-level concentration for a few days, the DO level decreases gradually till the end of the month. This situation happens when unexpected activities are happening, such as heavy rainfall, excessive algae, and phytoplankton growth. In these circumstances, the predictive models should capture the daily temporal pattern when forecasting future DO concentration. Moreover, the model should be robust so it can have stable prediction performance at different time steps.

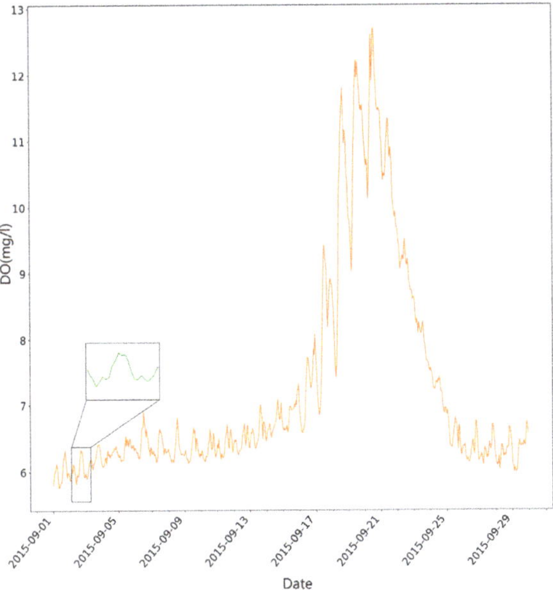

Figure 3. The concentration of DO measured in October 2015. The orange line describes the overall trend of DO concentration in this month. The green shape is an example of the daily pattern of DO concentration.

Figure 4 depicts the autocorrelation of DO for 48 hourly time steps over two days. The plot shows the correlation after the 26th lag is not statistical significance. It means that, in order to predict the trend of DO concentration, the information in the previous 26 h is the most important. This result is also supported by the fact that the concentration of DO follows a daily pattern (Figure 3).

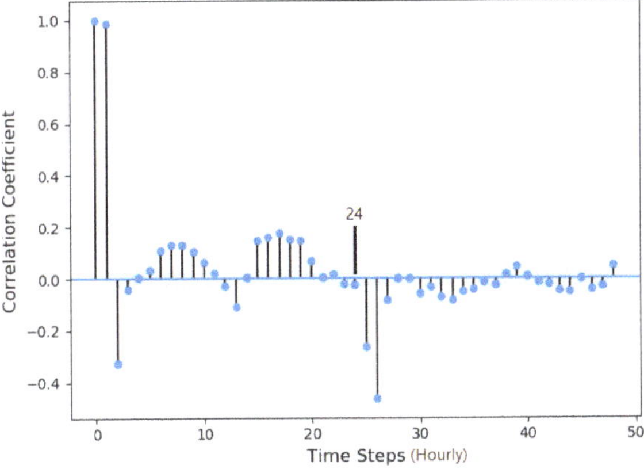

Figure 4. Partial autocorrelation of DO. The concentration of DO is collected hourly.

Hence, by both considering the partial autocorrelation results and the trends of DO under different time scales, we choose to use the data from 24 historical time steps as the input for our predictive model. In this way, the input can cover the information from the previous 24 h, which indicates the complete daily pattern of the DO concentration.

2.2. kPCA-RNN Model Description

2.2.1. Kernel PCA Based Input Abstraction

Principal component analysis (PCA) is routinely applied for linear dimensionality reduction and feature abstraction [23]. The diagonal of the correlation matrix transforms the original principal correlated variables into principal uncorrelated (orthogonal) variables called principal components (PCs), which are weighed as linear combinations of the original variables. The eigenvalues of the PCs are a measure of associated variances, and the sum of the eigenvalues coincides with the total number of variables.

The standard PCA only allows linear dimensionality reduction. However, the multivariate water quality data have a more complicated structure which cannot be easily represented in a linear subspace. In this paper, kernel PCA (kPCA) [24] is chosen as a nonlinear extension of PCA to implement nonlinear dimensionality reduction for water quality variables. The kernel represents an implicit mapping of the data to a higher dimensional space where linear PCA is performed.

The PCA problem in feature space F can be formulated as the diagonalization of an l-sample estimate of the covariance matrix [25], which can be defined as Equation (1):

$$\hat{C} = \frac{1}{l}\sum_{i=1}^{l} \Phi(x_i)\Phi(x_i)^T, \tag{1}$$

where $\Phi(x_i)$ are centred nonlinear mappings of input variables $x_i \in \mathbb{R}^n$. Then, we need to solve the following eigenvalue problem:

$$\begin{aligned} \lambda V &= \hat{C}V, \\ V &\in F, \lambda \geq 0. \end{aligned} \tag{2}$$

Note that all the solutions V with $\lambda \geq 0$ lie in the span of $\Phi(x_1), \Phi(x_2), ..., \Phi(x_l)$. An equivalently problem is defined below:

$$n\lambda\alpha = K\alpha, \tag{3}$$

where α denotes the column vector such that $V = \sum_{i=1}^{l} \alpha_i \Phi(x_i)$, and K is a kernel matrix which satisfies the following conditions:

$$\begin{aligned} &\iint K(x,y)g(x)g(y)dxdy > 0, \\ &\int g^2(x)dx < \infty, \end{aligned} \tag{4}$$

where $K(x,y) = \sum_{i=1}^{\infty} \alpha_i \psi(x)\psi(y), \alpha_i \geq 0$. Then, we can compute the kth nonlinear principal component of x as the projection of $\Phi(x)$ onto the eigenvector V^k:

$$\beta(x)_k = V^k \Phi(x) = \sum_{i=1}^{l} \alpha_i^k K(x_i, x). \tag{5}$$

Then, the first $p < l$ nonlinear components are chosen, which have the desired percentage of data variance. By doing this, the complexity of the original data series can be greatly reduced.

2.2.2. Recurrent Neural Network

Recurrent Neural Networks (RNN) have gained tremendous popularity over the last few years because of their capability in handling unstructured sequential data. In contradistinction to the feed-forward neural network, RNN has the information travelling in both directions. Computations derived from the earlier input are fed back into the network, which is critical in learning the nonlinear relationships between multiple water quality variables.

The general input to an RNN model is a variable-length sequence $x = \{x_1, x_2, ..., x_T\}$ where $x_i \in \mathbb{R}^d$ and d represents the dimention of x_i. At each time step, RNN maintains its internal hidden state h, which results in a hidden sequence of $\{h_1, h_2, ..., h_k\}$. The operation of an RNN at time step t can be formulated as:

$$h_t = f(w_{xh}x_t + w_{hh}h_{t-1}), \tag{6}$$

where $f()$ is an activation function, w_{xh} is the matrix of conventional weights between an input layer x and a hidden layer h, and w_{hh} is the matrix between a hidden layer h and itself at adjacent time steps.

The output of RNN is computed by:

$$y_t = w_{hy}h_t, \tag{7}$$

where w_{hy} is the matrix of weights between the hidden layer h and output y.

As exhibited in Figure 5, the structure of the RNN model across time can be expressed as a deep neural network with one layer per time step. Because this feedback loop occurs at every time step in the series, each hidden state contains traces not only of the previously hidden state, but also of all those that preceded h_{t-1} for as long as memory can persist.

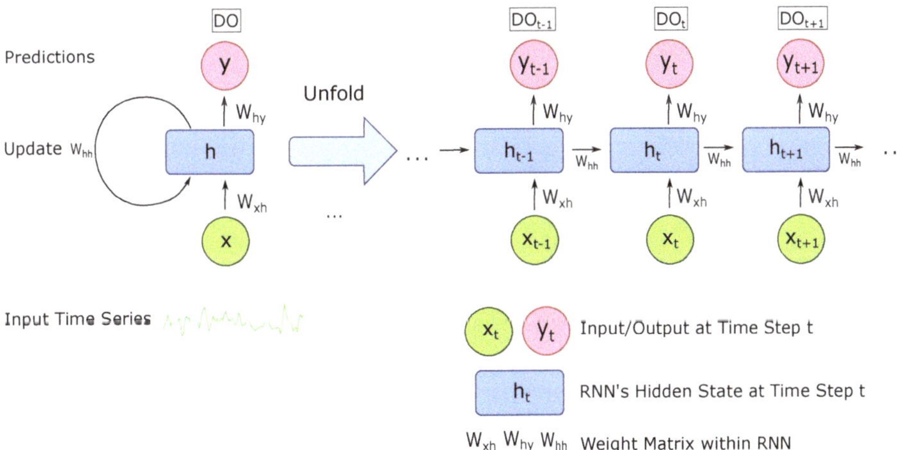

Figure 5. Recurrent neural network for predicting DO.

Compared to the transitional feed-forward neural network, the recurrent structure in RNN can preserve the sequential information in its hidden state. In this approach, the input information can be spanned many time steps as it cascades forward to affect the processing of each new example. The features of RNN networks are especially suitable for processing time series water quality data because of the following reasons: Firstly, water quality data are periodically collected from different sensors and the previous values have strong relationship with the following changing. Secondly, the pattern of many water quality variables can only be recognized when enough historical data are involved and analysed.

In the proposed water quality predictive model, we apply the RNN structure with the LSTM cell [26]. To predict the concentration of DO at time step $t+1$, the input time series include data in previous m time steps. Additionally, each time step has n water quality variables. Consequently, each input of the RNN model can be interpreted as a $m \times n$ matrix. The explicit hyperparameters of our RNN model will be outlined in the following Section 3.2.

2.3. Model Evaluation

We compared the kPCA-RNN model with the following three machine learning methods:

1. Feed-forward neural network (FFNN). FFNN has been broadly adopted for water quality analysis due to its capability in capturing nonlinear relationships within the short-term period [13].
2. General regression neural network (GRNN). GRNN [27] is a type of radial basis function neural network that has good nonlinear approximation ability and fast convergence speed. It has been widely applied in short-term water quality forecasting [14,28].
3. Support vector regression (SVR). SVR is a classic machine learning technique which can map inputs into higher dimensional space and interpret the problem as a linear regression [29].

The following performance indicators were applied to evaluate the predictive results. Those are the mean absolute error (MAE), the coefficient of determination (R^2), the root mean square error (RMSE), and the percent of prediction within a factor of 1.1 (FA1.1) [30]:

$$MAE = \frac{1}{n}\sum_{i=1}^{n}|f_i - \hat{f}_i| \tag{8}$$

$$R^2 = 1 - \frac{\sum_{i=1}^{n}(f_i - \hat{f}_i)^2}{\sum_{i=1}^{n}(f_i - \overline{f}_i)^2} \tag{9}$$

$$RMSE = \sqrt{\frac{1}{n}\sum_{i=1}^{n}(|f_i - \hat{f}_i|)^2} \tag{10}$$

$$FA1.1 = \frac{m}{n}, \qquad m = |0.9 < \frac{\hat{f}_i}{f_i} < 1.1| \tag{11}$$

where f_i, $\hat{f}_i$, n, and m represent the observed value, the predicted value, the number of observations, and the number of predictions within a factor of 1.1 of the observed values, respectively. Additionally, $\overline{f}_i = \frac{1}{n}\sum_{i=1}^{n} f_i$.

2.4. Workflow of Predicting DO

Figure 6 depicts the workflow of predicting the concentration of DO by using the kPCA-RNN model. There are two key steps in this workflow: applying the kPCA to denoise and reconstruct input data and implementing the RNN model to forecast the trend of dissolved oxygen in future time steps.

Firstly, the kPCA method is implemented on the tabulated water quality data (Table 1) to create corresponding principal components. The principal components with less importance are dropped to reduce the background noise in the original water quality dataset. Consequently, the remaining principal components are selected as new inputs for the predictive model.

Next, the input data are formed to $m \times n$ matrix as we explained in Section 2.2.2. After training and testing the RNN model, the concentration of DO in the upcoming time steps can be estimated.

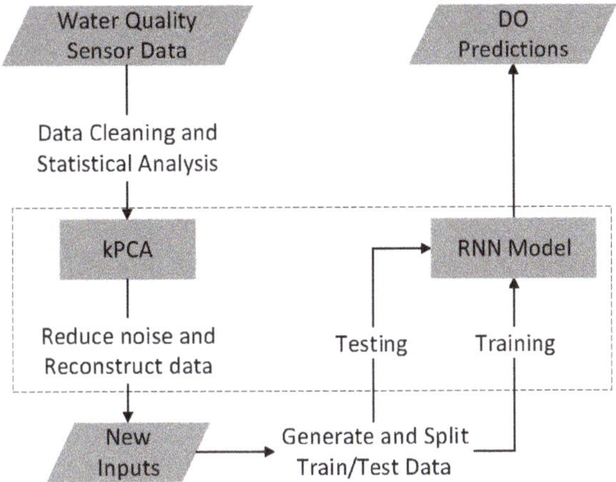

Figure 6. Workflow for predicting DO by applying the kPCA-RNN model. The dotted box highlights the key components of this proposed workflow.

The kPCA-RNN model described in Figure 6 differs from most existing DO forecasting models in the following aspects:

1. Instead of using the sensor data directly, the kPCA method is implemented to the water quality sensor data to construct new inputs based on principal components. This step can help reduce the background noise and keep the most useful information for DO forecasting tasks.
2. The recurrent neural network is applied to process the time series water quality data. The recurrent structure offers a powerful way of capturing the temporal patterns across a period of time, which is critical in forecasting the changing of DO concentration in the future.

3. Model Application

3.1. Applying kPCA on the Water Quality Data

We applied the kPCA method to the water quality dataset (Table 1) and obtained five principal components (Table 2).

Table 2. Descriptive statistics of five principal components.

Principal Components	Eigenvalue	Cumulative Variance Proportion (%)
PC1	466.6	44.4
PC2	285.3	71.5
PC3	129.8	83.8
PC4	114.8	94.8
PC5	55.1	100.0

Five principal components (Table 2) are ordered by their corresponded eigenvalue. The first principal component is the linear combination of all the variables that have a maximum variance, so it accounts for as much variation in the data as possible. After that, each succeeding component, in turn, has the highest variance possible under the constraint that it is orthogonal to the preceding components. The cumulative variance proportion of the first four principal components is 94.8%. This indicates that, retaining only the first four principal components, one can explain 94.8% of the full variance.

Figure 7 demonstrates the correlation scores between each water quality variable and a principal component (PC). The value at each cross point represents the correlation between two different items which are named on the left and bottom of the figure. This figure shows how each water quality variable contributes to each PC and also gives one the insight into how each PC can represent the information contained in different water quality variables.

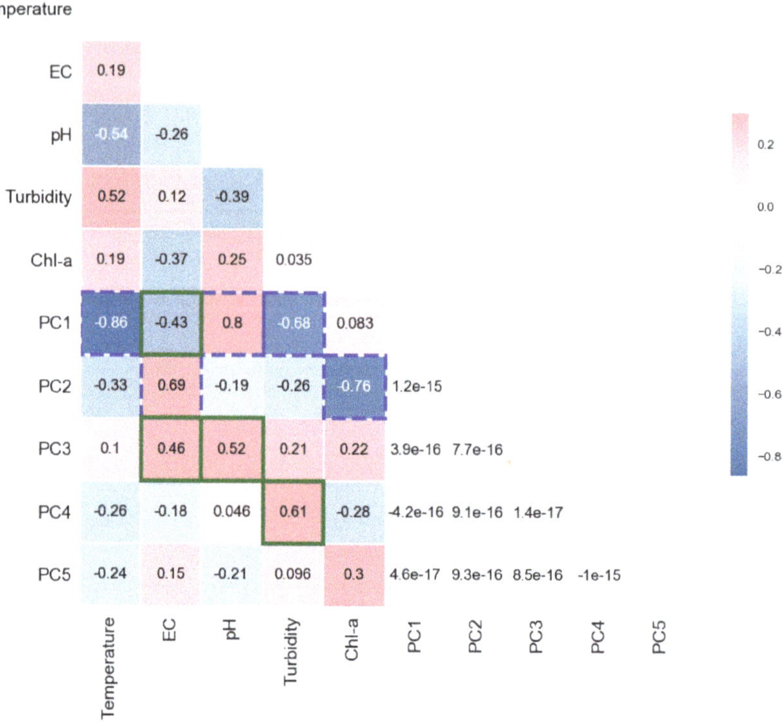

Figure 7. Correlations between water quality variables and principal components.

As has been pointed out, the first principal component (PC1) has the highest correlation (dotted box, Figure 7) with variables like temperature, pH and turbidity. The three dotted boxes in line 5 (PC1) highlight the highest correlation scores one got from the corresponding water quality variables (listed in the bottom axis). Furthermore, the second principal component (PC2) has the highest correlation (dotted box in line 6) with the remaining variables EC and Chl-a. This indicates that, by utilizing only principal components PC1 and PC2, most information involved in those five water quality variables can be presented. Furthermore, PC3 and PC4 also have a strong correlation with EC, pH, and Turbidity (solid box). On the contrary, PC5 has a low value of correlation coefficient to all water quality variables, which means it carries much noise information [31]. Accordingly, we accept the first four principal components as new inputs. The kPCA method can reduce the input size by 20% while still keeping the most valuable information.

3.2. RNN Hyperparameters Settings

One challenge of building a neural network model is optimizing the hyperparameters for predictive accuracy [32]. Generally, different neural network settings are required to achieve the promising results for different forecasting tasks. Hence, we need to choose proper neural network parameters for forecasting DO concentration in three different predictive horizons.

Three RNN models were designed to predict the next one, two, and three hours of DO concentration independently. Each RNN model has various parameters and they all accept four months of data (2928 samples) for training and one month of data (744 samples) for testing. Based on the partial autocorrelation analysis in Section 2.1.2, data from the previous 24 time steps were accepted as the model's input when predicting the concentration of DO in each future step.

The hyperparameters of the three RNN models were defined in Table 3.

Table 3. Experiments settings.

Model Settings	Experimental Cases		
	1 h Ahead	2 h Ahead	3 h Ahead
No. of Hidden Layers	1	2	3
No. of Hidden Units	40	30	20
Recurrent Cell	LSTM [1]	LSTM [1]	LSTM [1]
Optimizer	Adam [2]	Adam [2]	Adam [2]
No. of Historical Time Steps	24	24	24
No. of Training Data	2928	2928	2928
No. of Testing Data	744	744	744

[1] Long short-term memory [26]; [2] Adam [33].

3.3. Results and Discussion

Figure 8 illustrates the forecasting of DO concentration during October 2015 under three different predictive horizons. The upper part of each subfigure compares the actual measurements and predictions of the DO concentration at each time index. In all the three subfigures, over 90% of the predictions are located in the F1.1 range. This means that the proposed kPCA-RNN model can capture the moving average of the DO concentration. By learning information from the previous 24 time steps, the model can avoid most of the severe bias estimations.

The model yields predictions with R^2 value of 0.908 for 1 h ahead forecasting in Figure 8a. For 1 h ahead prediction, there is no time gap between the model's inputs and the prediction. In order to predict DO concentration at a specific time step, the model can learn all the historical measurements until the previous hour. Thus, the proposed kPCA-RNN reaches the highest accuracy for 1 h ahead forecasting.

Similarly, our proposed model achieves R^2 value of 0.823 for 2 h ahead forecasting in Figure 8b. When increasing the predictive horizon, the model does not predict what will happen after the last true measurement. Instead, the model needs to take a further step to generate the prediction. This usually happens when the model acts as an early warning system so there can be enough time for delivering management activities based on the forecasting results. In this circumstance, the model can only utilize what has been measured already to make the prediction. Hence, the prediction accuracy decreases slightly in this case.

In Figure 8c, the model obtains R^2 value of 0.671 for 3 h ahead forecasting. As we discussed above, it becomes more challenging when one increases the predictive horizon, while, in this case, around 93% prediction results are still within $\pm$ 10% range of the original observations (FA1.1). This gives us confidence that the proposed kPCA-RNN model can still yield promising estimations. As we discussed in Section 1, the rapidly changing of DO concentration in a few hours can put aquatic life under high stress. Hence, the promising predictions in a few hours ahead are significant in early warning and changing management activities.

In addition, we also listed the RMSE value at each time step for all the three experimental cases (lower part of each subfigure in Figure 8). This offers us a detailed insight into the prediction performance of the proposed kPCA-RNN model. The RMSE figures clearly indicate that our model has a stable performance accuracy at most of the time steps. This is critical in applying the model in processing the real-world water quality sensor data.

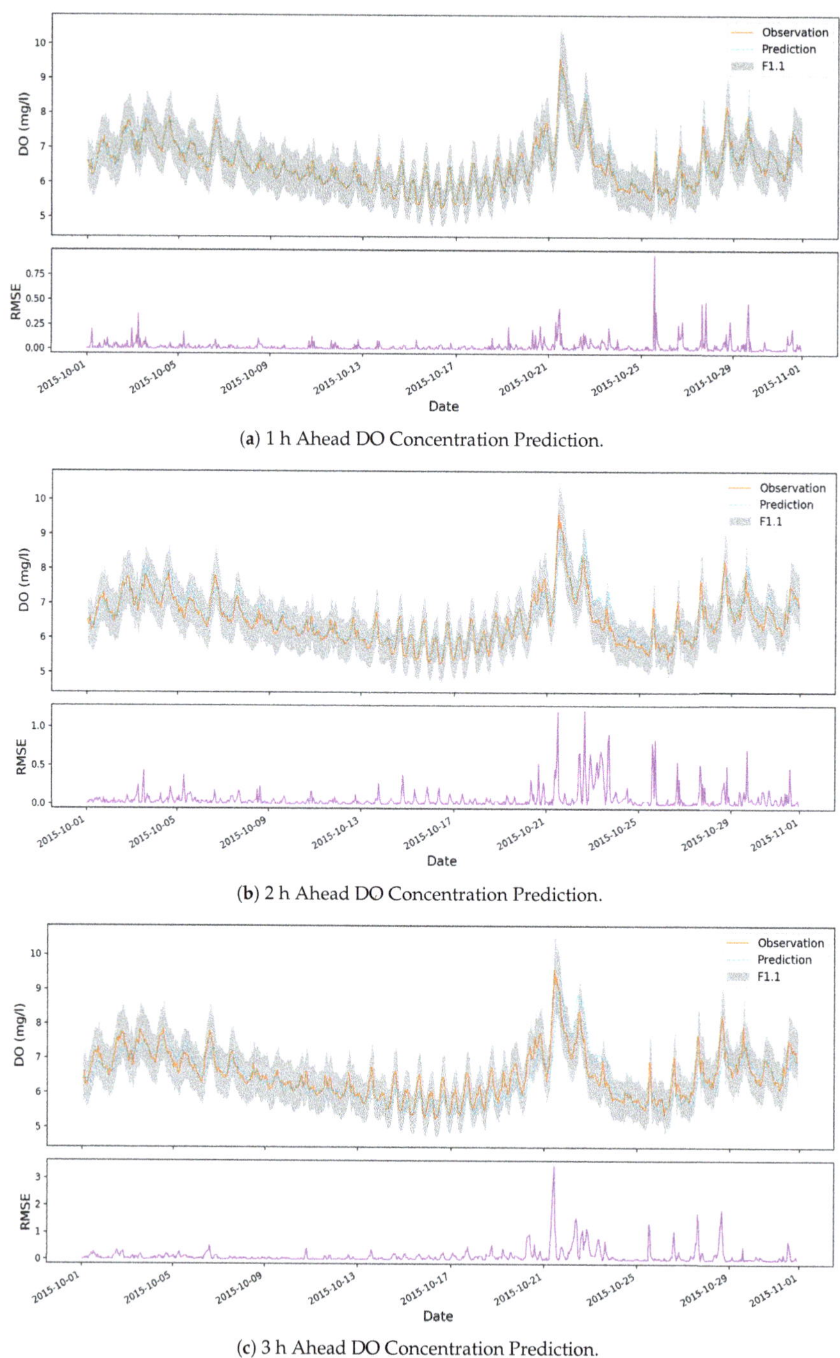

(**a**) 1 h Ahead DO Concentration Prediction.

(**b**) 2 h Ahead DO Concentration Prediction.

(**c**) 3 h Ahead DO Concentration Prediction.

Figure 8. 1, 2 and 3 h ahead predicting for the concentration of DO. In the upper part of each subfigure, the grey shadow, solid line and dotted line represent the FA1.1 range, DO observations and predictions, respectively. The lower part of each subfigure describes the RMSE value at each time step.

As can be seen, most biased estimations happened between 21 October 2015 and 29 October 2015, where there was a strong fluctuation of DO concentration. In the water monitoring reports published by the Queensland Government, there was a large amount of discharge for total nutrients, dissolved and particulate nutrients during that period of time. On the contrary, the discharge in the previous months was low. It indicates that the trend of concentration of DO was changing more frequently and heavily in October. However, the kPCA-RNN model is trained based on the concentration of DO obtained from historical months with regular DO change. Consequently, there are some predictions below the high points of the observations; for example, the predictions around 22 October 2015. Hence, it is necessary to involve extra water quality data to cover a longer time period.

We additionally compared the performance of the kPCA-RNN model with three models stated in Section 2.3. The same data set described in Section 2.1.1 was applied in all cases. For FFNN, we set the same neural network size as in the kPCA-RNN model. For GRNN, the standard deviation is set to 10 for the high dimensional inputs. For SVR, the Radial Basis Function kernel (RBF) is taken as the nonlinear kernel. The corresponding results are listed in Table 4.

Table 4. Performance comparison with the FFNN, SVR and GRNN.

Predictive Models	**Evaluation Criteria**			
	MAE	R^2	**RMSE**	**FA1.1**
1 h Ahead Prediction				
kPCA-RNN Model	**0.149**	**0.908**	**0.208**	**0.995**
FFNN	0.175	0.893	0.224	0.989
SVR	0.219	0.810	0.299	0.962
GRNN	0.263	0.727	0.355	0.944
2 h Ahead Prediction				
kPCA-RNN Model	**0.211**	**0.823**	**0.288**	**0.973**
FFNN	0.258	0.757	0.338	0.958
SVR	0.314	0.594	0.437	0.890
GRNN	0.295	0.648	0.403	0.926
3 h Ahead Prediction				
kPCA-RNN Model	**0.303**	**0.671**	**0.394**	**0.926**
FFNN	0.455	0.358	0.550	0.756
SVR	0.358	0.515	0.478	0.858
GRNN	0.320	0.562	0.450	0.910

The kPCA-RNN models offer the best performance in all three of the prediction cases (Table 4). For example, in the 1 h ahead prediction, 99.5% of the predictions are within the FA1.1 range, which demonstrates that the model has a stable accuracy for most predictions.

Specifically, the kPCA-RNN model has 8%, 17% and 40% improved performance on the RMSE than the FFNN in all three of the cases, respectively. Similarly, the kPCA-RNN model achieves 43%, 52% and 21% improved performance on the RMSE over the SVR. Compared to GRNN, our proposed model gains 41%, 29% and 12% performance improvement on the RMSE scores. The FFNN, SVR and GRNN are ineffective in predicting the changing of DO concentration with 2 or 3 h predictive horizon because their model structures are not designed to handle time series data and the temporal pattern cannot be efficiently captured.

Hence, the kPCA-RNN model can perform as an early warning predictor for DO in application areas such as aquaculture ponds. By providing the DO significant changing alarm, farmers can consider appropriate actions to maintain the DO on a suitable level for the health of the aquatic ecosystem.

4. Conclusions

To summarize, the kPCA-RNN model was able to successfully predict the trend of DO in the following 1 to 3 h. We evaluated our model based on water quality data from Burnett River, Australia and compared it with the FFNN, SVR and GRNN methods. The results demonstrate that our method is more accurate and stable to the alternative methods, especially when the predictive horizon is increasing. Furthermore, as a data-driven modeling method, the kPCA-RNN model is not limited to a specific hydrological area and can be extended to predict various water quality variables.

For future work, inputs can be improved to include extra information such as rail fall and cover more extended periods of time. In addition, the water quality predictive model can be extended to support predicting multiple water variables simultaneously.

Author Contributions: Methodology, writing—original draft preparation, Y.-F.Z.; writing—review and editing, P.F.; project administration, writing—review and editing, P.J.T. All authors have read and agreed to the published version of the manuscript.

Funding: This research received no external funding.

Acknowledgments: This work was conducted within the CSIRO Digiscape Future Science Platform.

Conflicts of Interest: The authors declare no conflict of interest.

References

1. Barzegar, R.; Adamowski, J.; Moghaddam, A.A. Application of wavelet-artificial intelligence hybrid models for water quality prediction: a case study in Aji-Chay River, Iran. *Stoch. Environ. Res. Risk Assess.* **2016**, *30*, 1797–1819. [CrossRef]
2. Hrachowitz, M.; Benettin, P.; van Breukelen, B.M.; Fovet, O.; Howden, N.J.; Ruiz, L.; van der Velde, Y.; Wade, A.J. Transit times-the link between hydrology and water quality at the catchment scale. *Wiley Interdiscip. Rev. Water* **2016**, *3*, 629–657. [CrossRef]
3. Jin, T.; Cai, S.; Jiang, D.; Liu, J. A data-driven model for real-time water quality prediction and early warning by an integration method. *Environ. Sci. Pollut. Res.* **2019**, *26*, 30374–30385. [CrossRef] [PubMed]
4. Tomić, A.Š.; Antanasijević, D.; Ristić, M.; Perić-Grujić, A.; Pocajt, V. A linear and nonlinear polynomial neural network modeling of dissolved oxygen content in surface water: Inter- and extrapolation performance with inputs' significance analysis. *Sci. Total. Environ.* **2018**, *610–611*, 1038–1046. [CrossRef]
5. King, A.J.; Tonkin, Z.; Lieshcke, J. Short-term effects of a prolonged blackwater event on aquatic fauna in the Murray River, Australia: considerations for future events. *Mar. Freshw. Res.* **2012**, *63*, 576–586. [CrossRef]
6. Zhang, Y.; Thorburn, P.J.; Fitch, P. Multi-Task Temporal Convolutional Network for Predicting Water Quality Sensor Data. In Proceedings of the 26th International Conference on Neural Information Processing (ICONIP2019), Sydney, Australia, 12–15 December 2019; Volume 1142, pp. 122–130.
7. Hawkins, C.P.; Olson, J.R.; Hill, R.A. The reference condition: predicting benchmarks for ecological and water-quality assessments. *J. North Am. Benthol. Soc.* **2010**, *29*, 312–343. [CrossRef]
8. Marcomini, A.; Suter II, G.W.; Critto, A. *Decision Support Systems for Risk-Based Management of Contaminated Sites*; Springer Science & Business Media: New York, NY, USA, 2008; Volume 763.
9. Chubarenko, I.; Tchepikova, I. Modelling of man-made contribution to salinity increase into the Vistula Lagoon (Baltic Sea). *Ecol. Model.* **2001**, *138*, 87–100. [CrossRef]
10. Chapra, S.; Pelletier, G.; Tao, H. *QUAL2K: A Modeling Framework for Simulating River and Stream Water Quality*; Documentation and user manual; Tufts University: Medford, MA, USA, 2003.
11. Ay, M.; Özgür Kişi. Estimation of dissolved oxygen by using neural networks and neuro fuzzy computing techniques. *KSCE J. Civ. Eng.* **2016**, *21*, 1631–1639. [CrossRef]
12. Zhang, Y.; Thorburn, P.J.; Wei, X.; Fitch, P. SSIM -A Deep Learning Approach for Recovering Missing Time Series Sensor Data. *IEEE Internet Things J.* **2019**, *6*, 6618–6628. [CrossRef]
13. Zhang, Y.; Fitch, P.; Vilas, M.P.; Thorburn, P.J. Applying Multi-Layer Artificial Neural Network and Mutual Information to the Prediction of Trends in Dissolved Oxygen. *Front. Environ. Sci.* **2019**, *7*, 46. [CrossRef]
14. Antanasijević, D.; Pocajt, V.; Perić-Grujić, A.; Ristić, M. Modelling of dissolved oxygen in the Danube River using artificial neural networks and Monte Carlo Simulation uncertainty analysis. *J. Hydrol.* **2014**, *519*, 1895–1907. [CrossRef]
15. Li, X.; Sha, J.; Wang, Z.l. A comparative study of multiple linear regression, artificial neural network and support vector machine for the prediction of dissolved oxygen. *Hydrol. Res.* **2017**, *48*, 1214–1225. [CrossRef]

16. Sundermeyer, M.; Oparin, I.; Gauvain, J.L.; Freiberg, B.; Schlüter, R.; Ney, H. Comparison of feedforward and recurrent neural network language models. In Proceedings of the 2013 IEEE International Conference on Acoustics, Speech and Signal Processing, Vancouver, BC, Canada, 26–31 May 2013; pp. 8430–8434.
17. Karami, J.; Alimohammadi, A.; Seifouri, T. Water quality analysis using a variable consistency dominance-based rough set approach. *Comput. Environ. Urban Syst.* **2014**, *43*, 25–33. [CrossRef]
18. Liu, S.; Che, H.; Smith, K.; Chang, T. A real time method of contaminant classification using conventional water quality sensors. *J. Environ. Manag.* **2015**, *154*, 13–21. [CrossRef]
19. chuan Wang, W.; wing Chau, K.; Qiu, L.; bo Chen, Y. Improving forecasting accuracy of medium and long-term runoff using artificial neural network based on EEMD decomposition. *Environ. Res.* **2015**, *139*, 46–54. [CrossRef]
20. Ambient Estuarine Water Quality Monitoring Data. Available online: https://data.qld.gov.au/dataset (accessed on 20 November 2017).
21. Great Barrier Reef Catchment Loads Monitoring Program. Available online: https://www.reefplan.qld.gov.au/measuring-success/paddock-to-reef/catchment-loads/ (accessed on 1 June 2018).
22. Macdonald, R.K.; Ridd, P.V.; Whinney, J.C.; Larcombe, P.; Neil, D.T. Towards environmental management of water turbidity within open coastal waters of the Great Barrier Reef. *Mar. Pollut. Bull.* **2013**, *74*, 82–94. [CrossRef]
23. Wold, S.; Esbensen, K.; Geladi, P. Principal component analysis. *Chemom. Intell. Lab. Syst.* **1987**, *2*, 37–52. [CrossRef]
24. Schölkopf, B.; Smola, A.; Müller, K.R. Kernel principal component analysis. In Proceeedings of the Artificial Neural Networks—ICANN'97: 7th International Conference, Lausanne, Switzerland, 8–10 October 1997; Springer Berlin Heidelberg: Berlin/Heidelberg, Germany, 1997; pp. 583–588.
25. Ince, H.; Trafalis, T.B. Kernel principal component analysis and support vector machines for stock price prediction. *IIE Trans.* **2007**, *39*, 629–637. [CrossRef]
26. Hochreiter, S.; Schmidhuber, J. Long short-term memory. *Neural Comput.* **1997**, *9*, 1735–1780. [CrossRef]
27. Specht, D.F. A general regression neural network. *IEEE Trans. Neural Netw.* **1991**, *2*, 568–576. [CrossRef]
28. Barzegar, R.; Moghaddam, A.A. Combining the advantages of neural networks using the concept of committee machine in the groundwater salinity prediction. *Model. Earth Syst. Environ.* **2016**, *2*, 26. [CrossRef]
29. Granata, F.; Papirio, S.; Esposito, G.; Gargano, R.; de Marinis, G. Machine learning algorithms for the forecasting of wastewater quality indicators. *Water* **2017**, *9*, 105. [CrossRef]
30. Roberts, W.; Williams, G.P.; Jackson, E.; Nelson, E.J.; Ames, D.P. Hydrostats: A Python package for characterizing errors between observed and predicted time series. *Hydrology* **2018**, *5*, 66. [CrossRef]
31. Verbanck, M.; Josse, J.; Husson, F. Regularised PCA to denoise and visualise data. *Stat. Comput.* **2013**, *25*, 471–486. [CrossRef]
32. Jaques, N.; Gu, S.; Turner, R.E.; Eck, D. Tuning recurrent neural networks with reinforcement learning. *arXiv* **2017**, arXiv:1611.02796v3.
33. Kingma, D.P.; Ba, J. Adam: A method for stochastic optimization. *arXiv* **2014**, arXiv:1412.6980.

Article

Conceptual Mini-Catchment Typologies for Testing Dominant Controls of Nutrient Dynamics in Three Nordic Countries

Fatemeh Hashemi [1,*], Ina Pohle [2], Johannes W.M. Pullens [3], Henrik Tornbjerg [1], Katarina Kyllmar [4], Hannu Marttila [5], Ahti Lepistö [6], Bjørn Kløve [5], Martyn Futter [7] and Brian Kronvang [1]

1. Department of Bioscience, Aarhus University, Vejlsøvej 25, 8600 Silkeborg, Denmark; hto@bios.au.dk (H.T.); bkr@bios.au.dk (B.K.)
2. Environmental and Biochemical Sciences, The James Hutton Institute, Craigiebuckler, Aberdeen AB15 8QH, Scotland, UK; ina.pohle@hutton.ac.uk
3. Department of Agroecology, Aarhus University, Blichers Allé 20, 8830 Tjele, Denmark; jwmp@agro.au.dk
4. Department of Soil and Environment, Swedish University of Agricultural Sciences, P.O. Box 7014, 750 07 Uppsala, Sweden; katarina.kyllmar@slu.se
5. Water Resources and Environmental Engineering Research Unit, Olulu University, 90014 Oulu, Finland; Hannu.marttila@oulu.fi (H.M.); bjorn.klove@oulu.fi (B.K.)
6. Finnish Environment Institute, Latokartanonkaari 11, 00790 Helsinki, Finland; ahti.lepisto@ymparisto.fi
7. Department of Aquatic Sciences and Assessment, Swedish University of Agricultural Sciences, P.O. Box 7050, 750 07 Uppsala, Sweden; martyn.futter@slu.se

* Correspondence: fh@bios.au.dk

Received: 3 May 2020; Accepted: 16 June 2020; Published: 22 June 2020

Abstract: Optimal nutrient pollution monitoring and management in catchments requires an in-depth understanding of spatial and temporal factors controlling nutrient dynamics. Such an understanding can potentially be obtained by analysing stream concentration–discharge (C-Q) relationships for hysteresis behaviours and export regimes. Here, a classification scheme including nine different C-Q types was applied to a total of 87 Nordic streams draining mini-catchments (0.1–65 km^2). The classification applied is based on a combination of stream export behaviour (dilution, constant, enrichment) and hysteresis rotational pattern (clock-wise, no rotation, anti-clockwise). The scheme has been applied to an 8-year data series (2010–2017) from small streams in Denmark, Sweden, and Finland on daily discharge and discrete nutrient concentrations, including nitrate (NO_3^-), total organic N (TON), dissolved reactive phosphorus (DRP), and particulate phosphorus (PP). The dominant nutrient export regimes were enrichment for NO_3^- and constant for TON, DRP, and PP. Nutrient hysteresis patterns were primarily clockwise or no hysteresis. Similarities in types of C-Q relationships were investigated using Principal Component Analysis (PCA) considering effects of catchment size, land use, climate, and dominant soil type. The PCA analysis revealed that land use and air temperature were the dominant factors controlling nutrient C-Q types. Therefore, the nutrient export behaviour in streams draining Nordic mini-catchments seems to be dominantly controlled by their land use characteristics and, to a lesser extent, their climate.

Keywords: water quality; concentration–discharge relationship; export behaviour; hysteresis; PCA

1. Introduction

The need for nutrient load reduction from both agricultural and non-agricultural lands to avoid harmful impacts on groundwater and surface water resources, including eutrophication and hypoxia in aquatic ecosystems, is widely recognised [1]. The Baltic Sea is among the most heavily degraded marine ecosystems worldwide, due in part to excessive nutrient loads [2]. Nordic countries in the Baltic region also suffer from nutrient pollution issues and their management requires better understanding of the nutrient sources [3].

Understanding the temporal and spatial mechanisms of nutrient dynamics as well as a diversity of factors controlling the transfer of nutrients to surface waters is essential for setting targets for water quality thresholds and providing catchment-scale non-point nutrient pollution mitigation options [4–6]. However, nutrient pollution is particularly difficult to manage, as controlling factors include complex hydrological and biogeochemical processes and relationships between nutrient sources, forms, and their transport to surface waters [6–11]. Among the relevant spatial controls of nutrient dynamics, we can mention the distribution and area of land use pressures on water quality, erodibility, and leaching sensitivity for sediment and nutrients, available water for transport, and hydrological connectivity which is related to geology, land use, topography, and climate [12–14]. Discharge and nutrient concentrations are basic stream measurements for quantifying nutrient export from catchments and are needed for obtaining an integrated understanding of stream hydrological and biogeochemical processes [15]. Moreover, assessments of the temporal variability of controls on nutrients transported by streams and their drivers, including agricultural land use/management, hydro-meteorological variables, and nutrient pathways [16–18] are important for designing water quality management strategies [19].

Understanding the complexity of catchment-scale spatial and temporal mechanisms of nutrient dynamics and loads is difficult and demanding [5,17]. Many studies have investigated the relationship between discharge (Q) and concentration (C) to characterise the temporal and spatial variability of nutrient loads and several indicators have been suggested [20–37]. Some earlier studies have proposed that relationships between C and Q can be described using log-linear relationships. These relationships can be chemo-static, i.e., constant C across all Q values, or chemo-dynamic with either a positive relationship defined as enrichment where C increases with Q or dilution when C decreases with increasing Q [20–22,25,26,32].

Given that a log-linear relationship between C and Q often does not sufficiently capture the variability of the data, some researches have suggested splitting concentration–discharge (C-Q) curves at the median flow to analyse the C-Q relationships during low and high flows separately [36,37]. Hysteresis patterns in C-Q relationships have also analysed to identify time lags between C and Q during runoff events [29,38]. Hysteresis patterns can either be clockwise (higher concentrations at the rising limb compared with the falling limb of the hydrograph) or anticlockwise (lower concentrations at the rising limb compared with the falling limb of the hydrograph).

When developing typologies describing similarities in catchment solute exports, there is a need for including both spatial and temporal controlling factors. To this end, a classification scheme for deriving typologies from low-frequency (e.g., monthly) water quality data distinguishing between nine different C-Q relationship types, defined as combinations of export behaviour and the rotational pattern of the hysteresis has been developed [35]. To be able to better represent catchment and substance-specific export characteristics, they used C-Q relationships with hydrograph segmentation where C-Q curves were split at an automatically derived optimal segmentation discharge, and introduced separate C-Q models for rising and falling hydrograph limbs. Assessing export behaviour and rotational pattern of the hysteresis based on water quality data in mini-catchments seems promising to gain more knowledge about C-Q dynamics and their spatial controls. The importance of mini-catchments is due to the intimate connection between terrestrial and aquatic ecosystems. This is of great significance for water resource management because small headwater streams markedly influence the downstream water quality [39].

Therefore, the main objectives of our study were to explore mini-catchment similarity in C-Q relationships for different N and P forms and to identify spatial and temporal controls on in-stream nutrient concentrations. We hypothesised that development of a mini-catchment typology can assist in informing stakeholders on the expected impacts of climate change, land use change, and nutrient management for the resulting nutrient losses and eutrophication impacts. To address these objectives, we investigated daily discharge and discrete nutrient (nitrate, organic N, dissolved reactive P, and particulate P) concentration data collected over an 8-year period from 87 small streams in Denmark, Sweden, and Finland.

2. Materials and Methods

2.1. Data and Study Sites

Discharge and nutrient concentration data were obtained from a total of 87 Nordic streams draining catchments with a size ranging from 0.1 km^2 to 65 km^2 in Denmark, Sweden, and Finland (Figure 1). An 8-year daily record (2010–2017) of stream discharge and weekly to bi-weekly concentration measurements for nitrate (NO_3^-), total organic N (TON calculated as total N minus NO_3^-), dissolved reactive phosphorus (DRP) and particulate P (PP defined as total P minus DRP) were available. We used 56 stations in Denmark, 13 in Sweden, and 18 in Finland (Table 1, Table S1).

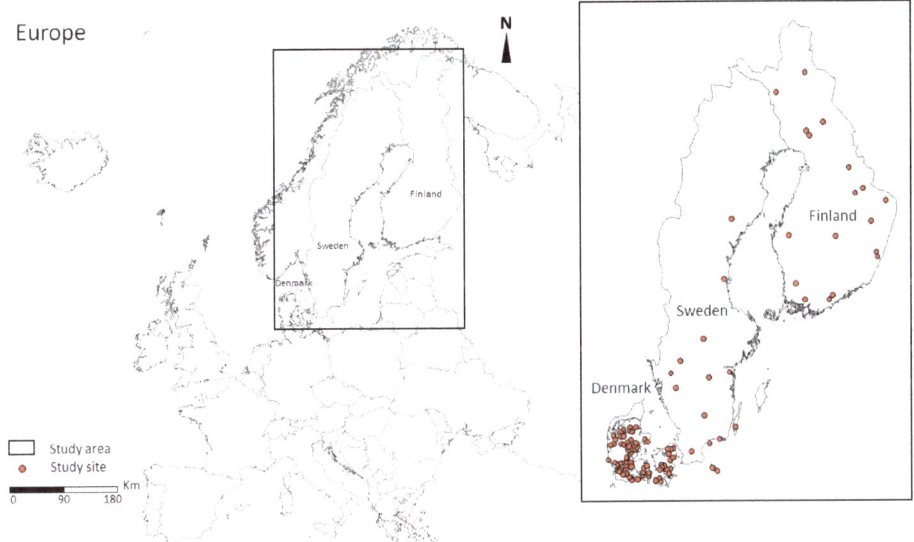

Figure 1. Locations of the mini-catchments investigated in the study.

Table 1. Numbers of stations and total number of nutrient monitoring concentrations for N (nitrate and total organic N) and P forms (dissolved reactive P and particulate P) per year for streams in Denmark, Sweden, and Finland, including mean, range, and standard deviation (SD).

Country	Number of Stations	Number of Nutrient Concentration Observations per Year per Station							
		N forms				P forms			
		Mean	Range		SD	Mean	Range		SD
			Max	Min			Max	Min	
Denmark	56	104	276	55	47	105	265	60	44
Sweden	13	186	209	143	16	186	209	143	16
Finland	18	123	200	57	37	120	192	52	36

The Danish data were extracted from the national monitoring database (ODA) at Aarhus University [40]. The Swedish data on agricultural catchments were extracted from the river flow monitoring programs for small agricultural catchments [41]. Data on forested (non-agricultural) catchments were obtained from the Swedish Integrated Monitoring programme [42]. The Finnish data were extracted from the hydrological database of the Finnish Environment Institute (SYKE) [43]. Mean annual precipitation in the study region ranges from 513 mm to 763 mm (standard reference period 1971–2000) and mean daily air temperature between −0.42 °C and 8.84 °C (standard reference period 1971–2000). The climatic characteristics with a resolution of a 0.5 degree grid (50 km) were derived from a database provided by the Swedish Meteorological and Hydrological Institute (SMHI). Land use data were reported as percentages of agriculture (including arable and grassland used for agriculture) and nature (including forested and naturally vegetated land) and were derived separately for each catchment in Denmark [44], Sweden [41,42], and Finland [45,46] (Table S2). As built up areas and open water areas (watercourses, ponds, lakes, etc.) are not included, land use values do not necessarily sum to 100%. Soil characteristics were derived from the 1:5000000 scale FAO Digital Soil Map of the World (DSMW). Soils were reclassified into 6 classes (Table S3) based on percentages of sand, till, moraine, sediment, gravel, and organic soil.

2.2. Analysis of Concentration–Discharge (C-Q) Relationships and Classification

We classified nine types of concentration–discharge (C-Q) relationships based on export regime and hysteresis [35] (Figure 2). We distinguished three export regimes: enrichment (increase of C with Q), constant (no significant relationship between C and Q) and dilution (decrease of C with Q), and three hysteresis patterns: clockwise, no hysteresis, and anticlockwise. An export regime showing enrichment is likely the result of solute mobilisation due to large element storage such as nitrate leaching from arable soils and mobilisation of legacy nutrients in the catchment [25,47]. A constant export regime may be indicative of a similar element distribution, fixed source areas, or concurrent hydrological and biogeochemical processes [26,37,48]. In contrast, dilution regimes are caused by source limitation such as a decrease in the number of point sources or exhaustion of catchment pools [25]. Clockwise hysteresis may be related to a fast response to flushing or erosional processes while anticlockwise hysteresis may be caused by delayed transport processes from, e.g., groundwater and upstream parts of the catchment [33].

C-Q relationships were specified according to a power-law model (Equation (1)):

$$C = aQ^b \tag{1}$$

where C is concentration, a a coefficient with units of concentration/discharge, Q discharge, and b is a unitless exponent representing the slope of the log-transformed C-Q relationship. All models were fit using R-3.6.1 [49] using the nlsLM function in minpack.lm [50]. To better represent specific export characteristics at catchment scale, we used an automatically derived optimal segmentation discharge and separate C- Q models for rising and falling hydrograph limbs [35]. We estimated parameters for four different C-Q models depending on whether the chemical measurement was made on a rising or falling limb of the hydrograph and if flow was higher or lower than the optimal segmentation value. We identified export regimes on the basis of modeled concentrations (N = 200, 100 each for rising and falling limb based on power-law models) considering high/low flow quantiles and hysteresis. Modeled rather than observed concentrations were used for classification and better comparability between catchments with differing chemical sampling frequencies.

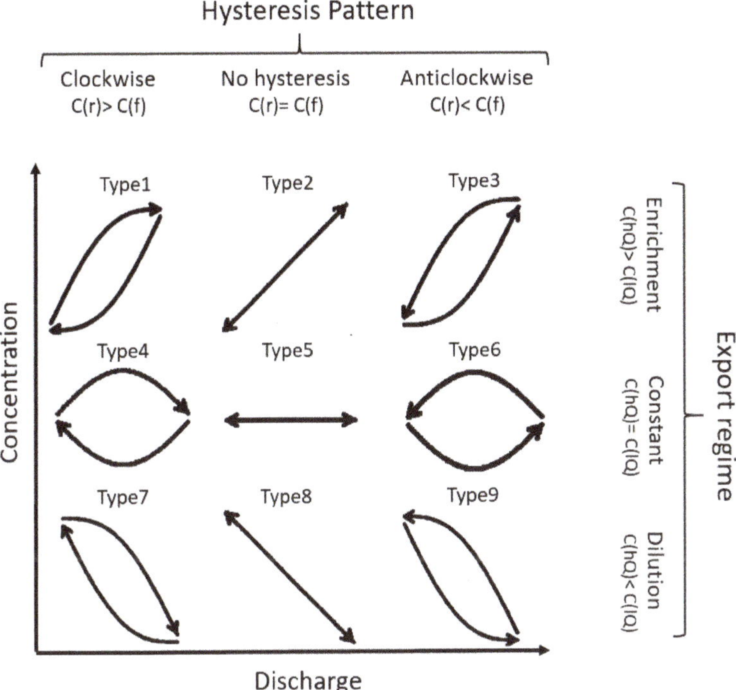

Figure 2. Schematic of the concentration–discharge (C-Q) classification considering both export regime and hysteresis pattern resulting in 9 types of C-Q relationships. Hysteresis patterns include: clockwise (concentration (C) at the rising limb (r) is higher than at the falling limb (f)), no hysteresis (no difference in concentrations at the rising and falling limbs), and anticlockwise (concentration at the falling limb is higher than at the rising limb). Export regime includes: enrichment (concentration increases with discharge) and constant (concentration does not change with discharge) and dilution (concentration decreases with discharge—lower C at high (hQ) than at low discharge (lQ)). Modified from [35].

The analysis of hysteresis patterns was based on discretization of runoff events in long-term daily discharge time series into rising and falling hydrograph limbs. Runoff events were defined as consecutive time periods when daily discharge exceeded base flow using the function base flows in the R package hydrostats [51]. To define rising and falling limbs of the hydrograph, we considered rising limbs as the periods between the day with the lowest discharge before an event until the peak flow of the event and falling limbs as the days after peak flow until the lowest flow after the discharge event. Clockwise hysteresis was identified by higher concentrations on the rising limb. Anticlockwise hysteresis was identified by higher concentrations on the falling limb. When there were no significant differences in concentration between the rising and falling limbs of the hydrograph, we assumed that there was no hysteresis. We classified C-Q types using these models (Equation 1). Hysteresis patterns were assessed by comparing concentrations on the rising and the falling limb using the Kruskal-Wallis test (p-value ≤ 0.05). To investigate whether hydrological regime has any significant importance for export regime, linear correlations (R^2) between base-flow index (BFI) and the exponent of the C-Q relationships for NO_3^-, TON, DRP, and PP were calculated. The BFI was defined as the sum of base flows divided by the total flow [52] using the function base flows in the R package hydrostats [51].

2.3. *Links between C-Q Types and Catchment Characteristics*

C-Q types and catchment behaviour were separately related to climatic characteristics (mean daily climate norms for precipitation and air temperature between 1971–2000), catchment area, land cover (agriculture and nature), and soil types (Table S2) in each catchment. To investigate similarity C-Q relationships and to explore dominant influences on the respective groupings, multivariate analyses of the C-Q relationships and catchment controls for each of the N and P forms were performed using Principal Component Analysis (PCA) [53]. PCA is a statistical technique for reducing the dimensionality of large datasets to increase interpretability while simultaneously minimising information loss [54]. It uses an orthogonal transformation to convert a set of observations of possibly correlated variables into a set of values of linearly uncorrelated principal components that successively maximise variance [55]. We used a PCA Biplot to simultaneously display information on the observations (C-Q types for each N and P form) and the variables (catchment characteristics) [53]. Two-dimensional Biplots represent the information contained in two of the principal components and are an approximation of the original multidimensional space [54]. Biplots typically display the first two principal component axes because the first principal component (PC1) always is the direction in space along which projections have the largest variance and the second principal component (PC2) is the direction, which maximizes variance among all directions orthogonal to the first [56].

3. Results

3.1. *Classification of C-Q Relationships for Nutrient Forms*

The results for NO_3^- in the Nordic region showed enrichment in 64% of the catchments (Figure 3 and Table 2), a constant relationship for 24% of the catchments, and dilution accounting for the remaining 12% (Figure 3 and Table 2). Country-specific analysis for NO_3^- showed enrichment dominated in both the Danish (70%) and the Swedish (77%) mini-catchments (Figure 3). In the Finnish mini-catchments, dilution was also of high importance (39%) (Figure 3). The dominant NO_3^- hysteresis patterns in the region were no hysteresis (48%) and clockwise (42%), while anti-clockwise occurred in only 10% of the catchments (Table 2). The Danish mini-catchments showed primarily clockwise hysteresis patterns (50%), whereas the no hysteresis pattern was most prevalent in the Swedish (62%) and Finnish (56%) mini-catchments (Figure 3). The regional results for TON differed from those of NO_3^-; a constant C-Q regime type was the dominant behaviour (60%), followed by dilution (24%) and enrichment (16%) (Figure 3 and Table 2).

Country specific analysis of TON showed constant behaviour dominated in both the Danish (73%) and the Swedish (46%) mini-catchments (Figure 3). However, in the Swedish mini-catchments, dilution also was of high importance (39%) (Figure 3). In the Finnish mini-catchments, enrichment (56%) was the dominant behaviour (Figure 3). The dominant regional TON hysteresis pattern was no hysteresis (66%), followed by clockwise (31%) and anti-clockwise (3%) (Table 2). The dominant hysteresis pattern was no hysteresis in Denmark (71%) and Sweden (85%), while the clockwise pattern dominated in Finland (56%) (Figure 3).

The regional DRP C-Q results showed dominance of constant behaviour for a total of 59% of the mini-catchments but with a high number of catchments showing dilution behaviour (33%) and very few showing enrichment (8%) (Figure 3 and Table 2). National patterns for DRP were similar with dominance of constant behaviour (Figure 3). The dominant hysteresis patterns for DRP in the region were clockwise (50%), followed by no hysteresis (47%) and anti-clockwise (3%) (Table 2). Clockwise hysteresis patterns prevailed in Denmark (54%), whereas no hysteresis prevailed in Sweden (54%) and Finland (56%) (Figure 3).

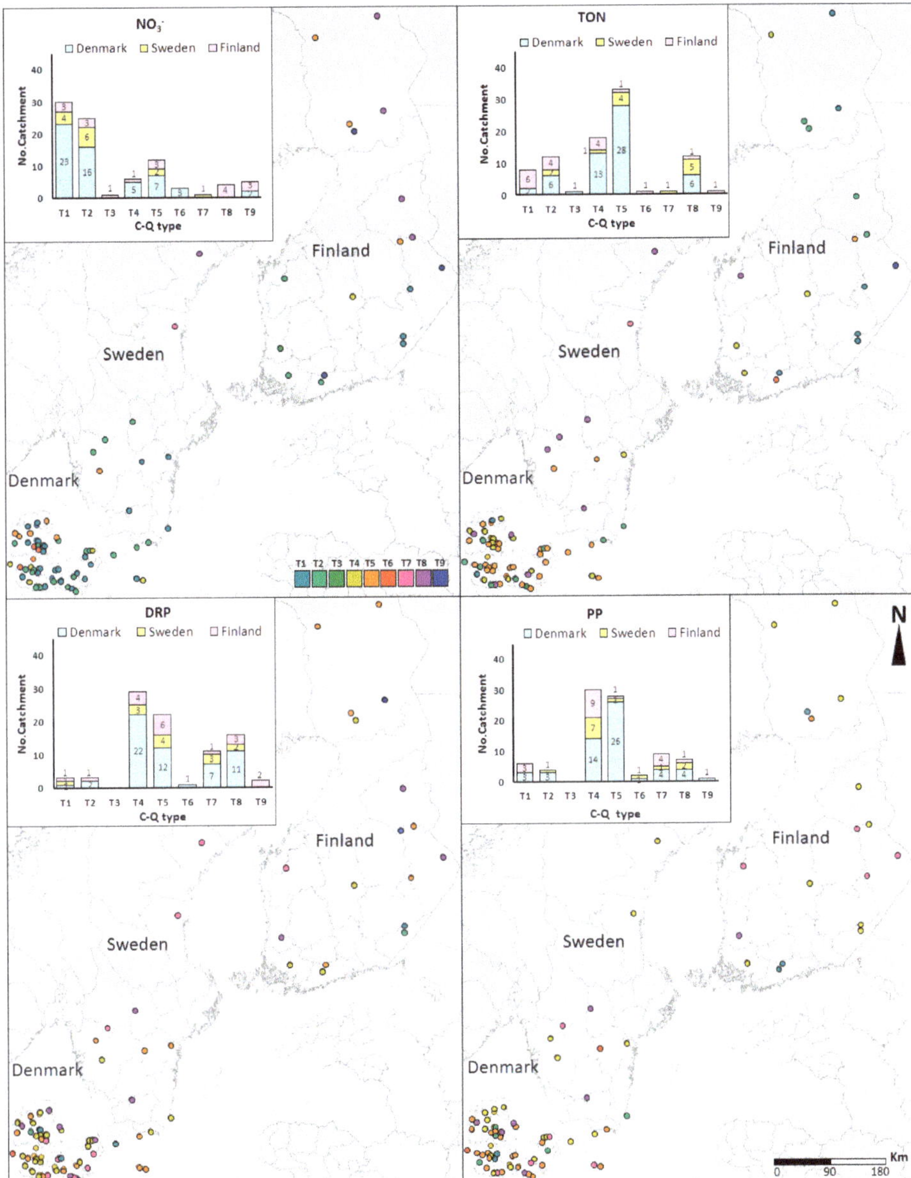

Figure 3. C-Q relationship types (T1–T9) characterising mini-catchments in three Nordic countries: Denmark, Sweden, and Finland, including nitrate (NO_3^-), total organic N (TON), dissolved reactive P (DRP), and particulate P (PP), considering export regime (enrichment, constant, and dilution) and hysteresis patterns (clockwise, anticlockwise, and no hysteresis).

Similar to DRP, C-Q regional relationships for PP were dominated by constant behaviour (69%), followed by dilution (20%) and enrichment (11%) (Figure 3 and Table 2). No major national differences could be seen for PP C-Q behaviour (Figure 3). The dominant regional hysteresis pattern for PP in the region was clockwise (53%) followed by no hysteresis (44%) and anti-clockwise (3%) (Table 2).

In the Danish mini-catchments, no hysteresis pattern was clearly the most prominent (59%), whereas clockwise hysteresis dominated in Sweden (54%) and Finland (89%) (Figure 3). Further information on C–Q relationship types for nutrient forms is provided in supporting material Table S4.

Table 2. Results on the percentage of mini-catchments in three Nordic countries: Denmark, Sweden, and Finland, considering export regime (enrichment, constant, and dilution) and hysteresis patterns (clockwise, anti-clockwise, and no hysteresis) including nitrate (NO_3^-), total organic N (TON), dissolved reactive P (DRP), and particulate P (PP).

Nutrient Form	Catchments Categorized Based on Export Regime or Hysteresis Pattern (%)					
	Export Regime			Hysteresis Pattern		
	Enrichment	Constant	Dilution	Clockwise	No Hysteresis	Anti-Clockwise
NO_3^-	64	24	12	42	48	10
TON	24	60	16	31	66	3
DRP	8	59	33	50	47	3
PP	11	69	20	53	44	3

The relationships between hydrological regime measured as BFI and export regime for NO_3^-, TON, DRP, and PP were also investigated (Figure S1). Export regime for all four nutrient forms shows significant relationships to the base-flow (BFI) index (Figure S1). Although statistically significant relationships were obtained for NO_3^-, TON, DRP, and PP, the explanatory power of the linear regression was often low. The influence of hydrological regime was strongest for TON ($R^2 = 0.45$) and less so for PP ($R^2 = 0.11$), DRP ($R^2 = 0.27$), and NO_{3-} ($R^2 = 0.13$).

3.2. Associations between C-Q Types and Catchment Characteristics

Bi-plots showing C-Q relationship types and catchment parameters for each of NO_3^-, TON, DRP, and PP are presented in Figure 4. Further information on each PCA is provided in supporting materials Table S5.

In the PCA analysis for NO_3^-, PC1 explained 45.2% of the variance, whereas PC2 explained 15.8% (Figure 4A). The most important catchment characteristics for NO_3^- are land use type and air temperature; for example, enrichment (T1, T2, and T3) is mainly related to agricultural percentage cover and higher air temperature (Figure 4A). In the case of constant and dilution behaviour (T4–T8) they are principally related to percentage natural land cover and lower air temperatures. However, soil type and annual precipitation also seem to be important parameters in case of the two dilution behaviours T7 and T9 (Figure 4A).

For TON, the PCA analysis showed that PC1 explains 41.9% of the variance (agriculture and air temperature) and PC2 explains 16.7% of the variance (catchment area and precipitation) (Figure 4B). Other important explanatory parameters for TON are low annual precipitation for enrichment types T1 and T2, whereas the constant (T6) and the dilution type (T9) are related to catchments having high annual precipitation. In case of the enrichment type (T3) larger catchment size is a dominant controlling parameter and the dilution types (T7 and T9) are related to catchment soil type (Figure 4B).

The most important catchment characteristics for DRP are land use types and air temperature, which were related to PC1 that explained 41% of the variance. PC2, which was related to precipitation and catchment area explained 16% of the variation (Figure 4C). However, precipitation also seems to influence enrichment types T1 and T2 and the constant type (T4) (Figure 4C).

As for PP, PC1 explained 40.8% and PC2 17.3% of the variance (Figure 4D). The most important catchment characteristics are land use types and air temperature for PC1 and catchment area for PC2. The regime types T1 and T2 are positively related and T8 is negatively related to catchment area. The constant regime types T4, T5, and T6 are related to a mixed-signal of catchment land use dominance.

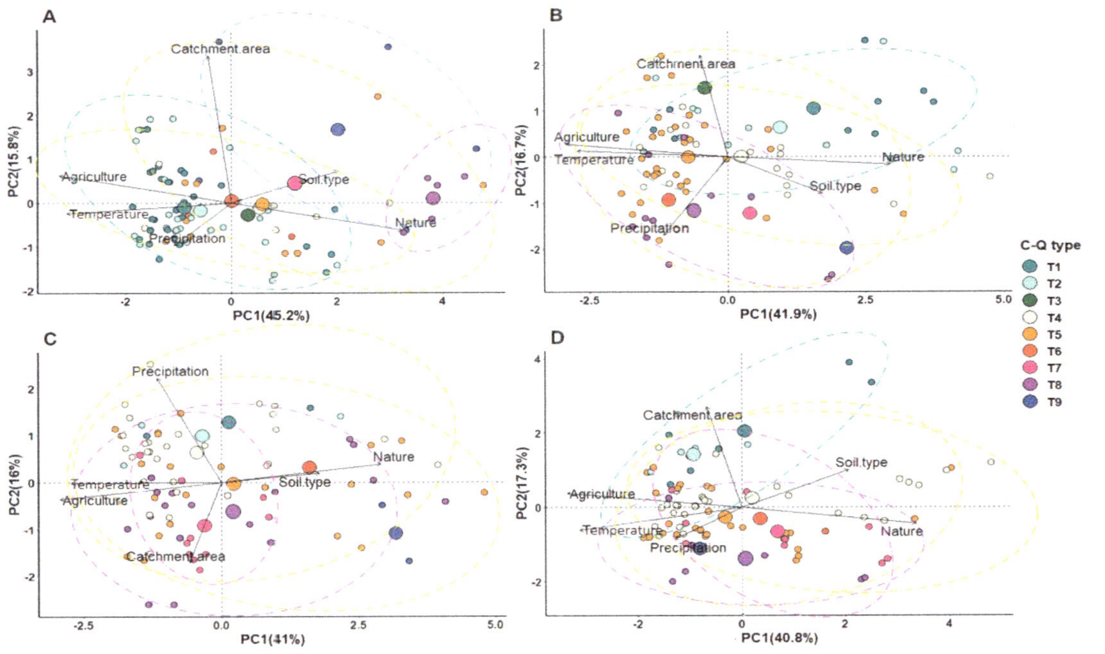

Figure 4. Principal Component Analysis (PCA)—Biplot for C-Q relationship types (T1-T9) of mini-catchments in Denmark, Sweden, and Finland for nitrate (NO_3^-) (**A**), total organic N (TON) (**B**), dissolved reactive P (DRP) (**C**), and particulate P (PP) (**D**). Variables are C-Q relationship types and catchment characteristics: land use (agriculture and nature), climate (mean daily precipitation and air temperature), soil type, and catchment area (km^2). PC 1 and 2 are the percentages of variance explained by principal component 1 and 2, respectively. Large individual points represent the centroid of the nine C-Q relationship types. Ellipses are drawn around the clusters.

4. Discussion

4.1. Classification of C-Q Relationships for Nutrient Forms

The classification of C-Q relationships obtained from low-frequency concentration data in Nordic mini-catchments according to export regime and hysteresis patterns provided insight into catchment behaviour and nutrient export patterns. We found a broad range of C-Q types in the catchments, whereby all C-Q types occur for all nutrient forms investigated with the exception of C-Q type 3 (i.e., enrichment in combination with anti-clockwise hysteresis) which does not occur for DRP and PP (Table 2). The complete range of C-Q types in our study might be ascribed to the fact that only statistically significant patterns of enrichment and dilution over the entire discharge range are considered, while other patterns, which might show non-statistically significant enrichment or dilution, are summarised as constant. Furthermore, the mini-catchments are very heterogeneous, including upland and lowland catchments with contrasting hydro-climatic conditions, topography, and land use and soil types.

Similar to an earlier Finnish study [25], enrichment behaviour (T1, T2, T3) occurs commonly in catchments with large and relatively homogeneous element stores for nutrients, indicating transport-limited export. As for NO_3^-, enrichment dominated in all countries, suggesting that high amounts of NO_3^- in agricultural catchment soils are readily available for leaching. This finding is consistent with other studies that showed enrichment for NO_3^- in catchments having a high proportion of agricultural areas [24,35,37,57]. The clockwise hysteresis pattern, which dominated the smaller Danish catchments, also points to a fast transport of NO_3^- from soils to streams, which might be linked to the dominance of tile-drained arable fields situated near to stream channels [58].

For TON, the differences in response behaviour—constant in Denmark (73%), constant (43%), and dilution (43%) in Sweden and enrichment in Finland (56%)—might be explained by differences in catchment sources and pathways. Finnish mini-catchments also had a clockwise hysteresis pattern, while no hysteresis characterised the majority of Danish and Swedish mini-catchments. This suggests that the source areas for TON in Finland might be tightly linked to riparian peatlands, whereas in Denmark and Sweden TON might be transported from more distant soil sources. Enrichment of TON (T3) can probably also be seen in streams draining dense livestock catchments with high manure and slurry inputs [35]. Constant behaviour of TON (T4, T5, T6) can result from reciprocal interactions between different drivers such as simultaneous enrichment and dilution in different parts of a catchment. Constant behaviour cannot be ascribed to the dominant spatial controls in the catchments [35]. Nevertheless, our results contrast with another study [47] evaluating variation in C-Q slopes derived from long-term low frequency N and P measurements that showed strong chemostatic behavior due to saturation and agricultural legacy effects.

Dilution behaviour (T7, T8, T9) commonly occurs in catchments with source-limited export, e.g., due to relatively low nutrient storage in non-agricultural catchments. For instance, a general association between bog (wetland) land cover and dilution caused by denitrification for NO_3^- is consistent with other studies [59,60]. Additionally, high volumes of available water for transport in wet catchments may cause dilution of TON [35,37]. Dilution behaviour can also be described by the dilution of downstream agricultural or natural or urban sources by runoff from upstream, more natural parts of the catchments.

For both DRP and PP, constant behaviour dominated in all mini-catchments in all countries, suggesting the existence of legacy P sources. For PP, the constant behaviour can possibly be explained by tight connection of legacy sources with the streams, e.g., stream banks can be a major source of PP [59,61]. For DRP, in one-third of the mini-catchments a dilution pattern was observed (Table 2). The dilution behaviour of DRP and PP can be ascribed to natural sources or (down-stream) point sources, which is consistent with other studies [35,37]. Dilution of DRP in catchments with a high percentage of agricultural land may indicate source-limited mobilisation of legacy stores in the catchments. The dilution pattern for DRP may also be attributed to point sources such as scattered households since these are often constant throughout the year and hence become diluted during high-flow periods.

A clockwise hysteresis pattern was found, principally for DRP and PP (Table 2). Clockwise hysteresis (T1, T4, T7) is a sign of fast responses of particulate and solute export to runoff events [33] and shows a close relationship or high connectivity between sources and receiving streams. This emphasises the importance of understanding the contribution of mini-catchments to P mobilisation [62,63]. Clockwise hysteresis of DRP and PP is associated mostly with a small percentage of natural land cover in the catchment points towards a strong influence of fast and highly connected transport pathways and point sources. The importance of artificial drainage as shortcuts of P exports has been demonstrated in many previous studies of water and nutrient losses in agricultural catchments [64,65]. Unfortunately, no detailed information about the occurrence of artificial drainage systems was available for this study. Therefore, the impact of artificial drainage on DRP and PP hysteresis patterns remains to be investigated. Anti-clockwise hysteresis (T3, T6, T9) can be explained by delayed transport processes due to the transport time within soils, the transport time between source areas and the catchment outlet as well as in-stream processing [33]. No significant association between the hysteresis patterns of N forms and catchment characteristics were found in this study. However, the hysteresis patterns documented for N forms might be explained by retardation of flow in soils, denitrification in wetlands, and the spatial differences in N losses within the catchment, triggering different response times between runoff events and N concentrations at the catchment outlets.

The contrasting behaviour of N and P can further be attributed to differences in nutrient sources such as a stronger influence of non-point sources on N than on P that generally originates from point sources [35]. Furthermore, differences in chemical properties and thus transport pathways might result in more homogeneous mobilisation of N from the entire catchment, P being predominantly mobilised from critical source areas with a high potential for surface runoff and a high connectivity with the river network [66,67].

4.2. Associations between C-Q Types and Catchment Characteristics

Associations between nutrient signals in different C-Q types and catchment characteristics are generally explained by a mixture of factors within the catchments related to source/pathways (i.e., extent and distribution of nutrient storage), soil properties (e.g., organic or mineral soils), and land use (cropping systems, agricultural intensity), hydrogeology (i.e., The available water for transport and hydrological connectivity), and climate (e.g., role of snow melt) and human impacts through drainage and point source discharges.

A conceptual understanding of the potential controls on in-stream nutrient concentrations underlying the C-Q classification (Figure 2) can be associated with catchment characteristics (Figure 4). Our multivariate analysis revealed an obvious association between C-Q classification and land use, which loaded strongly on PC1 for all nutrient forms. Catchment having the highest proportion of agricultural land showed dominance of enrichment types for NO_3^-, whereas dilution types were dominant for catchments having high proportions of natural land (forest or mixed types of natural vegetation) (Figure 4). These findings confirm the results of a study that aimed to elucidate the patterns and driving factors behind the N fluxes in a set of catchments in Uruguay and Denmark, which differ in land use and hydro-climatic conditions [68].

Average air temperature was found to be a co-driving parameter with land use for the different C-Q types and this is in line with air temperature being a strong controlling factor for mineralization of organic matter in soils, length of growing season, crop yield, and hence for removal of nutrients with harvest. Annual precipitation was important for TON, giving rise to constant or dilution types (T6 and T8) when it is high and enrichment types (T1–T3) when it is low (Figure 4B). This suggests different mechanisms for TON transport across the Nordic region as it can be supply-limited in wetter catchments but is transport-limited in drier catchments. Annual precipitation exerted an opposite control on DRP C-Q relationships as higher values were linked to enrichment types and lower values to dilution types

(Figure 4C). This pattern can possibly be explained by high legacy P stores in agricultural catchments and the impact of snow and soil frost in the drier, more northerly regions of the Nordic countries.

In Nordic catchments, snowmelt can contribute high amounts of water (up to 70%) to the annual runoff depending on the local annual climate. Snow and soil frost may affect C-Q relationships in several ways. In colder Nordic regions, snow accumulates from late autumn and melts in late spring and early summer. This leads to groundwater dominated discharge minima in February–March and snowmelt dominated runoff maxima in April–May. As the concentration of dissolved nutrients in streams depends partly on groundwater discharge, such conditions may strongly affect C-Q relationships and lead to a prevalence of dilution and anticlockwise C-Q types. Frozen soils often have low infiltration capacity [69]. Much of the snow melting on top of frozen soils discharges to streams with no or minimal soil contact, and this leads to low concentrations or dilution of all nutrient forms studied here [70]. On the other hand, snow melting on unfrozen soils can infiltrate and generate runoff of pre-event water. In such cases, concentrations of both dissolved and particulate bound nutrients might increase with runoff, just as in rain-generated hydrologic events.

Other studies have also looked at potential links between C-Q parameters and catchment characteristics. In most cases, correlations were poor [24,57] but these works evidenced relationships with catchment area, land use, and lithology. Similar to our study, a study in larger catchments in Scotland [35] did not identify any dominant spatial controls on stream nutrient concentration response with changes in flow. However, using C-Q typologies to classify catchments based on their export behaviour could still be used as part of a decision support system for the improvement of monitoring design and for spatially targeting of catchment scale mitigation measures. For example, in the spatial targeting of measures, application of transport mitigation measures (e.g., constructed wetlands) may be appropriate for catchment/nutrient combinations showing enrichment export regimes [71], while source mitigation measures may be more useful for catchment/nutrient combinations characterised by dilution export regimes [72]. Further, for enrichment export regimes, a higher frequency of water chemistry sampling may be appropriate compared with constant (medium-frequency) and dilution-type catchment/nutrient combinations [73].

The development of catchment typologies based on solute export behaviour and hysteresis could be useful for the transfer of information from data-rich to data-poor catchments [74], for impact assessment of climate, environmental and management changes on water quality and for parameterisation of water quality models [35]. However, finding the association between the typology presented here and catchment characteristics, e.g., topography, geology, and land use may be challenging owing to the complexity of different responses to both spatial and temporal mechanisms [5,14]. Within-catchment heterogeneity may also affect the links between catchment characteristics and C-Q relationships [23,32,75].

4.3. Evaluation of Methodology

Evaluating the limitations and uncertainties of water quality studies may contribute to increased awareness among watershed managers and policy makers, which in the end may contribute to more informed decisions. All catchment studies have limitations as well as technical and conceptual uncertainties that must be addressed. One potential limitation of this study is that catchments have a rapid response to precipitation inputs and sub-daily hydrographs were not available. However, the typology presented here is related to seasonal hysteresis patterns not the short terms brehavior during individual runoff events.

Uncertainties arise from many sources. These could include the way that export regime and hysteresis patterns are conceptualized and statistically tested, the mathematical functions relating concentrations to discharge may be an over-simplification of reality, the breakpoints assigned for modelling concentrations for different discharge quantiles, and how catchment characteristics are assumed to influence hydrological and nutrient flow processes. Within all of these issues there are also multiple interactions that may further contribute to uncertainties, and it is very difficult if not impossible to address all such issues. Further, uncertainties in input data such as soil map classifications, land use map classification, flow and concentration data could also contribute to uncertainties that will propagate to the final classification. In our study, we assumed that the reclassified soil, land use classes, and concentration data had no uncertainties, but in reality their uncertainty will affect the correlation between the different soil classes, land use classes, concentration data, and the C-Q types of the catchments. However, it may be possible to reduce the uncertainty in classification by obtaining a better understanding of the mini-catchments relative to soil types, land use types and hydrogeology (e.g., tile drainage) and their respective C-Q types. Thus, for a better representation of areas influencing in-stream water quality, it might be appropriate to investigate C-Q types in even smaller catchments (i.e., few hectares) with relatively homogeneous characteristics throughout.

PCA is a good data summary tool when the patterns of interest can be projected onto linear, orthogonal components. However, PCA also has limitations that must be considered when interpreting the output: the underlying structure of the data must be linear, patterns that are highly correlated may be unresolved because all PCs are uncorrelated, and the goal of PCA is to explain the maximum amount of variance and not necessarily to find clusters [76]. The method of selecting input data for PCA analysis is also an important factor to consider when interpreting the results.

5. Conclusions

Our work demonstrates the use of a novel nutrient concentration (C) and discharge (Q) (C-Q) classification [35] on 8 years of water quality data from 87 Nordic mini-catchments situated in Denmark, Finland, and Sweden. The dominant export regimes for nitrogen (N) were enrichment for nitrate and constant for total organic N. For phosphorus (P), both particulate P and dissolved reactive P were characterised by a constant export regime. Clockwise rotational hysteresis patterns dominated for nitrate, dissolved reactive P and particulate P, whereas a no hysteresis pattern dominated for total organic N.

As expected, our results showed that catchment land use exerted a dominant control on the C-Q types. Nutrient export behaviour in streams draining Nordic mini-catchments is determined by catchment land use characteristics and, to a lesser extent, by climate. All of these factors are important elements to be considered in future surface water nutrient management plans.

Supplementary Materials: The following are available online at http://www.mdpi.com/2073-4441/12/6/1776/s1.

Author Contributions: Conceptualisation, F.H. and B.K. (Brian Kronvang); methodology, F.H., I.P. and B.K. (Brian Kronvang); formal analysis, F.H.; investigation, all authors; resources, H.T., K.K., H.M. and A.L.; data curation, F.H.; writing – original draft preparation, F.H.; writing – review and editing, B.K. (Brian Kronvang), M.F., B.K. (Bjørn Kløve), K.K., H.M., I.P., J.W.M.P. and A.L; visualisation, F.H.; supervision, B.K. (Brian Kronvang); funding acquisition, B.K. (Brian Kronvang). All authors have read and agreed to the published version of the manuscript.

Funding: This study was part of the Nordic Centre of Excellence BIOWATER project and funded by Nordforsk (project number 82263).

Acknowledgments: We thankfully acknowledge valuable English language edits by Anne Mette Poulsen. The climatic characteristics with a resolution of a 0.5 degree grid (50 km) were derived from a database provided by the Swedish Meteorological and Hydrological Institute (SMHI) within the contract C3S_441_Lot1_SMHI of the Copernicus Climate Change Service (C3S), https://swicca.eu/). Further, Ina Pohle was supported by the Scottish Government's Rural and Environment Science and Analytical Services Division (RESAS).

Conflicts of Interest: The authors declare no conflict of interest.

References

1. Diaz, R.J.; Rosenberg, R. Spreading dead zones and consequences for marine ecosystems. *Science* **2008**, *321*, 926–929. [CrossRef] [PubMed]
2. Reusch, T.B.; Dierking, J.; Andersson, H.C.; Bonsdorff, E.; Carstensen, J.; Casini, M.; Czajkowski, M.; Hasler, B.; Hinsby, K.; Hyytiäinen, K.; et al. The Baltic Sea as a time machine for the future coastal ocean. *Sci. Adv.* **2018**, *4*, 8195. [CrossRef] [PubMed]

3. Smol, M.; Preisner, M.; Bianchini, A.; Rossi, J.; Hermann, L.; Schaaf, T.; Kruopienė, J.; Pamakštys, K.; Klavins, M.; Ozola-Davidane, R.; et al. Strategies for Sustainable and Circular Management of Phosphorus in the Baltic Sea Region: The Holistic Approach of the InPhos Project. *Sustainability* **2020**, *12*, 2567. [CrossRef]
4. Bouwman, A.F.; Bierkens, M.F.P.; Griffioen, J.; Hefting, M.M.; Middelburg, J.J.; Middelkoop, H.; Slomp, C.P. Nutrient dynamics, transfer and retention along the aquatic continuum from land to ocean: Towards integration of ecological and biogeochemical models. *Biogeosciences* **2016**, *10*, 1–23. [CrossRef]
5. Lundberg, C.J.; Lane, R.R.; Day, J.W., Jr. Spatial and temporal variations in nutrients and water-quality parameters in the Mississippi River-influenced Breton Sound Estuary. *J. Coast. Res.* **2014**, *30*, 328–336. [CrossRef]
6. Stutter, M.I.; Langan, S.J.; Cooper, R.J. Spatial and temporal dynamics of stream water particulate and dissolved N, P and C forms along a catchment transect, NE Scotland. *J. Hydrol.* **2008**, *350*, 187–202. [CrossRef]
7. Dong, Z.; Driscoll, C.T.; Campbell, J.L.; Pourmokhtarian, A.; Stoner, A.M.; Hayhoe, K. Projections of water, carbon, and nitrogen dynamics under future climate change in an alpine tundra ecosystem in the southern Rocky Mountains using a biogeochemical model. *Sci. Total Environ.* **2019**, *650*, 1451–1464. [CrossRef] [PubMed]
8. Lintern, A.; Webb, J.A.; Ryu, D.; Liu, S.; Waters, D.; Leahy, P.; Bende-Michl, U.; Western, A.W. What are the key catchment characteristics affecting spatial differences in riverine water quality? *Water Resour. Res.* **2018**, *54*, 7252–7272. [CrossRef]
9. Miller, M.P.; Tesoriero, A.J.; Hood, K.; Terziotti, S.; Wolock, D.M. Estimating discharge and nonpoint source nitrate loading to streams from three end-member pathways using high-frequency water quality data. *Water Resour. Res.* **2017**, *53*, 10201–10216. [CrossRef]
10. Wang, J.; Chen, G.; Kang, W.; Hu, K.; Wang, L. Impoundment intensity determines temporal patterns of hydrological fluctuation, carbon cycling and algal succession in a dammed lake of Southwest China. *Water Res.* **2019**, *148*, 162–175. [CrossRef] [PubMed]
11. Yang, S.; Büttner, O.; Kumar, R.; Jäger, C.; Jawitz, J.W.; Rao, P.S.C.; Borchardt, D. Spatial patterns of water quality impairments from point source nutrient loads in Germany's largest national River Basin (Weser River). *Sci. Total Environ.* **2019**, *697*, 134145. [CrossRef] [PubMed]
12. Bernal, S.; von Schiller, D.; Sabater, F.; Martí, E. Hydrological extremes modulate nutrient dynamics in Mediterranean climate streams across different spatial scales. *Hydrobiologia* **2018**, *719*, 31–42. [CrossRef]
13. Bracken, L.J.; Wainwright, J.; Ali, G.A.; Tetzlaff, D.; Smith, M.W.; Reaney, S.M.; Roy, A.G. Concepts of hydrological connectivity: Research approaches, pathways and future agendas. *Earth Sci. Rev.* **2013**, *119*, 17–34. [CrossRef]
14. Wood, M.E.; Macrae, M.L.; Strack, M.; Price, J.S.; Osko, T.J.; Petrone, R.M. Spatial variation in nutrient dynamics among five different peatland types in the Alberta oil sands region. *Ecohydrology* **2016**, *9*, 688–699. [CrossRef]
15. Jarvie, H.P.; Smith, D.R.; Norton, L.R.; Edwards, F.K.; Bowes, M.J.; King, S.M.; Scarlett, P.; Davies, S.; Dils, R.M.; Bachiller-Jareno, N. Phosphorus and nitrogen limitation and impairment of headwater streams relative to rivers in Great Britain: A national perspective on eutrophication. *Sci. Total Environ.* **2018**, *621*, 849–862. [CrossRef] [PubMed]
16. Hawtree, D.; Nunes, J.P.; Keizer, J.J.; Jacinto, R.; Santos, J.; Rial-Rivas, M.E.; Boulet, A.K.; Tavares-Wahren, F.; Feger, K.H. Time series analysis of the long-term hydrologic impacts of afforestation in the Águeda watershed of north-central Portugal. *Hydrol. Earth Syst. Sci.* **2015**, *19*, 3033–3045. [CrossRef]
17. Knapp, J.L.; Freyberg, J.V.; Studer, B.; Kiewiet, L.; Kirchner, J.W. Concentration-discharge relationships vary among hydrological events, reflecting differences in event characteristics. *Hydrol. Earth Syst. Sci. Discuss.* **2020**, *7*, 1–27.
18. Vale, S.S.; Dymond, J.R. Interpreting nested storm event suspended sediment-discharge hysteresis relationships at large catchment scales. *Hydrol. Process.* **2020**, *34*, 420–440. [CrossRef]
19. Scheffer, M.; Bascompte, J.; Brock, W.A.; Brovkin, V.; Carpenter, S.R.; Dakos, V.; Held, H.; Van Nes, E.H.; Rietkerk, M.; Sugihara, G. Early-warning signals for critical transitions. *Nature* **2009**, *461*, 53–59. [CrossRef]
20. Meybeck, M.; Laroche, L.; Dürr, H.H.; Syvitski, J.P.M. Global variability of daily total suspended solids and their fluxes in rivers. *Glob. Planet. Chang.* **2003**, *39*, 65–93. [CrossRef]
21. Moatar, F.; Meybeck, M.; Raymond, S.; Birgand, F.; Curie, F. River flux uncertainties predicted by hydrological variability and riverine material behaviour. *Hydrol. Process.* **2013**, *27*, 3535–3546. [CrossRef]

22. Moatar, F.; Floury, M.; Gold, A.J.; Meybeck, M.; Renard, B.; Chandesris, A.; Minaudo, C.; Addy, K.; Piffady, J.; Pinay, G. Stream solutes and particulates export regimes: A new framework to optimize their monitoring. *Front. Ecol. Evol.* **2019**, *7*, 516. [CrossRef]
23. Moatar, F.; Meybeck, M. Compared performances of different algorithms for estimating annual nutrient loads discharged by the eutrophic River Loire. *Hydrol. Process. Int. J.* **2005**, *19*, 429–444. [CrossRef]
24. Godsey, S.E.; Kirchner, J.W.; Clow, D.W. Concentration–discharge relationships reflect chemostatic characteristics of US catchments. *Hydrol. Process. Int. J.* **2009**, *23*, 1844–1864. [CrossRef]
25. Basu, N.B.; Destouni, G.; Jawitz, J.W.; Thompson, S.E.; Loukinova, N.V.; Darracq, A.; Zanardo, S.; Yaeger, M.; Sivapalan, M.; Rinaldo, A.; et al. Nutrient loads exported from managed catchments reveal emergent biogeochemical stationarity. *Geophys. Res. Lett.* **2010**, 37. [CrossRef]
26. Musolff, A.; Schmidt, C.; Selle, B.; Fleckenstein, J.H. Catchment controls on solute export. *Adv. Water Resour.* **2015**, *86*, 133–146. [CrossRef]
27. Musolff, A.; Fleckenstein, J.H.; Rao, P.S.C.; Jawitz, J.W. Emergent archetype patterns of coupled hydrologic and biogeochemical responses in catchments. *Geophys. Res. Lett.* **2017**, *44*, 4143–4151. [CrossRef]
28. Zhang, Q. Synthesis of nutrient and sediment export patterns in the Chesapeake Bay watershed: Complex and non-stationary concentration-discharge relationships. *Sci. Total Environ.* **2018**, *618*, 1268–1283. [CrossRef]
29. Minaudo, C.; Dupas, R.; Gascuel-Odoux, C.; Roubeix, V.; Danis, P.A.; Moatar, F. Seasonal and event-based concentration-discharge relationships to identify catchment controls on nutrient export regimes. *Adv. Water Resour.* **2019**, *131*, 103379. [CrossRef]
30. Aguilera, R.; Melack, J.M. Concentration-discharge responses to storm events in coastal California watersheds. *Water Resour. Res.* **2018**, *54*, 407–424. [CrossRef]
31. Hunsaker, C.T.; Johnson, D.W. Concentration-discharge relationships in headwater streams of the Sierra Nevada, California. *Water Resour. Res.* **2017**, *53*, 7869–7884. [CrossRef]
32. Zuecco, G.; Penna, D.; Borga, M.; van Meerveld, H.J. A versatile index to characterize hysteresis between hydrological variables at the runoff event timescale. *Hydrol. Process.* **2016**, *30*, 1449–1466. [CrossRef]
33. Bieroza, M.Z.; Heathwaite, A.L. Seasonal variation in phosphorus concentration–discharge hysteresis inferred from high-frequency in situ monitoring. *J. Hydrol.* **2015**, *524*, 333–347. [CrossRef]
34. Winnick, M.J.; Carroll, R.W.; Williams, K.H.; Maxwell, R.M.; Dong, W.; Maher, K. Snowmelt controls on concentration-discharge relationships and the balance of oxidative and acid-base weathering fluxes in an alpine catchment, East River, Colorado. *Water Resour. Res.* **2017**, *53*, 2507–2523. [CrossRef]
35. Pohle, I.; Glendell, M.; Baggaley, N.; Stutter, M. A classification scheme for concentration-discharge relationships based on long-term low-frequency water quality data. In *Geophysical Research Abstracts*; EGU2019-7425; EGU General Assembly: Vienna, Austria, 2019; Volume 21.
36. Meybeck, M.; Moatar, F. Daily variability of river concentrations and fluxes: Indicators based on the segmentation of the rating curve. *Hydrol. Process.* **2012**, *26*, 1188–1207. [CrossRef]
37. Moatar, F.; Abbott, B.W.; Minaudo, C.; Curie, F.; Pinay, G. Elemental properties, hydrology, and biology interact to shape concentration-discharge curves for carbon, nutrients, sediment, and major ions. *Water Resour. Res.* **2017**, *53*, 1270–1287. [CrossRef]
38. Dupas, R.; Jomaa, S.; Musolff, A.; Borchardt, D.; Rode, M. Disentangling the influence of hydroclimatic patterns and agricultural management on river nitrate dynamics from sub-hourly to decadal time scales. *Sci. Total Environ.* **2016**, *571*, 791–800. [CrossRef]
39. Alexander, R.B.; Boyer, E.W.; Smith, R.A.; Schwarz, G.E.; Moore, R.B. The role of headwater streams in downstream water quality 1. *JAWRA J. Am. Water Resour. Assoc.* **2007**, *43*, 41–59. [CrossRef]
40. NOVANA 2017-21. Se Det Nationale Overvågningsprogram for Vandmiljø og Natur (NOVANA)2017-21. Available online: http://mst.dk/service/publikationer/publikationsarkiv/2017/okt/novana-2017-21/ (accessed on 3 April 2019).
41. Kyllmar, K.; Forsberg, L.S.; Andersson, S.; Mårtensson, K. Small agricultural monitoring catchments in Sweden representing environmental impact. *Agric. Ecosyst. Environ.* **2014**, *198*, 25–35. [CrossRef]
42. Löfgren, S.; Aastrup, M.; Bringmark, L.; Hultberg, H.; Lewin-Pihlblad, L.; Lundin, L.; Karlsson, G.P.; Thunholm, B. Recovery of soil water, groundwater, and streamwater from acidification at the Swedish Integrated Monitoring catchments. *Ambio* **2011**, *40*, 836–856. [CrossRef]
43. Linjama, J.; Järvinen, J.; Kivinen, Y. The Finnish Environment. In *Hydrological Yearbook 2006–2010*; Korhonen, J., Haavanlammi, E., Eds.; Finnish Environmental Institute: Helsinki, Finland, 2012; p. 8.

44. Levin, G.; Iosub, C.I.; Jepsen, M.R. *Basemap02: Technical Documentation of a Model for Elaboration of a Land Use and Land-Cover Map for Denmark*; Aarhus University, DCE—Danish Centre for Environment and Energy: Aarhus, Denmark, 2017.
45. Vuorenmaa, J.; Rekolainen, S.; Lepistö, A.; Kenttämies, K.; Kauppila, P. Losses of nitrogen and phosphorus from agricultural and forest areas in Finland during the 1980s and 1990s. *Environ. Monit. Assess.* **2002**, *76*, 213–248. [CrossRef] [PubMed]
46. Tattari, S.; Koskiaho, J.; Kosunen, M.; Lepistö, A.; Linjama, J.; Puustinen, M. Nutrient loads from agricultural and forested areas in Finland from 1981 up to 2010—Can the efficiency of undertaken water protection measures seen? *Environ. Monit. Assess.* **2017**, *189*, 95. [CrossRef]
47. Bieroza, M.Z.; Heathwaite, A.L.; Bechmann, M.; Kyllmar, K.; Jordan, P. The concentration-discharge slope as a tool for water quality management. *Sci. Total Environ.* **2018**, *630*, 738–749. [CrossRef] [PubMed]
48. Li, L.; Bao, C.; Sullivan, P.L.; Brantley, S.; Shi, Y.; Duffy, C. Understanding watershed hydrogeochemistry: 2. Synchronized hydrological and geochemical processes drive stream chemostatic behavior. *Water Resour. Res.* **2017**, *53*, 2346–2367. [CrossRef]
49. R Core Team. *R: A Language and Environment for Statistical Computing*; R Foundation for Statistical Computing: Vienna, Austria, 2019. Available online: http://www.R-project.org/ (accessed on 12 December 2010).
50. Elzhov, T.V.; Mullen, K.M.; Spiess, A.N.; Bolker, B. Minpack. lm: R Interface to the Levenberg-Marquardt Nonlinear Least-Squares Algorithm Found in MINPACK, Plus Support for Bounds. R package version 1.2-1. 2016. Available online: https://cran.r-project.org/web/packages/minpack.lm/minpack.lm.pdf (accessed on 19 June 2010).
51. Bond, N. Hydrostats: Hydrologic Indices for Daily Time Series Data. R package version 0.2.7. 2019. Available online: https://cran.r-project.org/web/packages/hydrostats/hydrostats.pdf (accessed on 19 June 2010).
52. Gustard, A.; Bullock, A.; Dixon, J.M. *Low Flow Estimation in the United Kingdom*; IH report no. 108; Institute of Hydrology: Wallingford, UK, 1992.
53. Gabriel, K.R. The biplot graphic display of matrices with application to principal component analysis. *Biometrika* **1971**, *58*, 453–467. [CrossRef]
54. Jolliffe, I.T.; Cadima, J. Principal component analysis: A review and recent developments. *Philos. Trans. R. Soc. A Math. Phys. Eng. Sci.* **2016**, *374*, 20150202. [CrossRef]
55. Gower, J.C.; Lubbe, S.G.; Le Roux, N.J. *Understanding Biplots*; John Wiley & Sons: New York, NY, USA, 2011.
56. Abdi, H.; Williams, L.J. Principal component analysis. *Wiley Interdiscip. Rev. Comput. Stat.* **2010**, *2*, 433–459. [CrossRef]
57. Diamond, J.S.; Cohen, M.J. Complex patterns of catchment solute–discharge relationships for coastal plain rivers. *Hydrol. Process.* **2018**, *32*, 388–401. [CrossRef]
58. Grant, R.; Laubel, A.; Kronvang, B.; Andersen, H.E.; Svendsen, L.M.; Fuglsang, A. Loss of dissolved and particulate phosphorus from arable catchments by subsurface drainage. *Water Res.* **1996**, *30*, 2633–2642. [CrossRef]
59. Laubel, A.; Kronvang, B.; Hald, A.B.; Jensen, C. Hydromorphological and biological factors influencing sediment and phosphorus loss via bank erosion in small lowland rural streams in Denmark. *Hydrol. Process.* **2003**, *17*, 3443–3463. [CrossRef]
60. Stutter, M.I.; Graeber, D.; Evans, C.D.; Wade, A.J.; Withers, P.J.A. Balancing macronutrient stoichiometry to alleviate eutrophication. *Sci. Total Environ.* **2018**, *634*, 439–447. [CrossRef] [PubMed]
61. Kronvang, B.; Audet, J.; Baattrup-Pedersen, A.; Jensen, H.S.; Larsen, S.E. Phosphorus load to surface water from bank erosion in a Danish lowland river basin. *J. Environ. Qual.* **2012**, *41*, 304–313. [CrossRef] [PubMed]
62. Bol, R.; Gruau, G.; Mellander, P.E.; Dupas, R.; Bechmann, M.; Skarbøvik, E.; Bieroza, M.; Djodjic, F.; Glendell, M.; Jordan, P.; et al. Challenges of reducing phosphorus based water eutrophication in the agricultural landscapes of northwest Europe. *Front. Mar. Sci.* **2018**, *5*, 276. [CrossRef]
63. Dupas, R.; Musolff, A.; Jawitz, J.W.; Rao, P.S.C.; Jäger, C.G.; Fleckenstein, J.H.; Rode, M.; Borchardt, D. Carbon and nutrient export regimes from headwater catchments to downstream reaches. *Biogeosciences* **2017**, *14*, 4391. [CrossRef]
64. Koch, S.; Kahle, P.; Lennartz, B. Spatio-temporal analysis of phosphorus concentrations in a North-Eastern German lowland watershed. *J. Hydrol. Reg. Stud.* **2018**, *15*, 203–216. [CrossRef]

65. Laubel, A.; Jacobsen, O.H.; Kronvang, B.; Grant, R.; Andersen, H.E. Subsurface Drainage Loss of Particles and Phosphorus from Field Plot Experiments and a Tile-Drained Catchment. *J. Environ. Qual.* **1999**, *28*, 576–584. [CrossRef]
66. Sliva, L.; Williams, D.D. Buffer zone versus whole catchment approaches to studying land use impact on river water quality. *Water Res.* **2016**, *35*, 3462–3472. [CrossRef]
67. Sharpley, A.N.; Kleinman, P.J.; Jordan, P.; Bergström, L.; Allen, A.L. Evaluating the success of phosphorus management from field to watershed. *J. Environ. Qual.* **2009**, *38*, 1981–1988. [CrossRef]
68. Goyenola, G.; Graeber, D.; Meerhoff, M.; Jeppesen, E.; Mello, F.-D.; Vidal, N.; Fosalba, C.; Ovesen, N.B.; Gelbrecht, J.; Mazzeo, N.; et al. Influence of Farming Intensity and Climate on Lowland Stream Nitrogen. *Water* **2020**, *12*, 1021. [CrossRef]
69. Lundberg, A.; Ala-Aho, P.; Eklo, O.; Klöve, B.; Kværner, J.; Stumpp, C. Snow and frost: Implications for spatiotemporal infiltration patterns—A review. *Hydrol. Process.* **2016**, *30*, 1230–1250. [CrossRef]
70. Eskelinen, R.; Ronkanen, A.; Marttila, H.; Isokangas, E.; Kløve, B. Effects of soil frost on snowmelt runoff generation and surface water quality in drained peatlands. *Boreal Environ. Res.* **2016**, *21*, 556–570.
71. Carstensen, M.V.; Hashemi, F.; Hoffmann, C.C.; Zak, D.; Audet, J.; Kronvang, B. Efficiency of mitigation measures targeting nutrient losses from agricultural drainage systems: A review. *Ambio* **2020**. [CrossRef] [PubMed]
72. Hashemi, F.; Kronvang, B. Multi-functional benefits from targeted land use changes in a Danish catchment. *Ambio* **2020**, under review.
73. Kronvang, B.; Bruhn, A.J. Choice of sampling strategy and estimation method for calculating nitrogen and phosphorus transport in small lowland streams. *Hydrol. Process.* **1996**, *10*, 1483–1501. [CrossRef]
74. Krause, S.; Freer, J.; Hannah, D.M.; Howden, N.J.; Wagener, T.; Worrall, F. Catchment similarity concepts for understanding dynamic biogeochemical behaviour of river basins. *Hydrol. Process.* **2014**, *28*, 1554–1560. [CrossRef]
75. Ali, G.; Wilson, H.; Elliott, J.; Penner, A.; Haque, A.; Ross, C.; Rabie, M. Phosphorus export dynamics and hydrobiogeochemical controls across gradients of scale, topography and human impact. *Hydrol. Process.* **2017**, *31*, 3130–3145. [CrossRef]
76. Lever, J.; Krzywinsk, M.; Altman, N. Points of significance: Principal component analysis. *Nat. Methods* **2017**, *14*, 641–642. [CrossRef]

Article

Lag Time as an Indicator of the Link between Agricultural Pressure and Drinking Water Quality State

Hyojin Kim [1,*], Nicolas Surdyk [2], Ingelise Møller [1], Morten Graversgaard [3], Gitte Blicher-Mathiesen [4], Abel Henriot [2], Tommy Dalgaard [3] and Birgitte Hansen [1]

1 Department of Quaternary and Groundwater Mapping, Geological Survey of Denmark and Greenland (GEUS), C.F. Møllers Allé 8, Building 1110, 8000 Aarhus C, Denmark; ilm@geus.dk (I.M.); bgh@geus.dk (B.H.)
2 BRGM, 3, avenue C. Guillemin, BP 360009, 45060 Orléans CEDEX 2, France; n.surdyk@brgm.fr (N.S.); a.henriot@brgm.fr (A.H.)
3 Department of Agroecology, Aarhus University, Blichers Alle 20, 8830 Tjele, Denmark; morten.graversgaard@agro.au.dk (M.G.); tommy.dalgaard@agro.au.dk (T.D.)
4 Department of Bioscience, Aarhus University, Vejlsøvej 25, 8600 Silkeborg, Denmark; gbm@bios.au.dk
* Correspondence: hk@geus.dk

Received: 14 July 2020; Accepted: 19 August 2020; Published: 25 August 2020

Abstract: Diffuse nitrogen (N) pollution from agriculture in groundwater and surface water is a major challenge in terms of meeting drinking water targets in many parts of Europe. A bottom-up approach involving local stakeholders may be more effective than national- or European-level approaches for addressing local drinking water issues. Common understanding of the causal relationship between agricultural pressure and water quality state, e.g., nitrate pollution among the stakeholders, is necessary to define realistic goals of drinking water protection plans and to motivate the stakeholders; however, it is often challenging to obtain. Therefore, to link agricultural pressure and water quality state, we analyzed lag times between soil surface N surplus and groundwater chemistry using a cross correlation analysis method of three case study sites with groundwater-based drinking water abstraction: Tunø and Aalborg-Drastrup in Denmark and La Voulzie in France. At these sites, various mitigation measures have been implemented since the 1980s at local to national scales, resulting in a decrease of soil surface N surplus, with long-term monitoring data also being available to reveal the water quality responses. The lag times continuously increased with an increasing distance from the N source in Tunø (from 0 to 20 years between 1.2 and 24 m below the land surface; mbls) and La Voulzie (from 8 to 24 years along downstream), while in Aalborg-Drastrup, the lag times showed a greater variability with depth—for instance, 23-year lag time at 9–17 mbls and 4-year lag time at 21–23 mbls. These spatial patterns were interpreted, finding that in Tunø and La Voulzie, matrix flow is the dominant pathway of nitrate, whereas in Aalborg-Drastrup, both matrix and fracture flows are important pathways. The lag times estimated in this study were comparable to groundwater ages measured by chlorofluorocarbons (CFCs); however, they may provide different information to the stakeholders. The lag time may indicate a wait time for detecting the effects of an implemented protection plan while groundwater age, which is the mean residence time of a water body that is a mixture of significantly different ages, may be useful for planning the time scale of water protection programs. We conclude that the lag time may be a useful indicator to reveal the hydrogeological links between the agricultural pressure and water quality state, which is fundamental for a successful implementation of drinking water protection plans.

Keywords: lag times; link indicator; nitrate; agriculture; drinking water; DPLSIR framework

---

1. Introduction

The leaching of nitrate from intensive agricultural areas is one of the greatest threats to clean drinking water resources in Europe and the rest of the world. The groundwater and drinking water standard of nitrate is set to a maximum of 50 mg NO_3^-/L in the European Union (EU), following the recommendations of the World Health Organization (WHO) [1]. This level is set to protect infants from the acute condition called blue baby syndrome (methemoglobinemia) [2,3]. However, recent studies have reported that long-term exposure to drinking water containing nitrate below the current standard is suspected to have chronic adverse effects [3–6]. For instance, studies have consistently reported that risks of colorectal cancer, thyroid disease, and central nervous system birth defects are strongly associated with ingestion of drinking water nitrate [3]. The EU has issued a series of directives, guidelines, and policies over the last decades to set the requirements of the drinking water standards (i.e., Drinking Water Directive) and to protect the drinking water resources (e.g., Water Framework Directive, Groundwater Directive, Nitrate Directive, and Directive on the Sustainable Use of Pesticides). Each country in the EU has implemented these legislations in their national laws differently, due to adjustment to local politico-socio-economic-agri-hydrogeological conditions [7,8]. The Drinking Water Directive primarily focuses on large water supplies. Small water supplies in rural areas have been given less attention, resulting in worse cases of drinking water quality [9,10].

The Organization for Economic Co-operation and Development (OECD) has proposed a bottom-up approach that integrates multiple local and regional stakeholders, such as farmers, citizens, drinking water suppliers, policy-makers, and scientists, as an effective approach for addressing local issues in relation to drinking water security, rather than the most common top-down approach at the national and EU level [11]. Research has also shown that the success of water protection plans depends on the social, technical, financial, and political capacity of the stakeholders and institutional arrangements, such as policies and regulations [12–15]. In addition, non-regulatory tools, such as training, information campaigns, education, or voluntary programs, have been shown to be effective in increasing the engagement of stakeholders and consequently lead to the successful implementation of water protection programs [12].

Common understandings of the cause and effect relations between the agricultural pressure and drinking water state may be a fundamental step in finding common and achievable goals among stakeholders and in planning drinking water protection programs [8,16,17]. However, the cause–effect relations are often difficult to identify. A long lag time between the use of fertilizer and its effects on water quality is often considered to be one of the reasons for unclear correlations between the agricultural pressure and water quality state [8,18,19]. When authorities and farmers implement mitigation measures (e.g., catch crops, set-aside, and buffer zones) as part of water protection planning, they may have to wait multiple decades to observe the effects on the water quality in some geological settings. Consequently, it may be inevitable that farmers become less engaged in water protection programs, unless they are well-informed in advance. In addition, information that is key to evaluating the cause–effect relations is monitored and managed by different institutions and government bodies [8]. Therefore, it is often difficult to gather coherent information to understand the complexity of the system.

The hydrogeological system that links the agricultural and drinking water systems may be the first-order control on regulating the relationships between these two systems; however, its importance has often been neglected in communication among stakeholders and the development of water protection plans. For example, the Driving force–Pressure–State–Impact–Response (DPSIR) framework is one of the most widely used conceptual frameworks in integrated water resource management to explain the causal relationships between society and the environment [20–25]. The DPSIR describes the feedback among social and economic developments, including the driving forces (D); pressures (P) on the environment; state (S) of environmental changes; impacts (I) on ecosystems, human health, and society; and a societal response (R) [26,27]. Under this framework, however, there is no component to explain the relationships between pressure and state, which may vary from site to site. The DPSIR framework has previously been criticized for not addressing the dynamics of the system, and others have argued that there should be a focus on the links between the nodes of the DPSIR [22–25].

In this study, therefore, we proposed a new component, named "Link", for the DPSIR framework in order to better explain the interactions between pressure and state, called DPLSIR (Figure 1). Furthermore, we suggested the lag time between agricultural pressure (i.e., nitrate inputs from the agricultural system) and water state (i.e., nitrate concentrations in the hydrogeological system) as a key link indicator. The contaminant of interest is nitrate. The objectives of this study are to evaluate a cross-correlation function analysis as a simple methodology for estimating lag times using data from three case study site in Europe (two in Denmark and one in France) and then to investigate controls of the temporal scale and spatial heterogeneity of lag times to better characterize the hydrogeological link between agricultural pressure and water quality state.

2. Conceptual Framework

2.1. The DPLSIR Framework

The DPLSIR framework focuses on the understanding of underlying controls for the relationships between pressure and state (Figure 1). Pressure, in this study, is the amount of agricultural stress released from the agricultural system. We define the agricultural system as the zone where all agricultural activities occur, and this zone is physically defined by the zone above the bottom of the soil layer (e.g., plough layer; Figure 2). The state is the quality of the water in the hydrogeological system. The hydrogeological system is defined by the entire zone below the soil layer through which water and contaminants move (Figure 2). In the hydrogeochemical system, various biogeochemical reactions may change the concentrations of nitrate. Nitrate reduction can occur both in the soil layer and below the interface between the oxic and reduced zones, where the reduction capacity—the amount of nitrate-reducing material, such as pyrite and organic carbon—is high (Figure 2). In some places, this interface can be several meters thick, developing in an anoxic nitrate-reducing zone. Link provides information about these hydrological and biogeochemical processes that are responsible for the release, retention, and transport of water, as well as contaminants. In this study, we focused on transport of water and contaminants—how and how fast nitrate moves through the hydrogeological system. Further research may require reviewing and developing indicators to parameterize release and retention of contaminants in the hydrogeological system.

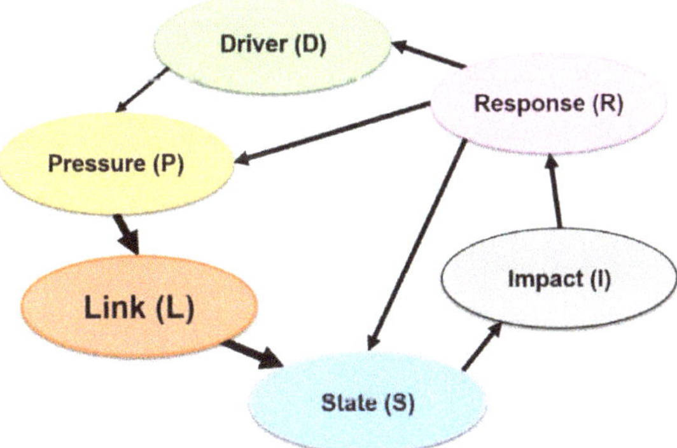

Figure 1. DPLSIR framework proposed in this study, including driving forces (D); pressures (P) on the environment; the link (L) between pressure and state; the state (S) of environmental changes; impacts (I) on ecosystems, human health, and society; and a societal response (R).

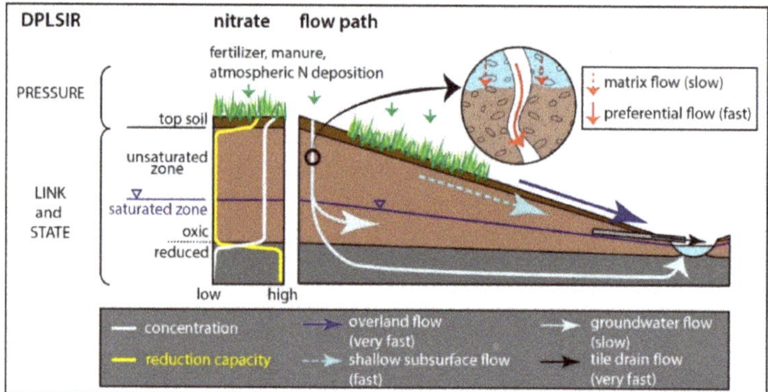

Figure 2. Conceptual understanding of the vertical distribution of nitrate from the agricultural system to the hydrogeological system and pathways in the hydrogeological system. The agricultural system is represented by the pressure indicators and the hydrogeological system is described by the link and state indicators. The reduction capacity is defined as the sum of solid nitrate-reducing compounds, e.g., pyrite and organic matter.

2.2. *Link: Flow Pathways and Lag Time*

Lag times not only quantify the delay time between the pressure stress and state response, but also reveal primary pathways of water and contaminants [28]. To recharge groundwater, water primarily flows vertically via matrix flow pathways and/or preferential flow pathways (Figure 2). This water eventually emerges in the surface water. Matrix flow is a pathway through pore spaces in the soil matrix. The lag times of matrix flow can be long (years to decades). Preferential flow is a pathway via macro-pores in soils and fractures in bedrock, bypassing a dense or less permeable matrix [29,30]. The macro porous spaces in soils can be created along root channels, soil fauna channels, cracks (i.e., freeze–thaw and wetting–drying), fissure, or soil pipes [29]. Preferential flow may be transiently active; however, it can deliver a significant quantity of contaminants with a very short time delay (hours to weeks) [31,32]. Therefore, the groundwater table and groundwater chemistry of matrix flow-dominated systems will exhibit relatively small variations and slow changes over time compared to those of preferential flow-dominated systems.

3. Materials and Methods

3.1. *Case Study Sites*

The three case study sites are intensive farming areas with groundwater-based drinking water abstraction. The first case study site is Tunø Island in Denmark (Figure 3). Tunø is a small island (area: 3.5 km^2) located in the Kattegat Sea, between the North Sea and the Baltic Sea (Table 1). The main part of the subsurface of the island consists of clay-rich glacial deposits of Quaternary origin. However, in the northeastern part, where drinking water abstraction wells are located, Quaternary sandy deposits dominate the near-surface layers and contain an aquifer vulnerable to pollution (Figure 3). The groundwater is the sole source of drinking water and there is only one public water supply system, which produces approximately 10,000 m^3/year of water for around 100 inhabitants and tourists.

During the 1970s, the farmers in the area around the drinking water abstraction wells changed from mixed conventional farming to intensive vegetable production, mainly the production of leeks, cabbages, and onions. This change in farming practice led to high inputs of mineral fertilizers. Since the mid-1980s, Denmark has adopted the EU directives on legislation at the national level to protect water resources, including groundwater [33]. On top of these national-level regulations, in Tunø, additional water protection plans have been implemented since 1989, mainly because high levels of nitrate were detected in the abstraction wells (≈150 mg NO_3^-/L). From 1989, the central part of the recharging area of the abstraction wells was set as a protection zone, and in 1991, this area was extended to more than double the size to cover the major parts of the recharge area (shaded with green in Figure 3). In this inner action area as well as one field at the western border of this area, all agricultural activity has been banned, and the land has been converted to set-aside with permanent grass. In the outer part of the recharge area (total protection area), fertilizer application has been regulated to the level of the economical optimum application of nitrogen for crop production (circled area in Figure 3).

Table 1. Overview of the basic characteristics of the case study sites.

	Tunø Island, Denmark	**Aalborg-Drastrup, Denmark**	**La Voulzie Catchment, France**
Study area (km^2)	0.25	9.92	115
Climate	Coastal temperate	Coastal temperate	Temperate
Geology	Quaternary glacial sediment	Quaternary glacial sediment and fractured limestone	Massive limestone
Source of drinking water	Groundwater	Groundwater	Groundwater (springs)
Drinking water production (m^3 $year^{-1}$)	10,000	1.5 million	20 million
Number of consumers	100	26,000	400,000
Water protection plans and activities	Permanent grass over the recharging area	Afforestation in some places and protection zone limiting nitrate leaching to 25 mg/L and no pesticide use	Various agricultural measures
Contaminant of concern	Nitrate	Nitrate and pesticides	Nitrate and pesticides

The second case study site is Aalborg-Drastrup in Denmark (Figure 3). The case study area is in northern Jutland. In the area, thin layers of glacial sand and clay till deposits of Quaternary origin overlie a fractured Maastrictian limestone reservoir. Like Tunø, groundwater is the only source of drinking water. There are well fields belonging to a public water supply system and several privately owned wells. The wells in the well field were constructed between 1958 and 2005, and the well field produces 1.5 mm^3 $year^{-1}$ of water distributed to around 26,000 consumers in the Aalborg area (Table 1).

The groundwater in the Aalborg-Drastrup case study area is one of the most vulnerable aquifers to nitrate pollution in Denmark because of absent or thin protecting clay layers above the chalk aquifer with a low nitrate reduction capacity in the majority of the recharging area, except in the northernmost part where thick layers of marine clay overlay the chalk (Figure 3). At the same time, there are intensive dairy and pig farms in the area, as well as arable farms. During the 1980s, the water authorities realized that the drinking water abstraction was threatened by pollution of nitrate from the agricultural production. Like Tunø, therefore, additional water protection plans were introduced. Since 2001, a series of mitigation measures were implemented, including setting a protection zone (afforestation) close to the abstraction wells, limiting nitrate leaching from zero to less than 50 mg/L of nitrate, and banning the use of pesticide.

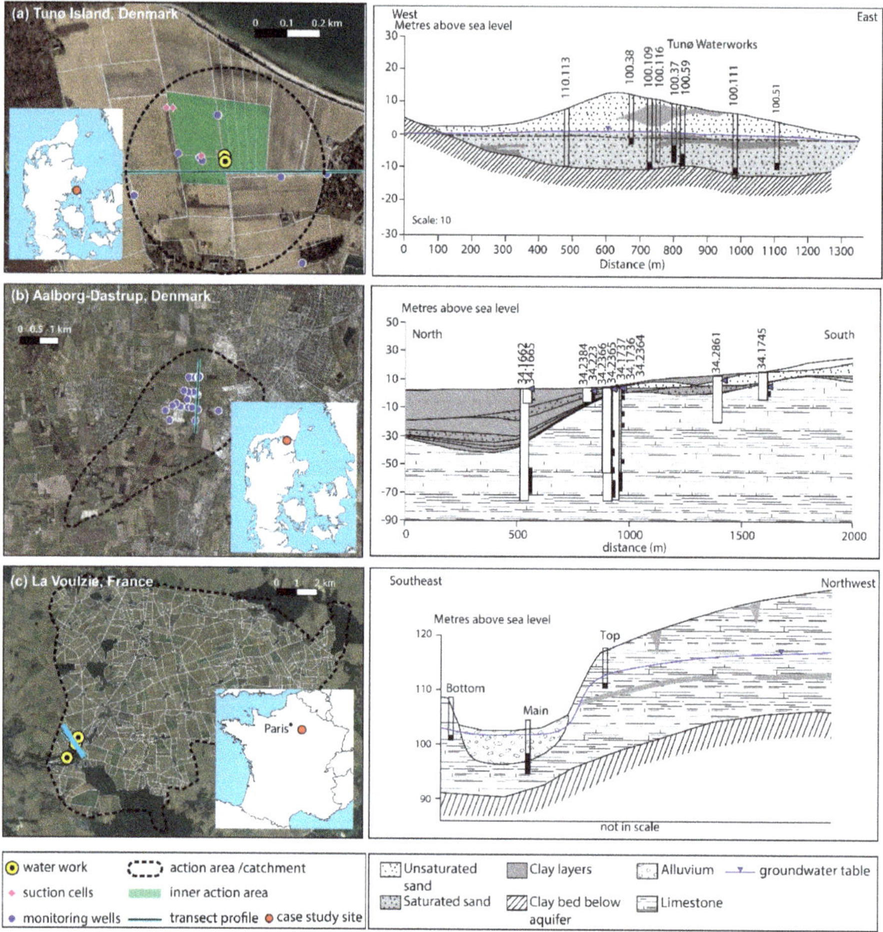

Figure 3. Maps of the three case study sites: Tunø Island, Denmark (**a**); Aalborg-Drastrup, Denmark (**b**); and La Voulzie, France (**c**).

The third case study site is the La Voulzie catchment area (drainage area: 115 km^2) in France (Table 1). The catchment is located about 70 km southeast of Paris. The catchment is primarily underlain by massive limestone (Figure 3). The aquifer is one of the main resources in the water supply system (Ile-de-France) and a very important drinking water source for Paris. Eau de Paris is a public organization in charge of managing water resources for Paris. The catchment provides 20 mm^3/year of water, which is equivalent to 10% of Eau de Paris's resources, to about 400,000 people. The catchment covers 15 rural municipalities with about 100 farmers (including one organic farmer). The area is intensively used by agriculture—90% of the surface area is cultivated, 40% of which is planted with wheat (in rotation with other cereals, rape seed, and sugar beet).

The highest potential nitrogen (N) leaching was achieved in the late 1980s. The agricultural action program was launched in 1991 for the watershed (Fertimieux) to limit high concentrations of nutrients. A change in fertilizer use occurred in the late 1990s in France. This change is attributed to the beginning of environmental awareness and anticipation of the Common Agricultural Policy in 1992. The nitrate concentrations in springs, which is raw water for drinking water production, have been relatively invariant at approximately 55–70 mg NO_3^-/L for the last two decades.

3.2. Materials

For all three study sites, three types of historical data were collected to carry out the study: (1) N surplus, (2) groundwater chemistry, and (3) groundwater age. N surplus (or N balance) is used as an environmental indicator [34,35] and it has also been applied in many EU member states including Denmark and France. Furthermore, studies have reported that N surplus is a good indicator to explain N leaching from the soil and nitrate pollution in groundwater [36–39]. Methods and input data to calculate N surplus differ among the EU member states [40]. This study compared the temporal variability of agricultural N pressure and groundwater state of each case study site, not across the sites; therefore, the methodological differences would not affect the lag time analysis. N surplus is calculated with respect to farm gate (i.e., farm gate N surplus) or to soil surface (i.e., soil surface N surplus). In this study, soil surface N surplus was the primary agricultural N pressure indicator (unit: kg N/ha/year). N surplus of each case study site was calculated according to the member state's standard method, but in principal, soil surface N surplus calculates the differences between the annual quantity of N entering the soil and the annual quantity of N leaving the soil surface by harvested crops. For Tunø and La Voulzie, the soil surface N surplus at the relevant catchment scale was used and for the Aalborg-Drastrup site; both soil surface and farm gate N surpluses at the regional level were used. The water chemistry data were time series of nitrate in any type of water, e.g., soil pore water and groundwater. The annual average concentration of nitrate in water was used as a water quality state indicator (unit: mg NO_3^-/L). The groundwater age data were measured or simulated values of residence times and/or water ages (unit: year). A few studies have reported chlorofluorocarbon (CFC) reduction in reduced groundwater [41,42]; hence, in this study, we only collected data of oxic and nitrate-reducing groundwater. In this study, we did not conduct extra measurements. Instead, we focused on existing data available from public databases and reports, where the data have been collected and reported following the national regulations and official guidelines at the time of collection.

3.2.1. Tunø Island, Denmark

For the Tunø study sites, input data to calculate soil surface N surplus at field scale were available; therefore, this was calculated for the period from 1975 to 2018 at the 345 m^2 outer protection area. A time series of modeled annual atmospheric N deposition is publicly available at the municipality scale (2006–2016) [43]. Time series of the main crop type and fertilization were constructed for Tunø for the period of 1975–2018, at the field scale. Recent data (2010–2018) on the main crop type are available from the Ministry of Environment and Food, Danish Agricultural Agency [44], while older data from the Tunø study site were reconstructed from information on crop rotations obtained from reports [45,46]. Information on the use of fertilizer was based on reported values [46].

In Denmark, all the groundwater chemistry data are registered in the national borehole database, JUPITER, and are managed by the Geological Survey of Denmark and Greenland. The data are publicly available. Therefore, nitrate data for both Danish case study sites were obtained from JUPITER [47]. In Tunø, groundwater ages were determined at seven monitoring points using the CFC method [48].

3.2.2. Aalborg-Drastrup, Denmark

Soil surface N surplus at regional scale (1990–2018) was used for the Aalbog-Drastrup site, and the detailed method and input data for the calculations are described in Windolf et al. [49]. Hansen et al. [50] reported farm gate N surplus of Denmark at the geo-region scale for the period of 1950–2007. Between 1990 to 2007, the soil surface N surplus was offset by 39 kg/ha/year on average compared to the farm gate N surplus; therefore, the soil surface N surplus for the Aalborg-Drastrup site before 1990 was estimated from the farm gate N surplus by subtracting the offset.

Within the action areas of the Aalborg-Drastrup site, monitoring wells that are part of the national groundwater monitoring program are located (Figure 3). The groundwater chemistry of these wells has been monitored regularly since 1989, and groundwater age was also determined using the CFCs method [51]. The groundwater chemistry and groundwater age data were extracted from JUPITER and the report [51].

3.2.3. La Voulzie, France

For the La Voulzie site, the N surplus calculated at the University of Tours was used, with detailed description of the calculation being found in Poivert et al. [52]. The methods for N surplus calculation of Denmark and France are similar. The N surplus for the study area was extracted from an online tool available for France (Cassis-N; https://geosciences.univ-tours.fr/cassis/login) [52,53]. The concentrations of nitrate were collected by the water company "Eau de Paris" at the "La Voulzie" springs. Water chemistry has been monitored since 1927. The water chemistry was monitored 1–4 times per year before 1986 and then every month after 1986. In this study, we focused on three sampling points: top, main, and bottom springs, which are naturally under pressure (Figure 3). The main spring is the primary source of drinking water produced at the water company, and its volume represents 45% of the total volume of the drinking water production. The top spring collects water in the top limestone layer over the thin clay layer that is situated about 5 m below the land surface (Figure 3). The bottom spring and the main spring collect water in a ≈10 m thick limestone layer situated below the thin clay layer (Figure 3). The average depth of the three wells is 14 m, and the groundwater table at the main spring is 1 m below the land surface.

3.3. Lag Time Estimations

The lag time between soil surface N surplus (x) and annual average concentrations of nitrate in water (y) was calculated using a cross-correlation method (CCF) using a correlation coefficient function of Matlab. The CCF method assumes a linear dependency of two variables, i.e., soil N surplus and nitrate concentrations, and the lag time is the time difference between a soil surface N surplus peak and a nitrate concentration peak. The lag times were calculated for a range of time lags (k; year); for example, a correlation between x at time t and y at time t + k was calculated. The range of k was from 0 to 50 years. The CCF provides two main results: (1) the strength of the correlations between x and y (correlation coefficient, r), and (2) the significance of the correlation (p-value). The strength of the correlation varies between 0 (no correlation) and ±1 (strong correlation). The highest r and statistically significant k-year was defined as lag time.

4. Results

4.1. Time-Series of Surface Soil N Surplus and Water Chemistry

Figure 4a shows the soil surface N surplus at the field scale calculated in this study and the farm gate N surplus at the geo-regional scale of Tunø Island [50]. When the protection plans were first introduced, the soil N surplus at the field scale from the agricultural system in the inner action area decreased to the input from the atmospheric deposition, as no other input and no harvest occurred from those fields (yellow circles in Figure 4a). No manure was applied in both the inner and the outer protection area. The soil N surplus at the field scale of the total protection area was averaged at 35 kg N/ha/year before the protection plan period (1975–1990) and was averaged 12 kg N/ha/year after that (1991–2018; gray bars in Figure 4a). The farm gate N surplus at the geo-regional scale peaked around the mid-80s (≈90 kg N/ha/year) and has decreased since then (≈40 kg N/ha/year), which is explained by the effects of the national N regulations [50]. The farm gate N surplus at the geo-regional scale was much higher than the soil surface N surplus at the field scale (Figure 4a). The discrepancy may be because of stricter regulations implemented at the local scale.

As a result of the intensified agriculture, the nitrate concentrations in groundwater and drinking water increased up to >100 mg NO_3^-/L in the 1980s (Figure 4d). As the protection plans set in, nitrate concentrations in soil pore water and groundwater started to respond. The annual average concentrations of nitrate of soil pore water (1.2 m) in the inner protection area decreased nearly to zero almost instantaneously

(Figure 4b). The annual average concentration of nitrate in shallow groundwater (<9 m deep) started to decrease a few years after the protection plans (Figure 4c) and those in the deeper groundwater slowly decreased down to ≈10 mg NO_3^-/L over the past two decades (Figure 4d). The annual average concentration of nitrate of groundwater in the outer protection area also decreased over time, but the concentrations were still relatively high (50–100 mg NO_3^-/L; Figure 4e).

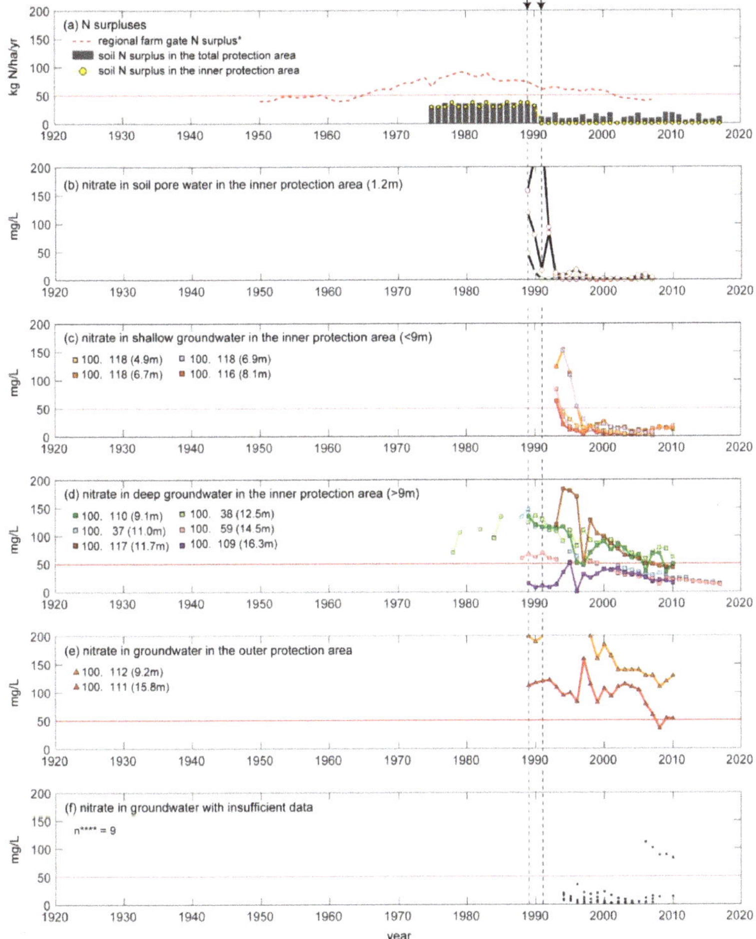

Figure 4. Overview of the nitrogen (N) surpluses and annual averages of nitrate concentrations in the soil pore water and groundwater of Tunø Island, Denmark. (**a**) The soil surface N surplus at the field level of the inner protection area and in the total protection area are shown in yellow circles and gray bars, respectively. The black arrows and dotted vertical lines show the years that the mitigation measures were implemented. The horizontal red line is the drinking water standard for nitrate (50 mg/L). The annual average concentrations of nitrate in soil pore water (circles), shallow groundwater (squares), and deep groundwater (squares) in the inner action area are shown in (**b**–**d**), respectively. Those of groundwater in the outer protection zone (triangles) are shown in (**e**). (**f**) shows the annual average nitrate concentrations of nine (**** n = 9; dots) monitoring points where the cross-correlation function (CCF) analysis could not be conducted due to either small variabilities of nitrate concentrations or an insufficient data length (* Hansen et al. [50]).

In terms of Aalborg-Drastrup, the two regional N surpluses, the soil surface and the farm gate, are shown in Figure 5a. These regional scale data do not precisely represent the impact of the mitigation measured implemented at the local level; however, they may be a good proxy to estimate the overall trend of agricultural N pressure of the study area. The soil surface N surplus decreased from 159 kg N/ha/year in 1990 to 83 kg N/ha/year in 2018. In this period, the annual input of manure was at a stable level of approximately 104 kg N/ha/year, while the input of mineral fertilizer decreased from 139 to 70 kg N/ha/year, as mitigation measures for higher utilization of N in manure were implemented, e.g., higher manure storage capacity and increased manure application in spring and summer rather than in the period with higher leaching in autumn and winter.

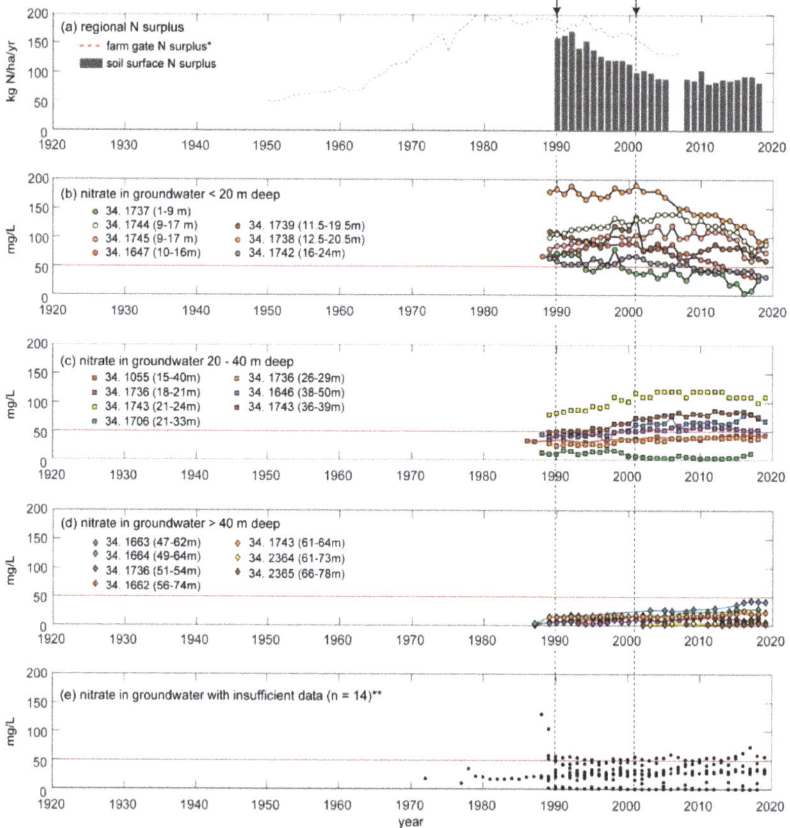

Figure 5. Overview of regional N surpluses and annual averages concentrations of nitrate in groundwater of Aalborg-Drastrup. The soil surface N surplus and farm gate N surplus at the regional scale (* Hansen et al. [50]) are shown in (**a**). The annual average concentrations of nitrate of shallow (filter interval < 20 m), middle (filter interval between 20 and 40 m), and deep groundwater (filter interval deeper than 40 m) are shown in (**b–d**), respectively. Those of groundwater with insufficient data and no temporal variability are shown in (**e**). The annotations in the figure are the same as those of Figure 4. ** number of groundwater wells.

In the monitoring site of the Aalborg-Drastrup (Figure 3), the annual average concentrations of nitrate of groundwater varied within a wide range (1–200 mg NO_3^-/L), and their temporal variability also was heterogeneous (Figure 5b–e). The groundwater in shallower depth (filter interval < 20 m) showed the greatest temporal variability, and the nitrate concentrations showed peaks around

2000–2010 and decreased over time (Figure 5b). The groundwater measured between 20–40 m intervals showed relatively damped responses (Figure 5c). In some monitoring points, the nitrate concentrations peaked around 2010 and slowly decreased over the past 10 years (e.g., 34. 1743 in Figure 5c). While nitrate of groundwater in the deeper intervals (>40 m) were relatively low, i.e., <10 mg NO_3^-/L, in some groundwater wells, the concentrations continuously increased over the entire monitoring period (e.g., 34. 1663 in Figure 5d). At 14 monitoring points, either the groundwater was in reduced conditions, the nitrate concentrations were temporally invariant, or the nitrate concentrations showed only increasing trends (Figure 5f).

In the La Voulzie catchment area, the soil surface N surplus at the catchment scale was highest around the mid-80s (≈77 kg N/ha/year) and then decreased to ≈55 kg N/ha/year around 1990 (Figure 6a). Over the past 30 years, the soil surface N surplus ranged between 15 and 66 kg N/ha/year (Figure 6a). The groundwater chemistry has been monitored since 1927. Until the 1960s, the annual average concentrations of nitrate of groundwater and spring waters ranged between 20 and 25 mg NO_3^-/L, but the level continuously increased up to 50–110 mg NO_3^-/L until 1993 (Figure 6b). The nitrate concentrations of the top spring decreased back to approximately 74 mg NO_3^-/L around 2000 and then were temporally invariant for the last 20 years at this level (Figure 6b). Those of the main and bottom springs also were relatively stable at around 55 mg NO_3^-/L for the last 30 years (Figure 6b). The monthly data revealed that the nitrate concentrations of all three springs did not show any seasonality—standard deviations of all the nitrate measurements between 2000 and 2014 were 2.8 (top spring), 2.2 (main spring), and 1.8 (bottom spring).

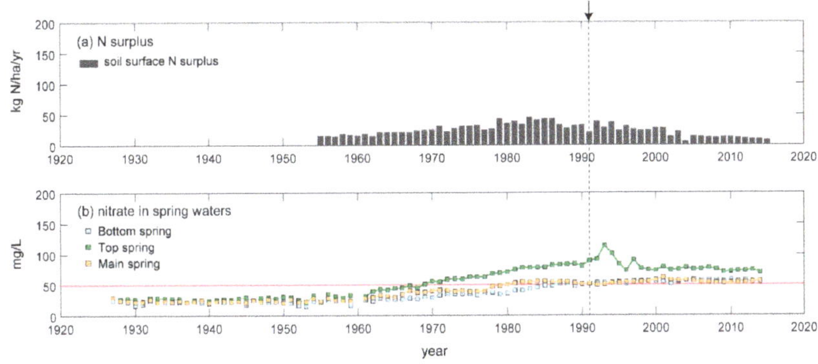

Figure 6. Overview of the soil surface N surplus (**a**) and annual average concentrations of nitrate of groundwater (spring) of La Voulzie (**b**). The figure annotations are the same as those of Figure 4.

4.2. Lag Time between Agricultural Pressure and Groundwater Quality State

Table 2 summarizes the results of the lag time analysis of the case study sites and water chemistry. For the Tunø site, at the 15 points with statistically significant results, the soil surface N surplus showed strong correlation with the annual average concentrations of nitrate in water. For example, except for the deepest groundwater, the correlation coefficients were higher than 0.76 and the results were statistically significant ($p < 0.005$) in all the cases (Table 2). For the deepest well (monitoring point 100. 109), which is in nitrate-reducing conditions, the correlation coefficient was 0.53, and it was statistically significant ($p < 0.05$; Table 2). The lag times progressively increased with an increasing depth (Figure 7a and Table 2). For instance, the soil pore water measured at 1.2 m below the land surface and showed lag times of 0–2 years, while the deep wells (deeper than 10 m) showed lag times of 5–16 years.

Table 2. Summary of water chemistry and the lag time estimation.

	Observation Period of Water Chemistry Data (Total Number of Observations)	Average of Nitrate in mg/L (Standard Deviation)	Average of Dissolved Oxygen in mg/L (Standard Deviation)	Lag Time in Years (Correlation Coefficient)	Groundwater Age (Year)
		Tunø Island, Denmark			
Soil pore water (3 points)	1989–2007 (130–142)	35 (69)	-	0–2 (0.75–0.85) **	
100. 118 (4.92 m)	1994–2007 (20)	16 (14)	7.3 (0.1)	5 (0.81) **	
100. 118 (6.72 m)	1993–2010 (30)	44 (54)	7.8 (1.9)	5 (0.85) **	
100. 118 (6.92 m)	1994–2007 (26)	49 (59)	6.8 (3.9)	5 (0.82) **	
100. 116 (8.07 m)	1993–2010 (28)	9.8 (13)	5.5 (0.2)	3 (0.76) **	
100. 110 (9.1 m)	1989–2010 (44)	87 (30)	6.7 (1.2)	4 (0.76) **	15
100. 112 (9.2 m) †	1989–2010 (40)	176 (38)	3.7 (1.1)	9 (0.86) **	15
100. 37 (11 m)	1989–2017 (50)	77 (43)	3.7 (1.8)	5 (0.87) **	
100. 117 (11.7 m)	1993–2010 (30)	100 (50)	4.6 (2.3)	6 (0.78) **	
100. 38 (12.5 m)	1978–2010 (53)	118 (34)	5.3 (1.1)	8 (0.78) **	21
100. 59 (14.5 m)	1985–2017 (76)	44 (19)	2.3 (1.9)	10 (0.88) **	
100. 111 (15.8 m) †	1989–2010 (40)	100 (26)	0.2 (0.2)	16 (0.77) **	22
100. 109 (16.25 m)	1989–2010 (40)	24 (13)	0.2 (0.2)	20 (0.53) *	26
		Aaborg-Drastrup, Denmark			
34. 1737 (1–9 m)	1989–2018 (45)	43 (19)	1.9 (1.7)	3 (0.78) **	21
34. 1744 (9–17 m)	1989–2019 (46)	119 (15)	8.2 (0.9)	16 (0.78) **	23
34. 1745 (9–17 m)	1989–2019 (44)	95 (18)	8.7 (0.7)	23 (0.48) *	20
34. 1647 (10–16 m)	1988–2018 (45)	69 (18)	0.4 (0.3)	12 (0.90) **	22
34. 1739 (11.5—19.5 m)	1989–2019 (45)	88 (18)	6.8 (0.8)	9 (0.73) **	16
34. 1738 (12.5–20.5 m)	1989–2019 (42)	158 (27)	6.8 (1.6)	12 (0.94) **	16
34. 1055 (15–40 m)	1986–2019 (18)	38 (4.5)	4.5 (1.1)	26 (0.84) **	
34. 1742 (16–24 m)	1989–2019 (45)	53 (8.4)	5.7 (0.6)	10 (0.83) **	30
34. 1736 (18–21 m)	1989–2018 (44)	50 (7.3)	4.7 (1.6)	28 (0.76) **	19
34. 1706 (21–33 m)	1988–2017 (46)	8.1 (4.8)	6.2 (2.2)	4 (0.78) **	8
34. 1743 (21–24 m)	1989–2019 (44)	106 (14)	8.6 (0.9)	25 (0.91) **	26
34. 1736 (26–29 m)	1989–2018 (44)	34 (5.3)	3.4 (0.4)	29 (0.92) **	30
34. 1743 (36–39 m)	1989—2018 (43)	69 (13)	7.5 (0.5)	29 (0.96) **	40
34. 1646 (38–50 m)	1988–2019 (45)	57 (12)	5.2 (0.3)	29 (0.88) **	28
34. 1663 (47–62 m)	1987–2019 (17)	26 (12)	3.2 (2.5)	28 (0.86) **	
34. 1664 (49–64 m)	1987–2018 (20)	16 (7.8)	2.4 (2.5)	35 (0.98) **	
34. 1736 (51–54 m)	1989–2018 (43)	9.3 (3.7)	0.7 (0.5)	29 (0.82) **	48
34. 1662 (56–74 m)	1987–2018 (15)	11 (6.2)	0.8 (0.3)	29 (0.86) **	
34. 1743 (61–64 m)	1989–2019 (42)	18 (2.6)	3.4 (0.3)	42 (0.91) **	46
34. 2364 (61–73 m)	2003–2019 (12)	5.5 (1.9)	1.2 (1.3)	38 (0.95) **	
34. 2365 (66–78 m)	2002–2019 (14)	3.0 (1.2)	1.1 (2.1)	38 (0.76) *	
		La Voulzie Catchment, France			
Top spring	1928–2014 (456)	66	-	8 (0.78) **	–
Main spring	1927–2014 (469)	47	9.2	15 (0.70) **	–
Bottom spring	1927–2014 (463)	43	9.9	24 (0.83) **	–

* Statistically significant at $p < 0.05$; ** $p < 0.005$; † outer protection zone.

In the Aalborg-Drastrup site, at 21 of the 35 monitoring points, the nitrate concentrations in groundwater showed significant correlations with the surface soil N surplus at the regional scale (Table 2). Except for one monitoring point (34. 1745 in Table 2), the correlation coefficients were higher than 0.73 and they were statistically significant ($p < 0.005$). Like Tunø, the lag times roughly increased with an increasing depth; however, they showed a greater variability (Figure 7 and Table 2). For example, at the filter interval between 21–33 m (34. 1706), 4 years of lag year was estimated. In the deeper part of the groundwater (>40 m), the lag times also varied between 29 to 42 years.

In the La Voulzie catchment area, the soil surface N surplus and the annual average concentrations of nitrate of all three springs showed strong ($r = 0.70$–0.83) and statistically significant correlations ($p < 0.005$). The lag times estimated for the top, main, and bottom springs were 8, 15, and 24 years (Table 2).

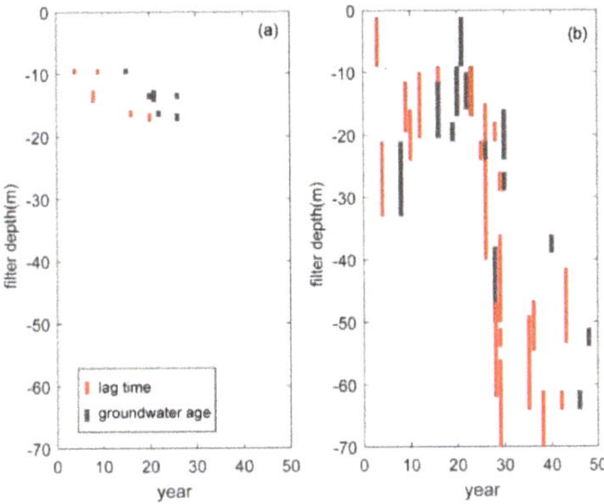

Figure 7. Groundwater ages (chlorofluorocarbon (CFC) measurements) and lag times (CCF statistical estimates) of groundwater in (**a**) Tunø and (**b**) Aalborg-Drastrup. The intervals of the intake filters are on the *y*-axis.

4.3. Groundwater Age

For the Danish case study sites, groundwater age data were collected to compare them with the lag time analysis results. In Tunø, groundwater age data are available at five monitoring points in the inner action area and at two points outside the inner action area [48]. Like the lag time, the groundwater age of the Tunø groundwater linearly increased with an increasing depth (Figure 7a). However, the groundwater ages for the same depth were 6–14 years longer than the lag times estimated in this study (Figure 7a). In the Aalborg-Drastrup site, like lag times, groundwater ages varied widely with depth (Figure 7b). For instance, young groundwater (<10 years old) was found at 21–33 m below the land surface, while old groundwater (≈20 years old) was found in the near surface groundwater (Figure 7b). Like in Tunø, the groundwater ages were relatively longer than the lag times calculated in this study.

5. Discussion

5.1. Methodological Evaluation of the Lag Time Estimation and Data Requirements

Lag times may be one of the key pieces of information for developing water protection programs and setting achievable goals [54–57]. Various models have been developed to estimate the lag times between mitigation measures and the water quality changes [55–59]. Most models are process-based and provide comprehensive views of the system. Modelling is applicable both data-rich and data-poor sites [55]. However, these models often require scientific and technical training to use them. Therefore, these tools may not be readily available for many stakeholders. Furthermore, such models are often based on various assumptions; therefore, building trust on the modeling results may require considerable efforts. Studies have reported that involving the stakeholders in the modelling processes improve the acceptable of the modeling results [55]. This study selected the CCF analysis, which is based on direct measurements. This method is a simple, intuitive, and easy-to-use method for the stakeholders. Such measurement-based methods require good quality and quantity of data [55], and we also found that the CCF analysis may provide meaningful and statistically significant results if the following conditions are satisfied.

First, continuous, long-term monitoring data on both the agricultural pressure and groundwater state should be available. Obviously, the time series of the data must be sufficiently long to document the cause and effect relations between the pressure and state. For example, in La Voulzie, both agricultural pressure and groundwater state data were available for nearly a century and could thus fully cover the changes in agricultural pressure and the groundwater state responses (Figure 6). For the two Danish sites, regional long-term farm gate N surplus data indicated the trends in agricultural pressure, confirmed by overlapping more recent soil surface balances. These long-term data allowed statistical analysis and the results of the CCF analysis were statistically significant. However, links between regional N-surplus pressure data and effects on resulting levels of nitrate leaching should be subject to further research [40]. Supplementary analyses of more detailed historical, agricultural pressure data [60] for the local Parishes covering both of the two Danish sites showed that the farm gate N surplus levels for both Tunø (represented by the Parish of Tunø) and Aalborg-Drastrup (represented by the two Parishes of Dall and Navling) were about 20% and 10% lower than the regional balances shown in Figures 4 and 5, respectively, but with similar decadal trends. This confirms the validity of the current analyses. Furthermore, the local data confirmed the high decrease in N surplus pressure for Tunø after the 1980s, as well as historical trends back in time similar to the regional results. For Aalborg-Drastrup, the agricultural pressure according to local agricultural N-surplus followed the regional trend after 1980 but was significantly higher before 1980. Thereby, the value of localized data about links between agricultural pressures and lag time in groundwater quality was supported.

Second, particularly for evaluating the effectiveness of the mitigation measures, the time series should capture the changes in pressure and state. For instance, in Tunø, at some sampling points, groundwater monitoring had started a few years after the protection plans were implemented (Figure 4f). The nitrate concentrations in those wells already decreased below 10 mg NO_3^-/L, and they were relatively invariant over time. With these time series, the CCF analysis cannot produce statistically meaningful results.

Third, in cases of the groundwater state, only oxic groundwater should be considered because nitrate is redox-sensitive. In nitrate-reducing and/or -reduced conditions, nitrate concentrations decrease not only because of the changes in the inputs from the agricultural system, but also because of biogeochemical reactions (denitrification to gaseous N compounds) in the subsurface. If the denitrification reactions dominantly control the nitrate concentrations in groundwater, the correlations between the agricultural pressure and groundwater state might not be strong or statistically significant. For example, in Tunø, the deepest groundwater (100. 109) was nitrate-reducing, the CCF analysis estimated weak correlations, and one of the pressure indicators showed an insignificant result (Table 2). These results are attributed to its nitrate-reducing condition. In this well, oxygen was nearly depleted, but its nitrate concentrations were much lower (approximately 25 mg NO_3^-/L) than the overlying part of the groundwater (44–170 mg NO_3^-/L), indicating that nitrate had begun to be reduced. Therefore, to evaluate the effects of agricultural activities, including mitigation measures, on the groundwater quality, the role of nitrate reduction reactions in the hydrogeological system should be excluded.

5.2. Hydrogeological Control of Lag Times

The hydrological system determines the pathways between a source (i.e., agricultural soil) and a sink (i.e., groundwater) and the travel speed of the contaminants through the pathways. All the hydrogeological systems, however, have complex networks of flow pathways, and this complexity has been a major challenge in communication among stakeholders, water authorities, and scientists. To develop a water protection program with local stakeholders, it may be more important to identify the most dominant pathways of water and nitrate transport, rather than to fully understand the complexity of the hydrogeological system. By targeting the primary pathways, the efficiency of the mitigation measures may be increased [19,61]. The main role of the link indicators is to reveal the primary pathways of the site so that the stakeholders can develop action plans and set achievable goals. Tunø is a sandy aquifer underlain by Quaternary glacial deposits, where heterogeneous geology and flow pathway network are expected. However, our data showed that lag times and groundwater ages linearly increased with an increasing depth at this site (Figure 7a). These results imply that the signals of the agricultural pressure are propagated downward, predominantly via matrix flow (i.e., diffusion and dispersion through the porous medium) to groundwater. In a matrix flow-dominated system, the groundwater state may respond to the agricultural pressure slowly, which is consistent with the relatively long lag times of the Tunø site.

The Aalborg-Drastrup site is underlain by limestone. Limestone aquifers are also well known for their complexity and anisotropy of flow networks because both matrix flow via porous medium and preferential flow via fractures play a key role in transporting water and contaminants [62]. This complexity may explain why the lag time and residence times heterogeneously varied with depth. When preferential flow via large fractures is the dominant pathway, the agricultural pressure signals will rapidly propagate to groundwater, and consequently young groundwater and short lag times can be found at deep depths in this case (Figure 7b). Conversely if matrix flow via porous medium or macroscale fractures plays a more important role in delivering water and nitrate, the groundwater state responses will be significantly delayed. At the Aalborg-Drastrup site, we observed both patterns and these observations imply that both pathways may be active. In this case study site, mitigation measures that target rapid (i.e., preferential flow) and slow (i.e., matrix flow) pathways will be needed.

La Voulzie is also underlain by limestone. A previous study using multiple tracers in this area revealed that, at this site, nitrate transport mainly occurs through matrix flow on a long-term timescale [63]. Consistent with this, the lag times of nitrate were 8–24 years (Table 2) and the monthly water chemistry data showed no seasonal variations. At this site, therefore, a long-term water protection program will be necessary. Furthermore, the results also show that the lag time variability among the springs may be explained by the hydrogeological structure of the subsurface. At the top spring, the estimated lag time is shorter than those of the other two springs. These differences may be because the top spring is located above the relatively impermeable clay layer (Figure 3), while the other two are located below it. Therefore, the top spring may receive water that circulates only within the top limestone layer. In sum, we conclude that the link indicators may be useful for better understanding of how water and nitrate move through the hydrological system and selecting the best mitigation measures. The underlying hydrogeology is the primary control of the relations between the agricultural pressure and groundwater state. Therefore, the hydrogeological system should be well-characterized in the context of the objectives and the spatial and temporal scales of the water protection programs.

5.3. Link Indicators: Groundwater Age vs. Lag Time

The time lags between pressure and state can also be estimated through groundwater age measurements using environmental tracers such as CFCs, $^{3}H/^{3}He$, and noble gases (Ar, He, Kr, Ne, and Xe) if the contaminant is highly soluble [64–68]. Nitrate is primarily transported as solutes and behaves conservatively in the oxic zone; therefore, the groundwater age may be a useful indicator for estimating the time lag of groundwater state responses. Indeed, previous studies of groundwater ages and groundwater chemistry have successfully documented the changes of nitrate concentrations in groundwater chemistry as responses to the changes of regulations or agricultural practices at decadal scale [64–68].

Our results show that the lag times estimated using the statistical CCF method were shorter than the measured CFC groundwater ages in Tunø. Such differences could be attributed to uncertainties associated with each measurement; however, in principle, lag time and groundwater age represent two different hydrological timescales. Lag time measures the time difference between a pressure peak and a state peak; therefore, it quantifies how fast a water parcel or an input signal moves through a groundwater system [69]. In other words, lag time is transit time. Comparatively, groundwater age is the mean residence time of the water; therefore, it is related to water mass movement and the volume of the groundwater. If the lengths of pathways of a hydrogeological system are identical, its lag time is half of the mean residence time [70]. However, such hydrogeological systems do not exist. The lengths of pathways are greatly different, even for a relatively homogeneous system; therefore, some signals can propagate through the system via short pathways (i.e., short lag time), while the rest take longer pathways (i.e., long residence time). In other words, water can be contaminated quickly, but it will take much longer to remediate it.

When the lag time and groundwater age are used as link indicators in communications with stakeholders, it is important to make a distinction between these two link indicators and to assess the limitations of each indicator. The CCF method is a statistical analysis tool and it does not provide process-based understandings of the lag time between the agricultural pressure and groundwater state. The lag time is controlled by the interplay between hydrological and biogeochemical processes [71], which potentially results in an unclear correlation between the agricultural pressure and groundwater state and in highly uncertain estimations of lag times. Therefore, lag time results should be carefully examined on the basis of good understandings of the hydrogeological system.

The groundwater age is the mean age of the exiting waters at the monitoring point in the aquifer, and it does not provide information about the age distribution by itself. Information about the age distribution, however, may be critical to identifying the sources and timing of the contamination [72] and to estimating the time span of the water protection plans. If the ages of a water body show a normal distribution, for example, then it will take twice as long as the mean residence time to replace all the waters in the water body. However, it is well documented that groundwater is a mixture of waters with vastly different ages, often displaying a skewed distribution with a long tail or a bimodal distribution (e.g., [72,73]). Several models and multiple tracers methods are available to estimate the distribution of water ages, and different methods may predict different distribution patterns (e.g., [66,72,74]). Therefore, when using the groundwater age as a link indicator, the age distribution and limitations associated with the employed method should be provided.

5.4. *Lag Time as a Criterion for the Selection of Pressure and State Indicators*

Because different stakeholders have different interests, they focus on different aspects of drinking water protection plans. For example, farmers may be interested in the effects of the implemented measures on their fields. For the field level evaluation, the spatial scale of the agricultural N pressure (i.e., N surplus) will be essential. For example, the farm gate N surplus at the geo-regional scale of the Tunø case study site did not show the effects of the local mitigation measures, whereas soil surface N surplus at the field scale showed instantaneous reduction (Figure 4a). Indeed, many EU member states have adopted N surplus (or N balance, N budget) at the field level as an advisory or regulatory tool [40]. For instance, the Danish agriculture and Food Council works on a fertilizer planning system that precisely calculates the soil surface N balance and surplus on individual fields of the farmer, giving the farmer the opportunity to take this soil surface N surplus into account when the farmer plans the application of mineral and manure nitrogen to his or her crops. Therefore, farmers can readily calculate N surplus at the field scale.

In combination with the soil surface N surplus, soil pore water chemistry may be useful to evaluate the effects of mitigation measures at the field scale because of its short lag time. Our results show that nitrate in soil pore water responds to the agricultural pressure almost instantaneously (lag time between 0 and 2 years; Table 2) and the strength of their correlation was strong (0.75–0.85; Table 2) in Tunø. Indeed, the soil pore water chemistry, which is generally referred to as N leaching, is one of the most widely monitored parameters for scientific, monitoring, and regulatory purposes in order to quantify N losses from the agricultural system to the hydrogeological system [36,75–77]. Several methods have been developed, and the advantages and limitations of each method are well documented (e.g., [75,78,79]). Because of the spatial heterogeneity and temporal variability of soil pore water chemistry, establishing monitoring networks and obtaining representative values for N leaching will be costly and labor-intensive; however, nitrate in soil pore water collected below the root zone directly measures the amount of nitrate loss after the complex interplay between agricultural and biogeochemical N cycles in the soil layer. Furthermore, farmers or citizens can be part of the monitoring procedure; therefore, this may contribute to increasing the transparency of the indicator. We strongly recommend developing a sampling protocol to account for the spatial and temporal variability of the soil pore water chemistry that can document the link between N surplus and nitrate concentration in soil pore water.

On the other hand, water companies and authorities may be more interested in the changes in the quality of drinking water resources that likely integrate signals from the entire catchment or recharging area. In order to do this, N surplus may be calculated for the area of interest, i.e., catchment or recharging area, and nitrate concentrations in oxic groundwater can be used as a water quality state indicator. To specifically evaluate the effects of the mitigation measures, oxic groundwater should be focused upon because the effects of naturally occurring N reduction reactions are negligible in oxic conditions. In addition, our results of all three case study sites showed that the lag times generally increased with an increasing distance from the N source (e.g., depth or flow pathways). Therefore, oxic groundwater close to the N source may be most effective for rapid communication with the stakeholders. It is equally important to select representative locations for oxic groundwater monitoring because the groundwater chemistry and lag time responses can be spatially heterogenous. Therefore, monitoring programs should be carefully established to document the spatial and temporal variability of the site.

6. Conclusions

This study investigated lag times as a link indicator to better understand the cause–effect relations between agricultural N pressure and groundwater quality state using data from the three case study sites. We showed that the cross-correlation (CCF) analysis of soil surface N surplus and annual average concentrations of nitrate in water can be a simple and useful method to determine the lag times between agricultural N pressure and groundwater quality state if long-term pressure and state data are available. Furthermore, the spatial patterns of lag time distribution revealed the dominant pathways of water and nitrate, which then provide valuable information for the selection of mitigation measures. Groundwater age is another useful link indicator to estimate the timescale of the water quality state responses to the agricultural pressure, particularly in data-limited circumstances. However, lag time represents transit time while groundwater age measures residence time; thus, these two should be distinguished in communication with the stakeholders.

Altogether, we concluded that the link indicators investigated under the DPLSIR framework can illustrate some of the potential for enhancing the communication of uncertainties and complexity between the agricultural pressure and groundwater state. When dealing with groundwater and agriculture-related issues, the link between the pressure and state is important for the responses (in terms of measures and policies) adopted by authorities that are to be accepted and used by farmers. If there is a common understanding of the link among the stakeholders, then decisions and water protection plans might be easier to implement. Knowledge exchange on the link between decision-takers (farmers) and decision-makers (policymakers and water work) should be one of the first steps when developing water protection plans in the future.

Author Contributions: For research articles with several authors: conceptualization, H.K., A.H., N.S., and B.H.; methodology, H.K. and A.H.; validation, H.K., N.S., and A.H.; formal analysis, H.K., N.S., and A.H.; data curation, H.K., N.S., I.M., G.B.-M., and A.H.; writing—original draft preparation, H.K.; writing—review and editing, H.K., N.S., I.M., M.G., G.B.-M., T.D., and B.H.; visualization, H.K.; supervision, B.H.; funding acquisition, B.H. All authors have read and agreed to the published version of the manuscript.

Funding: This project has received funding from the European Union's Horizon 2020 research and innovation program under grant agreement no. 727984.

Acknowledgments: We would like to thank the reviewers for the comments. This paper has been produced in the context of the project Fairway (farm system management and governance for producing good water quality for drinking water supplies; grant no. 727984). The authors thank "Eau de Paris" for providing ground state data of the La Voulzie catchment. The authors also thank Lærke Thorling, who took part in collecting the original Tunø data, for discussing the Tunø case study site.

Conflicts of Interest: The authors declare no conflict of interest.

References

1. WHO. *WHO Guidelines for Drinking-Water Quality*; World Health Organization: Geneva, Switzerland, 2011; Volume 38.

2. Knobeloch, L.; Salna, B.; Hogan, A.; Postle, J.; Anderson, H. Blue babies and nitrate-contaminated well water. *Environ. Health Perspect.* **2000**, *108*, 675–678. [CrossRef] [PubMed]
3. Ward, M.; Jones, R.; Brender, J.; de Kok, T.; Weyer, P.; Nolan, B.; Villanueva, C.; van Breda, S. Drinking water nitrate and human health: An updated review. *Int. J. Environ. Res. Public Health* **2018**, *15*, 1557. [CrossRef] [PubMed]
4. Espejo-Herrera, N.; Gràcia-Lavedan, E.; Boldo, E.; Aragonés, N.; Pérez-Gómez, B.; Pollán, M.; Molina, A.J.; Fernández, T.; Martín, V.; La Vecchia, C.; et al. Colorectal cancer risk and nitrate exposure through drinking water and diet. *Int. J. Cancer* **2016**, *139*, 334–346. [CrossRef] [PubMed]
5. Schullehner, J.; Hansen, B.; Thygesen, M.; Pedersen, C.B.; Sigsgaard, T. Nitrate in drinking water and colorectal cancer risk: A nationwide population-based cohort study. *Int. J. Cancer* **2018**, *143*, 73–79. [CrossRef]
6. Temkin, A.; Evans, S.; Manidis, T.; Campbell, C.; Naidenko, O.V. Exposure-based assessment and economic valuation of adverse birth outcomes and cancer risk due to nitrate in United States drinking water. *Environ. Res.* **2019**, *176*, 108442. [CrossRef]
7. European Environmental Agency. Main Drinking Water Problems. Available online: https://www.eea.europa.eu/data-and-maps/figures/main-drinking-water-problems (accessed on 20 May 2020).
8. Glavan, M.; Železnikar, Š.; Velthof, G.; Boekhold, S.; Langaas, S.; Pintar, M. How to enhance the role of science in European Union policy making and implementation: The case of agricultural impacts on drinking water quality. *Water* **2019**, *11*, 492. [CrossRef]
9. European Commission. *Framework for Action for the Management of Small Drinking Water Supplies*; European Union: Brussel, Belgium, 2014.
10. *European Commission Synthesis Report on the Quality of Drinking Water in the EU Examining the Member States' Reports for the Period 2008–2010 under Directive 98/83/EC EN*; European Union: Brussel, Belgium, 2014.
11. OECD. OECD Principles on Water Governance. Available online: https://www.oecd.org/governance/oecd-principles-on-water-governance.htm (accessed on 20 May 2020).
12. Young, S.; Plummer, R.; FitzGibbon, J. What can we learn from exemplary groundwater protection programs? *Can. Water Resour. J.* **2009**, *34*, 61–78. [CrossRef]
13. De Loë, R.C.; Di Giantomasso, S.E.; Kreutzwiser, R.D. Local capacity for groundwater protection in ontario. *Environ. Manag.* **2002**, *29*, 217–233. [CrossRef]
14. Ivey, J.L.; de Loë, R.; Kreutzwiser, R.; Ferreyra, C. An institutional perspective on local capacity for source water protection. *Geoforum* **2006**, *37*, 944–957. [CrossRef]
15. Graversgaard, M.; Hedelin, B.; Smith, L.; Gertz, F.; Højberg, A.L.; Langford, J.; Martinez, G.; Mostert, E.; Ptak, E.; Peterson, H.; et al. Opportunities and barriers for water co-governance—A critical analysis of seven cases of diffuse water pollution from agriculture in Europe, Australia and North America. *Sustainability* **2018**, *10*, 1634. [CrossRef]
16. Serrat-Capdevila, A.; Valdes, J.; Gupta, H. Decision support systems in water resources planning and management: Stakeholder participation and the sustainable path to science-based decision making. In *Efficient Decision Support Systems—Practice and Challenges from Current to Future*; InTechOpen: London, UK, 2011; pp. 1–16.
17. Jacobs, K.; Lebel, L.; Buizer, J.; Addams, L.; Matson, P.; McCullough, E.; Garden, P.; Saliba, G.; Finan, T. Linking knowledge with action in the pursuit of sustainable water-resources management. *Proc. Natl. Acad. Sci. USA* **2016**, *113*, 4591–4596. [CrossRef] [PubMed]
18. Van Meter, K.J.; Basu, N.B. Time lags in watershed-scale nutrient transport: An exploration of dominant controls. *Environ. Res. Lett.* **2017**, *12*, 084017. [CrossRef]
19. Meals, D.W.; Dressing, S.A.; Davenport, T.E. Lag time in water quality response to best management practices: A review. *J. Environ. Qual.* **2010**, *39*, 85–96. [CrossRef]
20. European Environment Agency (EEA). EEA Glossary. Available online: https://www.eea.europa.eu/help/glossary/eea-glossary/dpsir (accessed on 1 May 2020).
21. Bagordo, F.; Migoni, D.; Grassi, T.; Serio, F.; Idolo, A.; Guido, M.; Zaccarelli, N.; Fanizzi, F.P.; De Donno, A. Using the DPSIR framework to identify factors influencing the quality of groundwater in Grecìa Salentina (Puglia, Italy). *Rend. Lincei* **2016**, *27*, 113–125. [CrossRef]
22. Mattas, C.; Voudouris, K.; Panagopoulos, A. Integrated groundwater resources management using the DPSIR approach in a GIS environment context: A case study from the Gallikos river basin, North Greece. *Water* **2014**, *6*, 1043–1068. [CrossRef]

23. Song, X.; Frostell, B. The DPSIR framework and a pressure-oriented water quality monitoring approach to ecological river restoration. *Water* **2012**, *4*, 670–682. [CrossRef]
24. Rekolainen, S.; Kämäri, J.; Hiltunen, M.; Saloranta, T.M. A conceptual framework for identifying the need and role of models in the implementation of the water framework directive. *Int. J. River Basin Manag.* **2003**, *1*, 347–352. [CrossRef]
25. Carr, E.R.; Wingard, P.M.; Yorty, S.C.; Thompson, M.C.; Jensen, N.K.; Roberson, J. Applying DPSIR to sustainable development. *Int. J. Sustain. Dev. World Ecol.* **2007**, *14*, 543–555. [CrossRef]
26. Smeets, E.; Weterings, R. *Environmental Indicators: Typology and Overview*; Technical Report No. 25; European Environment Agency: Copenhagen, Denmark, 1999.
27. Gabrielsen, P.; Bosch, P. *Environmental Indicators: Typology and Use in Reporting*; European Environment Agency: Copenhagen, Denmark, 2003.
28. Koh, E.-H.; Lee, E.; Kaown, D.; Green, C.T.; Koh, D.-C.; Lee, K.-K.; Lee, S.H. Comparison of groundwater age models for assessing nitrate loading, transport pathways, and management options in a complex aquifer system. *Hydrol. Process.* **2018**, *32*, 923–938. [CrossRef]
29. Beven, K.; Germann, P. Macropores and water flow in soils. *Water Resour. Res.* **1982**, *18*, 1311–1325. [CrossRef]
30. Hendrickx, J.M.H.; Flury, M. Uniform and preferential flow mechanisms in the vadose zone. In *Conceptual Models of Flow and Transport in the Fractured Vadose Zone*; National Academies Press: Washington, DC, USA, 2001; pp. 149–187. ISBN 978-0-309-07302-8.
31. Rosenbom, A.E.; Therrien, R.; Refsgaard, J.C.; Jensen, K.H.; Ernstsen, V.; Klint, K.E.S. Numerical analysis of water and solute transport in variably-saturated fractured clayey till. *J. Contam. Hydrol.* **2009**, *104*, 137–152. [CrossRef] [PubMed]
32. Rosenbom, A.E.; Ernstsen, V.; Flühler, H.; Jensen, K.H.; Refsgaard, J.C.; Wydler, H. Fluorescence imaging applied to tracer distributions in variably saturated fractured clayey till. *J. Environ. Qual.* **2008**, *37*, 448–458. [CrossRef] [PubMed]
33. Hansen, B.; Thorling, L.; Kim, H.; Blicher-Mathiesen, G. Long-term nitrate response in shallow groundwater to agricultural N regulations in Denmark. *J. Environ. Manag.* **2019**, *240*, 66–74. [CrossRef] [PubMed]
34. Parris, K. Agricultural nutrient balances as agri-environmental indicators: An OECD perspective. *Environ. Pollut.* **1998**, *102*, 219–225. [CrossRef]
35. Organisation for Econonomic Cooperation and Development. *Environmental Indicators for Agriculture. Methods and Results*; OECD Publishing: Paris, France, 2001; ISBN 9789264186149.
36. Blicher-Mathiesen, G.; Andersen, H.E.; Larsen, S.E. Nitrogen field balances and suction cup-measured N leaching in Danish catchments. *Agric. Ecosyst. Environ.* **2014**, *196*, 69–75. [CrossRef]
37. De Notaris, C.; Rasmussen, J.; Sørensen, P.; Olesen, J.E. Nitrogen leaching: A crop rotation perspective on the effect of N surplus, field management and use of catch crops. *Agric. Ecosyst. Environ.* **2018**, *255*, 1–11. [CrossRef]
38. Wick, K.; Heumesser, C.; Schmid, E. Groundwater nitrate contamination: Factors and indicators. *J. Environ. Manag.* **2012**, *111*, 178–186. [CrossRef]
39. Hansen, B.; Thorling, L.; Schullehner, J.; Termansen, M.; Dalgaard, T. Groundwater nitrate response to sustainable nitrogen management. *Sci. Rep.* **2017**, *7*, 8566. [CrossRef]
40. Klages, S.; Heidecke, C.; Osterburg, B.; Bailey, J.; Calciu, I.; Casey, C.; Dalgaard, T.; Frick, H.; Glavan, M.; D'Haene, K.; et al. Nitrogen surplus—A unified indicator for water pollution in Europe? *Water* **2020**, *12*, 1197. [CrossRef]
41. Sebol, L.A.; Robertson, W.D.; Busenberg, E.; Plummer, L.N.; Ryan, M.C.; Schiff, S.L. Evidence of CFC degradation in groundwater under pyrite-oxidizing conditions. *J. Hydrol.* **2007**, *347*, 1–12. [CrossRef]
42. Hinsby, K.; Højberg, A.L.; Engesgaard, P.; Jensen, K.H.; Larsen, F.; Plummer, L.N.; Busenberg, E. Transport and degradation of chlorofluorocarbons (CFCs) in the pyritic Rabis Creek aquifer, Denmark. *Water Resour. Res.* **2007**, *43*, W10423. [CrossRef]
43. Ellermann, T.; Bossi, R.; Nygaard, J.; Christensen, J.; Løfstrøm, P.; Monies, C.; Grundahl, L.; Geels, C.; Nielsen, I.E.; Poulsen, M.B. *Atmosfærisk Deposition 2017*; Videnskabelig Rapport Fra DCE—Nationalt Center for Miljø og Energi nr. 304; Aarhus Universitet: Aarhus, Denmark, 2019.
44. Ministry of Environment and Food, Danish Agricultural Agency. Available online: https://lbst.dk/landbrug/kort-og-markblokke/hvordan-faar-du-adgang-til-data/ (accessed on 1 May 2020).
45. Aarhus Amt. *Vandforsyning På Tunø*; Teknisk Rapport; Aarhus Amt: Højbjerg, Denmark, 1989; p. 87.

46. Jensen, J.C.S.; Thirup, C. *Nitratudvaskning i indsatsområde Tunø*; Report; Aarhus Amt: Højbjerg, Denmark, 2006; p. 31.
47. Geological Survey of Denmark and Greenland (GEUS). Jupiter. Available online: https://www.geus.dk/ (accessed on 1 May 2020).
48. Thorling, L.; Thomsen, R. *Tunø Status Report 1989–1999*; Aarhus Amt: Højbjerg, Denmark, 2001.
49. Windolf, J.; Blicher-Mathiesen, G.; Carstensen, J.; Kronvang, B. Changes in nitrogen loads to estuaries following implementation of governmental action plans in Denmark: A paired catchment and estuary approach for analysing regional responses. *Environ. Sci. Policy* **2012**, *24*, 24–33. [CrossRef]
50. Hansen, B.; Dalgaard, T.; Thorling, L.; Sørensen, B.; Erlandsen, M. Regional analysis of groundwater nitrate concentrations and trends in Denmark in regard to agricultural influence. *Biogeosciences* **2012**, *9*, 3277–3286. [CrossRef]
51. Thorling, L.; Ditlefsen, C.; Ernstsen, V.; Hansen, B.; Johnsen, A.R.; Troldborg, L. *Grundvandsovervågning Status Og Udvikling 1989–2018*; Teknisk Rapport, GEUS 2019; GEUS: Copenhagen, Denmark, 2019.
52. Poisvert, C.; Curie, F.; Moatar, F. Annual agricultural N surplus in France over a 70-year period. *Nutr. Cycl. Agroecosyst.* **2017**, *107*, 63–78. [CrossRef]
53. Poisvert, C.; Curie, F.; Gassama, N. *Evolution des Surplus Azotés (1960–2010): Déploiement National, Étude des Temps de Transfert et de L'impact du Changement des Pratiques Agricoles*; Université de Tours—UFR Sciences et Techniques: Tours, France, 2016; p. 45.
54. McGonigle, D.F.; Harris, R.C.; McCamphill, C.; Kirk, S.; Dils, R.; MacDonald, J.; Bailey, S. Towards a more strategic approach to research to support catchment-based policy approaches to mitigate agricultural water pollution: A UK case-study. *Environ. Sci. Policy* **2012**, *24*, 4–14. [CrossRef]
55. Cherry, K.A.; Shepherd, M.; Withers, P.J.A.; Mooney, S.J. Assessing the effectiveness of actions to mitigate nutrient loss from agriculture: A review of methods. *Sci. Total Environ.* **2008**, *406*, 1–23. [CrossRef]
56. Bouraoui, F.; Grizzetti, B. Modelling mitigation options to reduce diffuse nitrogen water pollution from agriculture. *Sci. Total Environ.* **2014**, *468–469*, 1267–1277. [CrossRef]
57. Vero, S.E.; Healy, M.G.; Henry, T.; Creamer, R.E.; Ibrahim, T.G.; Richards, K.G.; Mellander, P.E.; McDonald, N.T.; Fenton, O. A framework for determining unsaturated zone water quality time lags at catchment scale. *Agric. Ecosyst. Environ.* **2017**, *236*, 234–242. [CrossRef]
58. Bourgeois, C.; Jayet, P.-A. Regulation of relationships between heterogeneous farmers and an aquifer accounting for lag effects. *Aust. J. Agric. Resour. Econ.* **2016**, *60*, 39–59. [CrossRef]
59. Vervloet, L.S.C.; Binning, P.J.; Børgesen, C.D.; Højberg, A.L. Delay in catchment nitrogen load to streams following restrictions on fertilizer application. *Sci. Total Environ.* **2018**, *627*, 1154–1166. [CrossRef]
60. Odgaard, B.V.; Rømer, J.R. *Danske Landbrugslandskaber Gennem 2000 år: Fra Digevoldninger Til Støtteordninger*; Odgaard, B.V., Ed.; Aarhus University Press: Aarhus, Denmark, 2000; ISBN 978-87-7934-420-4.
61. Carter, A. How pesticides get into water—And proposed reduction measures. *Pestic. Outlook* **2000**, *11*, 149–156. [CrossRef]
62. Ghasemizadeh, R.; Hellweger, F.; Butscher, C.; Padilla, I.; Vesper, D.; Field, M.; Alshawabkeh, A. Review: Groundwater flow and transport modeling of karst aquifers, with particular reference to the North Coast Limestone aquifer system of Puerto Rico. *Hydrogeol. J.* **2012**, *20*, 1441–1461. [CrossRef] [PubMed]
63. Megnien, C. Transits multiples dans l'aquifère des sources de Provins (Ile-de-France) et stratification des écoulements souterrains; Multiple transits in the aquifer supplying the Provins springs (Ile-de-France) and stratified groundwater flow (en). *Hydrogéologie (Orléans)* **1998**, *4*, 63–71.
64. Gourcy, L.; Baran, N.; Vittecoq, B. Improving the knowledge of pesticide and nitrate transfer processes using age-dating tools (CFC, SF6, 3H) in a volcanic island (Martinique, French West Indies). *J. Contam. Hydrol.* **2009**, *108*, 107–117. [CrossRef] [PubMed]
65. Hansen, B.; Thorling, L.; Dalgaard, T.; Erlandsen, M. Trend reversal of nitrate in Danish groundwater—A reflection of agricultural practices and nitrogen surpluses since 1950. *Environ. Sci. Technol.* **2011**, *45*, 228–234. [CrossRef]
66. Böhlke, J.K.; Michel, R.L. Contrasting residence times and fluxes of water and sulfate in two small forested watersheds in Virginia, USA. *Sci. Total Environ.* **2009**, *407*, 4363–4377. [CrossRef]
67. Wells, M.; Gilmore, T.; Mittelstet, A.; Snow, D.; Sibray, S. Assessing decadal trends of a nitrate-contaminated shallow aquifer in Western Nebraska using groundwater isotopes, age-dating, and monitoring. *Water* **2018**, *10*, 1047. [CrossRef]

68. Böhlke, J.K.; Denver, J.M. Combined use of groundwater dating, chemical, and isotopic analyses to resolve the history and fate of nitrate contamination in two agricultural watersheds, Atlantic Coastal Plain, Maryland. *Water Resour. Res.* **1995**, *31*, 2319–2339. [CrossRef]
69. Manga, M. On the timescales characterizing groundwater discharge at springs. *J. Hydrol.* **1999**, *219*, 56–69. [CrossRef]
70. McGuire, K.J.; McDonnell, J.J. A review and evaluation of catchment transit time modeling. *J. Hydrol.* **2006**, *330*, 543–563. [CrossRef]
71. Hamilton, S.K. Biogeochemical time lags may delay responses of streams to ecological restoration. *Freshw. Biol.* **2012**, *57*, 43–57. [CrossRef]
72. Jakobsen, R.; Hinsby, K.; Aamand, J.; van der Keur, P.; Kidmose, J.; Purtschert, R.; Jurgens, B.; Sültenfuss, J.; Albers, C.N. History and sources of co-occurring pesticides in an abstraction well unraveled by age distributions of depth-specific groundwater samples. *Environ. Sci. Technol.* **2020**, *54*, 158–165. [CrossRef] [PubMed]
73. Weissmann, G.S.; Zhang, Y.; LaBolle, E.M.; Fogg, G.E. Dispersion of groundwater age in an alluvial aquifer system. *Water Resour. Res.* **2002**, *38*, 1198. [CrossRef]
74. Botter, G.; Bertuzzo, E.; Rinaldo, A. Catchment residence and travel time distributions: The master equation. *Geophys. Res. Lett.* **2011**, *38*, 1–6. [CrossRef]
75. U.S. Environmental Protection Agency Pore Water Sampling. 2013, 1–16. Available online: https://www.epa.gov/quality/pore-water-sampling (accessed on 1 May 2020).
76. Yeshno, E.; Arnon, S.; Dahan, O. Real-time monitoring of nitrate in soils as a key for optimization of agricultural productivity and prevention of groundwater pollution. *Hydrol. Earth Syst. Sci.* **2019**, *23*, 3997–4010. [CrossRef]
77. Velthof, G.L.; Hoving, I.E.; Dolfing, J.; Smit, A.; Kuikman, P.J.; Oenema, O. Method and timing of grassland renovation affects herbage yield, nitrate leaching, and nitrous oxide emission in intensively managed grasslands. *Nutr. Cycl. Agroecosyst.* **2010**, *86*, 401–412. [CrossRef]
78. Singh, G.; Kaur, G.; Williard, K.; Schoonover, J.; Kang, J. Monitoring of water and solute transport in the vadose zone: A review. *Vadose Zone J.* **2018**, *17*, 160058. [CrossRef]
79. Weihermüller, L.; Siemens, J.; Deurer, M.; Knoblauch, S.; Rupp, H.; Göttlein, A.; Pütz, T. In situ soil water extraction: A review. *J. Environ. Qual.* **2007**, *36*, 1735–1748. [CrossRef]